CURR
S-SC5
G484
2000
Gr.9-12TL

Chemistry

CONCEPTS AND APPLICATIONS

LABORATORY MANUAL
TEACHER EDITION

Author: Tom Russo

McGraw-Hill
New York, New York   Columbus, Ohio   Woodland Hills, California   Peoria, Illinois

Chemistry
Concepts and Applications

Student Edition
Teacher Wraparound Edition
Teacher Classroom Resources:

- Laboratory Manual SE and TE
- Study Guide SE and TE
- Problems and Solutions Manual
- Supplemental Practice Problems
- Chapter Review and Assessment
- Section Focus Transparency Package
- Basic Concepts Transparency Package
- Problem Solving Transparency Package
- ChemLab and MiniLab Worksheets
- Critical Thinking/Problem Solving
- Chemistry and Industry
- Consumer Chemistry
- Tech Prep Applications
- Applying Scientific Methods in Chemistry
- Spanish Resources
- Lesson Plans
- Microcomputer-based Labs

Technology

- Computer Test Bank: DOS and Macintosh
- CD-ROM Multimedia System: Windows and Macintosh Versions
- Videodisc Program
- Chapter Summaries, English and Spanish Audiocassettes
- MindJogger Videoquizzes
- Mastering Concepts in Chemistry - Software

The Glencoe Science Professional Development Series

- Cooperative Learning in the Science Classroom
- Alternate Assessment in the Science Classroom
- Lab and Safety Skills in the Science Classroom
- Performance Assessment in the Science Classroom

Glencoe/McGraw-Hill

A Division of The McGraw-Hill Companies

Copyright © by The McGraw-Hill Companies, Inc. All rights reserved. Permission is granted to reproduce the material contained herein on the condition that such material be reproduced only for classroom use; be provided to students, teachers, and families without charge; and be used solely in conjunction with the *Chemistry: Concepts and Applications* program. Any other reproduction, for use or sale, is prohibited without prior written permission of the publisher.

Send all inquiries to:
Glencoe/McGraw-Hill
936 Eastwind Drive
Westerville, Ohio 43081

ISBN 0-02-827460-1
Printed in the United States of America.

3 4 5 6 7 8 9 024 05 04 03 02 01 00 99

Contents

To the Teacher iv

	Page
Laboratory Safety	T1
Chemical Storage	T2
Waste Disposal	T3
Disposal of Chemicals	T4
Microchemistry	T5
Microplate Templates and Data Forms	T6
Copy Masters for Laboratory 5-1	T10
Student and Class Equipment Lists	T14
Chemical Suppliers	T17
Chemical List	T17
Post Lab Quizzes	T21
Answer Key	T43
Student Edition Contents	iii
Laboratory Activities	1

To the Teacher

A well-rounded laboratory program is an integral part of the high school curriculum. The ***Laboratory Manual Teacher Edition*** provides students with a variety of laboratory experiences that serve to reinforce the principles of chemistry encountered in ***Chemistry: Concepts and Applications.*** Each experiment involves the manipulation of apparatus, observation, data gathering, data analysis, and interpretation to form conclusions.

MICROCHEMISTRY

Chemistry: Concepts and Applications has taken the lead in providing the latest development in laboratory techniques—microchemistry. Microchemistry uses smaller amounts of chemicals than do conventional methods of chemistry lab instruction. Much more information about microchemistry and implementing these techniques in the laboratory program is provided on page T5.

Included in this teacher edition on pages T6–T9 are blackline copy masters for use with microchemistry experiments. Pages T6 and T8 have actual-size templates for 24-well and 96-well microplates, respectively. Students can place copies of these masters beneath microplates to help keep track of the contents of wells. Students can tailor the template for each experiment by labeling the columns and rows. Provided on pages T7 and T9 are Microplate Data Forms for 24-well and 96-well plates, respectively. These are grids that students can use to record data and observations. Although the appropriate Microplate Data Form is included in each microchemistry activity of the student edition, the copy masters on pages T7 and T9 should be used for those activities where more than one form is required.

TEACHER GUIDE PAGES

This teacher edition is organized so that all relevant teacher information for each experiment appears immediately following that experiment. In the student edition, two pages for student data and lab reports follow the experiment. In the teacher edition, those pages are used for teacher information and have the heading **Teacher Guide.** The following is a description of the **Teacher Guide** pages.

- **Title and Number:** The laboratory activities correlate to the text, so the experiment number corresponds to the textbook chapter to which the experiment is relevant. If there is more than one experiment associated with a chapter, a hyphenated number is used as an identifier.
- **Introduction:** The introduction appears in a block immediately following the title of the experiment. This discussion provides teacher background, additional theory, an overall purpose, and, often, a **Misconception** that students frequently hold regarding some aspect of the activity.
- **Process Skills:** This is a listing of expected learning outcomes for the experiment. These skills are stated in behavioral terms for the teacher's benefit.
- **Preparation Hints:** This section contains hints for setting up the experiment, helpful shortcuts, and suggestions for handling or storing specific reagents. Presented here also are safety precautions that apply to the preparation of reagents and to the overall activity rather than to certain procedure steps.
- **Materials:** This is the same list of materials as that in the student edition except that reagents and, in come cases, suggested sizes of equipment have been added. It is felt that this method gives teachers better control over the dispensing of chemicals. Complete instructions are given for the preparation of solutions.
- **Procedure Hints:** The hints presented here are based on classroom experience. **Troubleshooting** hints will aid you in anticipating procedural problems and in deciding which techniques should be demonstrated to students in advance. **Safety Precautions** deal with specific substances or manipulations. **Disposal** hints give advice on proper disposal of wastes resulting from the activity. Letter codes refer to disposal instructions on pages T4 and T5.
- **Data and Observations:** This section provides sample data typical of the data that students are likely to obtain in the experiment.
- **Analysis, Conclusions, Extension and Application:** The **Teacher Guide** pages provide detailed answers to the questions posed in the student pages.

POST LAB QUIZZES

A complete set of post lab quizzes is furnished in this teacher edition for all the experiments. The quizzes test student understanding of the purpose of the lab as well as specific procedures, skills, and conclusions.

Outlined here are some considerations on laboratory safety that are intended primarily for teachers and administrators. Safety awareness must begin with the principal, be supervised by the department head, and be important to the individual teacher.

Principals and supervisors should be familiar with the guidelines for laboratory safety and provide continual supervision to ensure compliance with those guidelines. Teachers have the ultimate responsibility for enforcing safety standards in the laboratory. They must set the proper example in the laboratory by observing all rules themselves. This behavior applies to such duties as wearing goggles and protective clothing, and not working alone in the laboratory. Planning is essential to laboratory safety, and that planning must include what to do in an emergency, as well as how to prevent accidents.

Numerous books and pamphlets are available on laboratory safety with detailed instructions on planning and preventing accidents. However, much of what they present can be summarized in the phrase, "Be prepared!" Know the rules and what common violations occur. Know where emergency equipment is stored and how to use it. Practice good laboratory housekeeping by observing these guidelines.

1. Store chemicals properly.
   a. Segregate chemicals by reaction types.
   b. Label all chemical containers; include purchase date, special precautions, and expiration date.
   c. When outdated, discard chemicals according to appropriate disposal methods.
2. Prohibit eating, drinking, and smoking in the laboratory.
3. Refuse to tolerate any behavior other than a serious, businesslike approach to laboratory work.
4. Require that eye and clothing protection be worn.
5. Expect proper use of equipment.
   a. Do not leave equipment unattended.
   b. Shield systems under pressure or vacuum.
   c. *NEVER* use open flames when a flammable solvent is in use in the same room. Use open flames only when necessary; substitute hot plates whenever possible.
   d. Instruct students in the proper handling of glass tubing.
   e. Instruct students in the proper use of pipets, cylinders, and balances.
   f. See that laboratory benches do not become catchalls for books, jackets, and so on.
6. Dispose of wastes correctly.

Waste disposal is subject to various federal and state regulations. The agencies charged with administering disposal of wastes are the U.S. Environmental Protection Agency (EPA) and its equivalent state agency. The regulations promulgated by the EPA are found in the Code of Federal Regulations (CFR), Title 40. Since some wastes may have to be disposed of away from the school, transportation of hazardous material becomes a problem. Regulations concerning the movement of hazardous materials are established by the Department of Transportation (DOT) in CFR Title 49. Both 40 CFR and 49 CFR are published annually. Changes are published as they occur in the Federal Register. A practical guide to waste disposal for laboratories is *Prudent Practices for Disposal of Chemicals from Laboratories,* published by the National Research Council, Washington: National Academy Press, 1983.

Other resources on waste disposal and other aspects of laboratory management include the following: N. Irving Sax, *Dangerous Properties of Industrial Materials,* 7th ed., New York: Van Nostrand-Reinhold, 1988; the chemical catalog and reference manual published by Flinn Scientific, Inc., P.O. Box 231, Batavia, IL 60510, (630) 879-6900; a handbook, *Less Is Better,* and two pamphlets, "Hazardous Waste Management" and "Chemical Risk: A Primer," available free in single copies from the Office of Federal Regulatory Programs, ACS Department of Government Relations and Science Policy, 1155 16th St. NW, Washington, DC 20036; and the ACS Chemical Health and Safety Referral Service at the same address, which provides referrals to literature, films, educational courses, or organizations that can provide safety information, (202) 872-4511.

DISCLAIMER

Glencoe Publishing Company makes no claims to the completeness of this discussion of laboratory safety and chemical storage. The material presented is not all-inclusive, nor does it address all of the hazards associated with handling, storage, and disposal of chemicals, or with laboratory practices and management.

The basis for all schemes of chemical storage is the minimizing of any chance of reaction, both between the contents of two containers and within a single container. The possibility for reaction between two containers is minimized by storing chemicals in groups that contain materials nonreactive with each other and that are well separated from other groups with which they will react. The principal method for reducing reaction within a single container is frequent housecleaning to remove chemicals nearing the end of their shelf life.

The starting point for proper chemical storage is a well-designed stockroom that meets local building code, OSHA, and NFPA regulations and recommendations. The stockroom should be cool, dry, and provide both adequate space and adequate ventilation. In general, chemicals should not be stored above eye level. Wood shelving with wood supports is preferable to the metal adjustable type. Shelf assemblies should be firmly attached to walls. Antiroll lips should be placed on all shelves. When storing chemicals, make sure bottle covers are secure, the outsides of the bottles have been wiped clean, and labels face forward.

Special attention needs to be given to the storage of flammable substances, toxic substances, water-sensitive substances, and compressed gases. The amounts of hazardous materials present in the laboratory itself should always be minimized. A good rule of thumb for flammable liquids is not more than 500 mL in the laboratory at one time. The following is one storage scheme for separating incompatible chemicals. Materials in Column A must be widely separated from those in Column B.

INCOMPATIBLE CHEMICALS

	Column A	Column B
I.	acids	bases
	A. nitric acid	metals, sulfuric acid, sulfides, nitrites and other reducing agents, chromic acid and chromates, permanganates silver, mercury
	B. oxalic acid	
	C. sulfuric acid	metals, chlorates, perchlorates, permanganates, nitric acid
II.	alkali and alkaline earth compounds, metals and their carbides, hydrides, hydroxides, oxides, peroxides	water, acids, halogenated organics, oxidizing agents*
III.	ammonia	halogens, halogenating agents, silver, mercury
IV.	carbon, activated	oxidizing agents*
V.	hydrogen peroxide	metals and their salts
VI.	inorganic azides	acids, heavy metals and their salts, oxidizing agents*
VII.	inorganic cyanides	acids, strong bases
VIII.	inorganic nitrates	acids, metals, nitrites, sulfur
IX.	inorganic nitrites	acids, oxidizing agents*
X.	inorganic sulfides	acids
XI.	organic compounds	oxidizing agents*
	A. acetylenes and monosubstituted acetylenes (R-CCH)	halogens, Group IB and IIB metals and their salts
	B. organic acyl halides	bases, organic hydroxy compounds
	C. organic anhydrides	bases, organic hydroxy compounds
	D. organic halogen compounds	aluminum metal
	E. organic nitro compounds	strong bases
XII.	phosphorus (yellow)	oxygen, oxidizing agents*, strong bases
XIII.	phosphorus pentoxide	water, halogenating agents
XIV.	powdered metals	acids, oxidizing agents*

*Oxidizing agents include chromates, dichromates, chromium(VI) oxide, halogens, halogenating agents, hydrogen peroxide and peroxides, nitric acid, nitrates, perchlorates and chlorates, permanganates, and persulfates.

The number of lawsuits against schools and the number of regulations by OSHA, EPA, and DOT are occurring at an increasing rate. All of these complications make life more difficult for chemistry teachers.

One major problem that teachers face is the prohibition by a principal of students working with any substance even remotely considered hazardous. In that circumstance, the teacher can always perform an experiment with students acting as observers, note takers, and data recorders. Even if the demonstration-type experiment is done behind a safety shield, at least the students get to see chemical phenomena really occurring.

Another problem that many teachers face is that of chemical waste disposal. One way to reduce waste is to conduct microscale experiments. Many of the ChemLabs in the textbook and experiments in the *Laboratory Manual* employ microscale equipment and techniques. Others use traditional techniques and apparatus. Thus you can tailor your laboratory chemistry program to the facilities and equipment that you have available.

A frequent question concerns what liquids may safely be disposed of in the sanitary drains. First, the teacher must be certain that the sewer flows to a wastewater treatment plant and not to a stream or other natural water course. Second, any substance from a laboratory should be flushed with *at least* 100 times its own volume of tap water. Third, the suggestions given below should be checked with local authorities because local regulations are often more stringent than federal requirements. The National Research Council book *Prudent Practices for Disposal of Chemicals from Laboratories* lists many substances that can be disposed of in the sanitary drain. Below are listed some positive and negative ions from the NRC lists. It is important to note that both the positive and negative ion of a salt must be listed in order for its drain disposal to be considered safe.

Positive Ions: aluminum, ammonium, bismuth, calcium, copper, hydrogen, iron, lithium, magnesium, potassium, sodium, strontium, tin, titanium, zinc

Negative Ions: borate, bromide, carbonate, chloride, hydrogen sulfite, hydroxide, iodide, nitrate, phosphate, sulfate, sulfite, tetraborate, thiocyanate

Note that, although hydrogen and hydroxide ions are listed, acids and bases should be neutralized before disposal. A good rule of thumb is that nothing of pH less than 3 or greater than 8 should be discarded without neutralizing it first.

Of the organic compounds most often found in high school laboratories, the following can be disposed of in the drain: methanol, ethanol, propanols, butanols, pentanols, ethylene glycol, glycerol, sugars, formaldehyde, formic and acetic acids, oxalic acid, sodium and potassium salts of carboxylic acids, esters with fewer than five carbon atoms, and acetone. More [illegible]ive lists are given in *Prudent Practices*.

What happens to substances that do not fall into one of the categories discussed? There are three possibilities. One is to treat waste chemically to convert it to a form that is drain-disposable. A good example is iodate ion, which is too strong an oxidizing agent to go untreated. However, it is readily reduced to iodide (disposable) by acidified sodium hydrogen sulfite. Many procedures for processing laboratory waste to a discardable form are found in *Prudent Practices*.

A second possibility is the recycling of waste. Good examples here are the recovery of valuable metals such as mercury and silver. Another is the recovery of solvents through distillation.

If waste cannot be recycled or processed to disposable form, then it must be packed and shipped by a Department of Transportation-approved shipper to a landfill designated to receive chemical and hazardous waste. Since that method of disposal is very expensive, it pays to reduce such waste to a minimum. Microscale experiments, already mentioned, aid in that task. There are also numerous processes in *Prudent Practices* for reducing both the bulk and hazard of wastes. An example is the reduction of chromate and dichromate waste solutions to chromium(III) solutions, which are then made basic, precipitating chromium(III) oxide. The oxide is filtered off, dried, and crushed, resulting in a significant mass and volume saving. As hazardous wastes are accumulated awaiting shipment, it is important to observe the proper storage procedures for separating incompatible substances.

Most high schools qualify as Small Quantity Hazardous Waste Generators under EPA regulations. However, it is usually wise to register with the state or local environmental protection agency. That agency is a good source of information on disposal, particularly on the availability of approved packers, shippers, and landfill operators. Even though wastes are packed, shipped, and interred by licensed or approved firms, generators of waste are responsible for their waste—FOREVER! Therefore, we should examine carefully the credentials of any firm we hire to handle our waste.

DISPOSAL OF CHEMICALS

Local, state, and federal laws regulate the proper disposal of chemicals. Before using these disposal instructions, you should confirm them with local regulators so that they do not violate your local or state laws. *No representation, warranty, or guarantee is made by Glencoe Publishing Company or the authors as to the completeness or accuracy of these disposal procedures.*

The proper disposal of chemicals should be done only by the teacher who must be wearing a laboratory apron, goggles, and rubber gloves. The disposal procedures should be done in an operating fume hood. Even teachers should never be alone in the laboratory when disposing of chemicals. They should have the proper type of fire extinguisher and a telephone or intercom nearby.

The disposal procedures listed are for relatively small amounts of waste. If a larger amount is produced, the stoichiometry of the chemical reactions in the disposal procedure will have to be calculated. Do your disposal by reacting very small amounts of the chemical at one time. Many reactions are exothermic, so the reaction vessel will need to be in an ice water bath.

Disposal advice given in the *Teacher Wraparound Edition* is keyed to the following lettered disposal procedures.

DISPOSAL A: These materials can be packaged and sealed in separate plastic containers and buried in a landfill approved for chemical and hazardous waste disposal. Local soil and water conditions cause local regulatory agencies to restrict what is permitted in the landfill. Contact your local regulatory agency before assuming it is acceptable to place these materials in your school's trash dumpster. Do not mix chemicals; place each one in its own container. Place the containers in a cardboard box and separate with vermiculite. Seal the closed box before disposing of it.

DISPOSAL B: Decant the water layer into a separate beaker. Then, discard the water layer down the drain. Label the other chemical and save it for future activities, following proper storage directions.

DISPOSAL C: Rinse the chemical down the drain with at least a 100-fold increase in the volume of water. Rinse only one chemical down the drain at a time, using plenty of water to dilute it. Do not mix chemicals in the sink or drain. Dissolve any solids in a beaker before placing them down the drain. *Do not rinse these chemicals down the drain if your school's drains go into a septic system.* This disposal procedure is intended for schools whose sanitary sewers go to a waste water treatment plant. Due to local soil and water conditions, local regulations may prohibit the disposal of these chemicals in this manner. Check with the local government regulatory agency before using this procedure.

DISPOSAL D: While being stirred, the acidic solution should be slowly added to a large beaker of cold water. Prepare a 1 *M* $Na_2CO_3 \cdot 10H_2O$ solution by dissolving 62 g in 500 mL of water. Then, slowly add the 1 *M* Na_2CO_3 solution to the diluted acid. Carbon dioxide gas will be evolved. When there is no more evolution of gas as more Na_2CO_3 solution is added, the solution can be tested with pH paper to verify that it is neutral. Rinse the neutralized solution down the drain with a large volume of water.

DISPOSAL E: Filter the solution through filter paper. Open the filter paper and allow it to dry. Place the dried solid and filter paper in a separate, plastic container surrounded by vermiculite in a box. Seal the box and dispose of it in a landfill approved to receive chemical and hazardous waste. Dilute the filtrate by adding it slowly, while stirring, to a large beaker of cold water. Verify that it is neutral by using pH paper. Rinse the neutral solution down the drain with a large volume of water.

DISPOSAL F: Place the substance in a plastic container and seal it. Completely label the container and, following proper storage directions, save it for future activities.

DISPOSAL G: For the iodine: add 18 g of sodium thiosulfate to the solution. Stir while warming the solution to 50°C. After the iodine is consumed, check with pH paper and add enough 3*M* NaOH solution (12 g NaOH in 100 mL water) to neutralize the solution. Then for the manganese or cobalt: add 6 g of sodium sulfide. After one hour of stirring, neutralize the solution with 3*M* NaOH. Verify with pH paper that the solution is neutral. Filter and place the MnS solid in a plastic bottle for disposal in an approved landfill. Treat the filtrate for excess sulfide by adding 12 g of iron(III) chloride with constant stirring. Filter the precipitate, Fe_2S_3. Place the solid in a plastic container, and seal it for disposal in an approved landfill. Rinse the neutralized solution down the drain with a large volume of water.

DISPOSAL H: Use an operating fume hood to disperse the small volume of gas produced.

DISPOSAL I: Slowly add the substance to a large beaker of cold water while stirring. Then in an ice bath, slowly add 6*M* HCl until the solution is neutralized. Verify by using pH paper. Rinse the neutralized solution down the drain, using a large volume of water.

MICROCHEMISTRY TECHNIQUES

Chemistry teaching at the high school level is presently facing several problems at once. Concern for student safety, environmental questions, cost of materials, and the necessity of adhering to a prescribed curriculum have all worked to make the laboratory program difficult to carry out. Still, the "lab" is the most tangible, best remembered, and most visible aspect of chemistry instruction. Without direct observation of chemical phenomena and the hands-on manipulation of the substances referred to in the classroom and the textbook, chemistry can become no more than a survey of chemical theory with little real meaning to students, especially students who are visual learners.

Glencoe Publishing Company has taken the lead in answering these problems by introducing microchemistry techniques in ***Chemistry: Concepts and Applications***—both in the textbook in the form of ChemLabs and in the *Laboratory Manual*. When you have limited time and a limited budget, microchemistry provides a safe, inexpensive, and time-efficient means to conduct a laboratory program. It gives you, the chemistry teacher, the opportunity to keep chemistry the investigative science that it should be and that students expect it to be.

Laboratory activities and experiments that use micro-amounts of chemicals provide a way to involve students in observation and manipulation of some substances that might otherwise be regarded as hazardous. Because drastically reduced amounts of chemicals are used, safety is greatly increased. Likewise, the expense of running a laboratory chemistry program is cut. Because of the savings in both time and expense with microchemistry, students are afforded the opportunity to do much more experimental chemistry than was possible with conventional labs. As with traditional chemistry, all processes should be conducted with safety goggles and protective clothing.

WASTE DISPOSAL IN MICROCHEMISTRY

Waste disposal for nearly all microchemistry activities is coded as Disposal J

DISPOSAL J: Collect all chemical solutions, precipitates, and rinse solutions in a polyethylene dishpan or similar container devoted to that purpose. Retain the solutions until the end of the period or day. In the fume hood, set up a hot plate with a l-L or 2-L beaker. Pour the collected solutions into the beaker. Turn the hot plate on low and allow the beaker to heat with the hood running and the hood door closed. The liquids and volatiles in the mixture will evaporate, leaving dried chemicals. Allow the beaker to cool. Continue to add solutions and waste until the beaker is $2/3$ full. Treat the waste as heavy metal waste. ***Dispose of the beaker and its dry contents in an approved manner.***

Copyright © Glencoe/McGraw-Hill, a division of The McGraw-Hill Companies, Inc.

24-well template

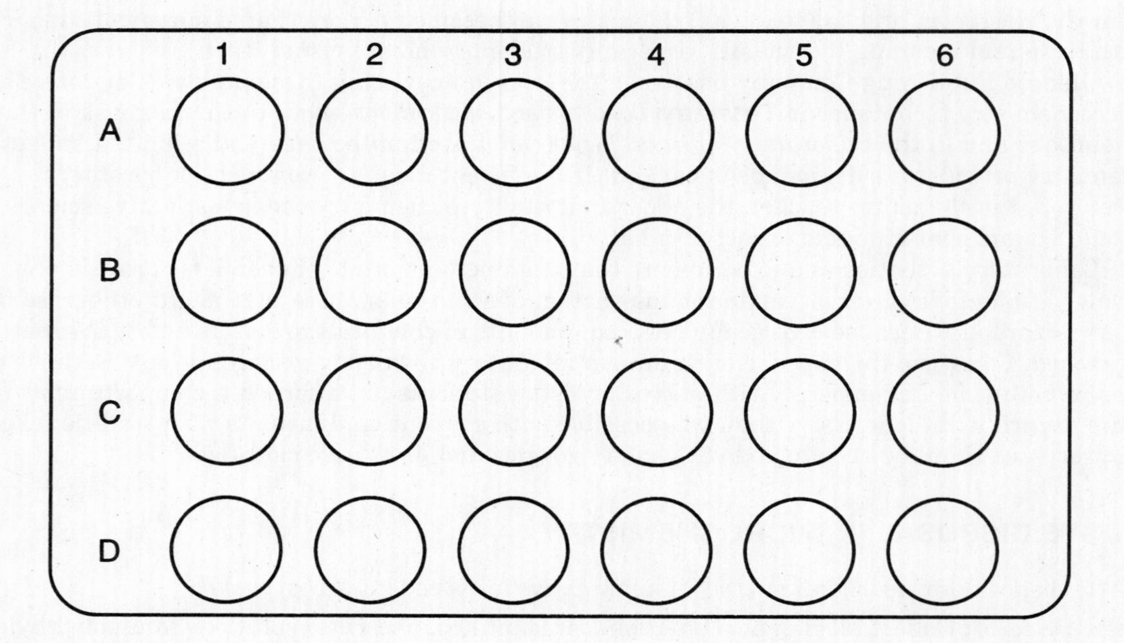

Name Date Class

Microplate data form

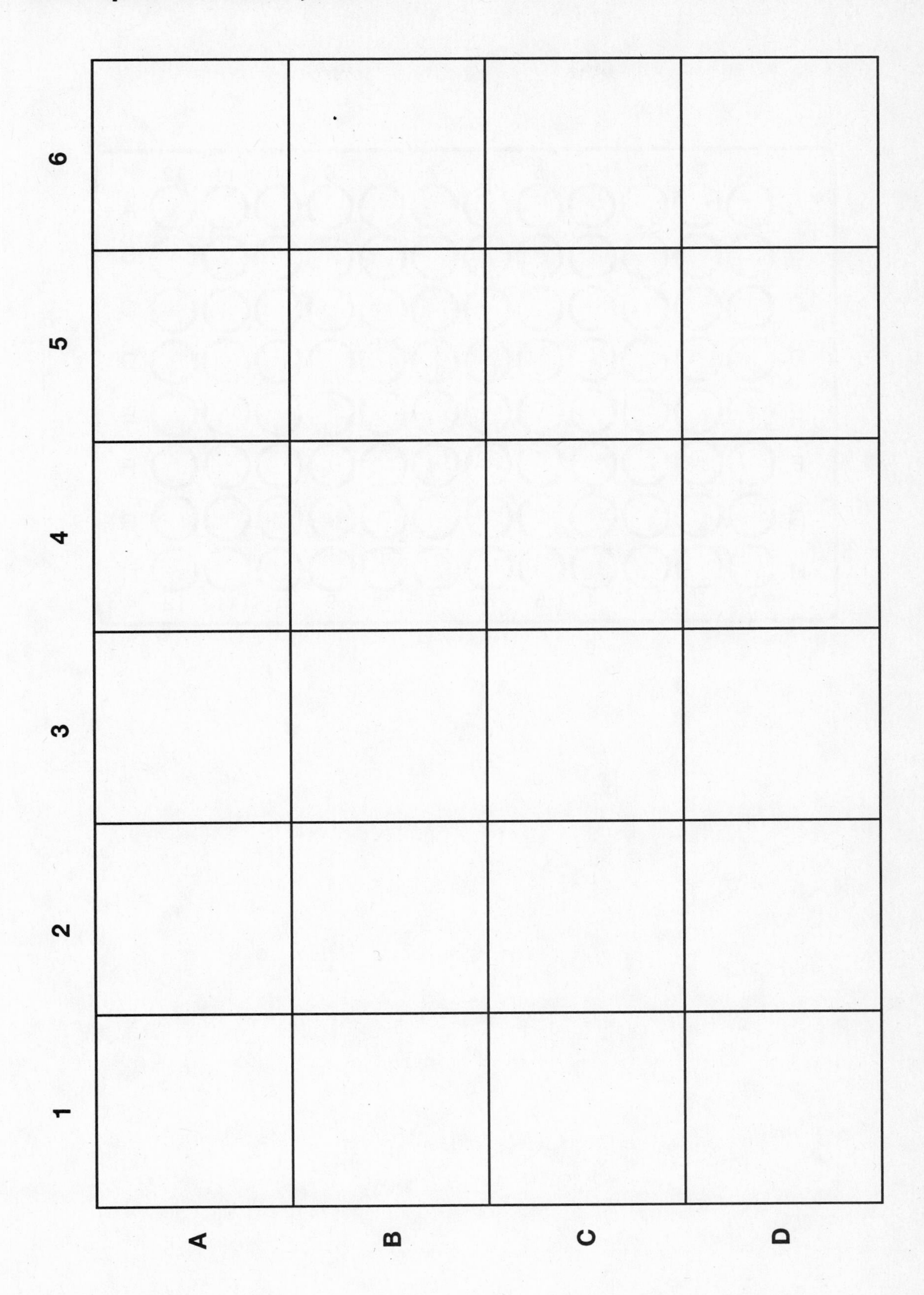

96-well template

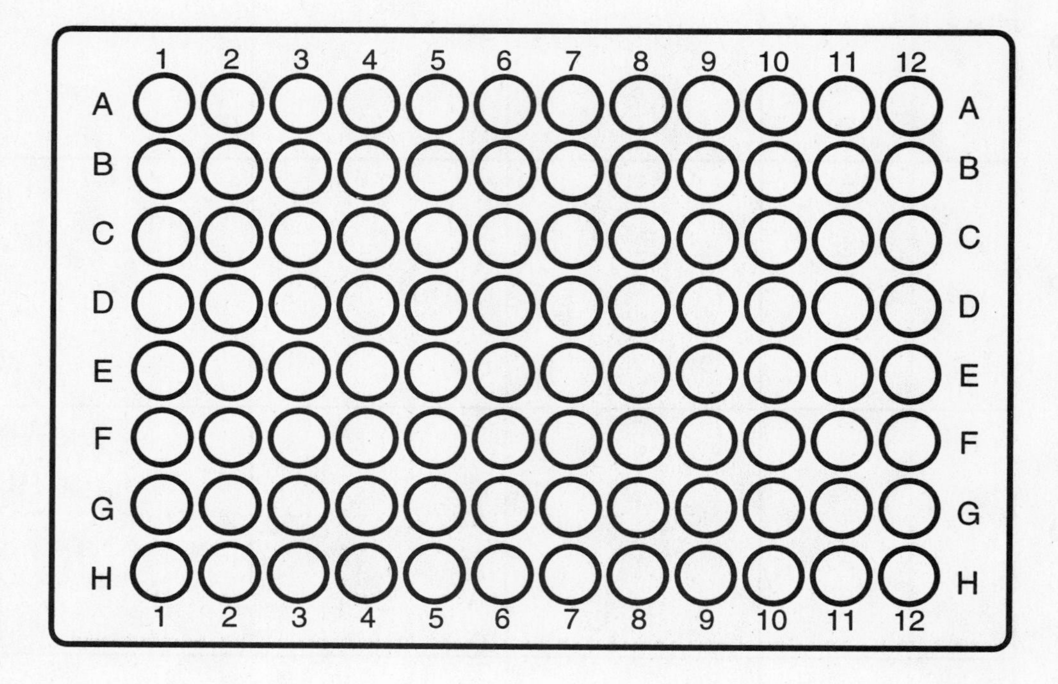

Name Date Class

Microplate data form

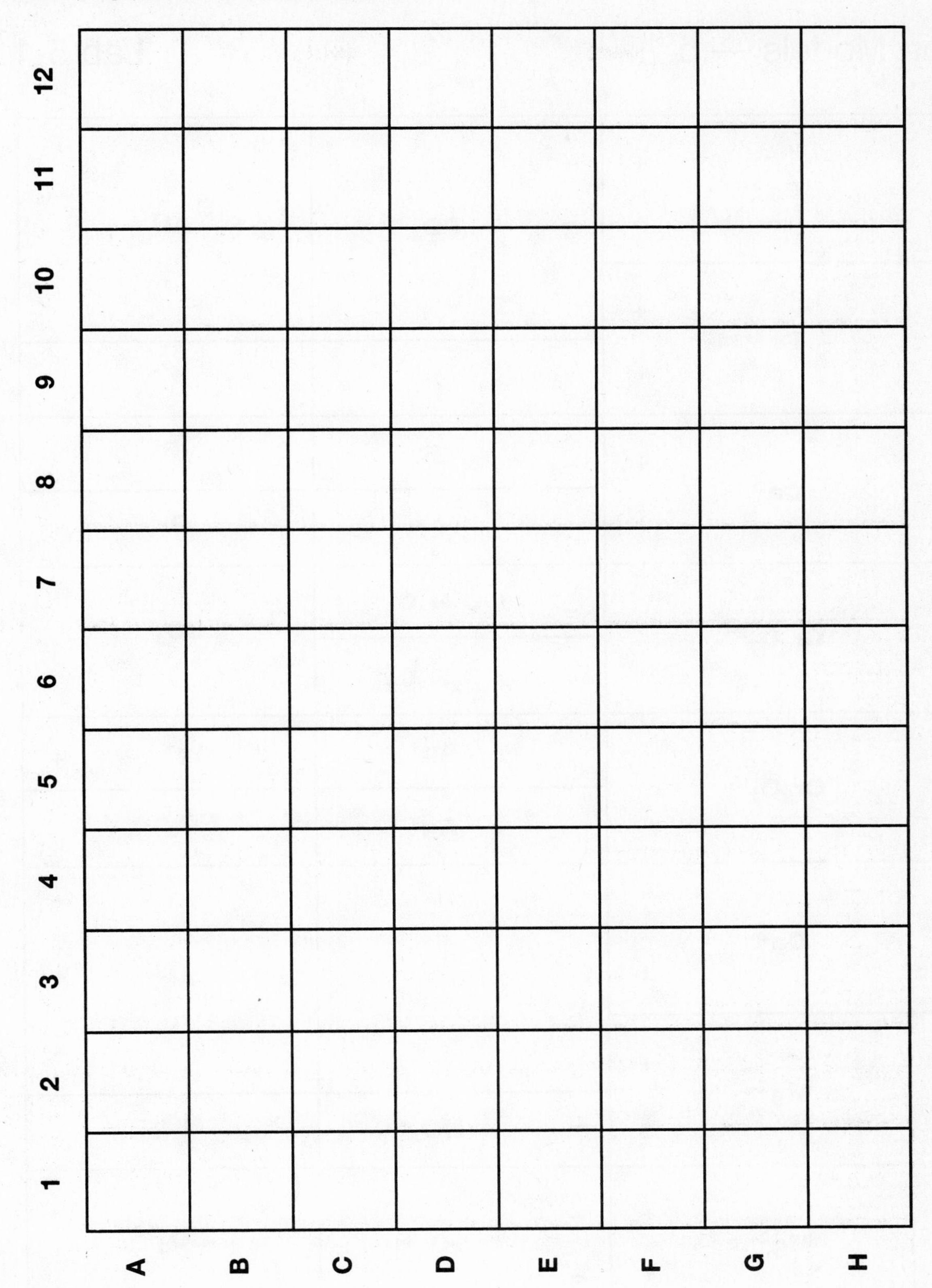

Copyright © Glencoe/McGraw-Hill, a division of The McGraw-Hill Companies, Inc.

1. Ion Models for Laboratory 5-1

Ion Models — p. 1 Lab 5-1

+ + Ba^{2+}	− − − PO_4^{3-}	Al^{3+}
+ + Fe^{2+}	− F^-	− F^-
+ + Ca^{2+}	− F^-	− F^-
	− Cl^-	− Br^-
− − $Cr_2O_7^{2-}$	− $C_2H_3O_2^-$	− − CO_3^{2-}
	− $C_2H_3O_2^-$	
− − $Cr_2O_7^{2-}$	Ag^+ +	Ag^+ +
	Ag^+ +	− NO_3^-
+ + Ca^{2+}	− − − PO_4^{3-}	Al^{3+} + + +
+ + Mg^{2+}	− NO_3^-	− NO_3^-
+ + Mg^{2+}	− − SO_4^{2-}	− − CO_3^{2-}

Copyright © Glencoe/McGraw-Hill, a division of The McGraw-Hill Companies, Inc.

1. Ion Models for Laboratory 5-1

Ion Models — p. 2 Lab 5-1

Ca^{2+} + +	– – SO_4^{2-}	– – CO_3^{2-}
Ba^{2+} + +	NH_4^+ +	NH_4^+ +
	Na^+ +	Na^+ +
Ba^{2+} + +	K^+ +	K^+ +
	Li^+ +	Li^+ +
Mg^{2+} + +	– F^-	– F^-
	Li^+ +	Li^+ +
Fe^{2+} + +	– NO_3^-	– $C_2H_3O_2^-$
	– NO_3^-	– $C_2H_3O_2^-$
Fe^{2+} + +	– Br^-	– Cl^-
	– Br^-	– Br^-
– – $Cr_2O_7^{2-}$	– Cl^-	– Cl^-
	Ag^+ +	Ag^+ +
– – $Cr_2O_7^{2-}$	– – SO_4^{2-}	– – SO_4^{2-}

Copyright © Glencoe/McGraw-Hill, a division of The McGraw-Hill Companies, Inc.

1. Ion Models for Laboratory 5-1

Ion Models — p. 3

Lab 5-1

Ca^{2+}	SO_4^{2-}	CO_3^{2-}
Ba^{2+}	Cl^-	Na^+
	Cl^-	Na^+
$Cr_2O_7^{2-}$	Br^-	K^+
	Br^-	K^+
Mg^{2+}	$C_2H_3O_2^-$	$C_2H_3O_2^-$
	$C_2H_3O_2^-$	Cl^-
Fe^{2+}	Br^-	Na^+
	Li^+	K^+
PO_4^{3-}	Al^{3+}	NH_4^+
		NH_4^+
		NH_4^+
NO_3^-	Al^{3+}	PO_4^{3-}
CO_3^{2-}		

Copyright © Glencoe/McGraw-Hill, a division of The McGraw-Hill Companies, Inc.

Table T1

Individual Student Equipment			
Quantity	**Description**	**Quantity**	**Description**
1	beaker, 50-mL	1	graduated cylinder (50-mL)
1	beaker, 100- or 150 mL	1	laboratory apron
2	beaker, 250-mL	1	24-well microplate
4	beaker, 400-mL	3	96-well microplate
1	chemical scoop	10	pipets, microtip
3	crucible	15	pipets, thin-stem
2	droppers	1	scissors
1	evaporating dish	1	spatula
1	flask, 125-mL Erlenmeyer	6	test tube (13 x 100-mm)
1	flask, 500-mL Erlenmeyer	25	test tube (18 x 150-mm)
1	funnel	1	test-tube brush
1	goggles	1	test-tube holder
1	graduated cylinder, 10-mL	2	test-tube rack

Table T2*

Class Equipment		
Equipment	**Quantity**	**Labs Used In**
alligator clips	90	4-2, 17-1
aluminum block	30	2-1
aluminum foil	19 200-sq-ft rolls	2-1, 13-1
baby bottle, plastic	30	12-3
balance	30	2-1, 2-2, 12-1, 12-2, 13-1, 13-2, 15-2, 20-2, 20-3, 21-1
balloon, round	150	9-2
balloon, small round	30	11-2
balloon, pear-shaped	120	9-2
barometer	1	10-2, 12-1, 12-3
basin, plastic	30	19-2
battery, 9-volt	30	4-2
battery clip	30	4-2
beaker, 50-mL	30	9-1, 15-2, 17-1
beaker, 100-mL	30	1-1, 1-2, 4-1, 6-1, 14-2
beaker, 250-mL	60	2-2, 20-1
beaker, 400-mL	120	6-2, 10-2, 11-1, 12-3, 13-1, 15-2, 19-1, 19-2
beaker, 1-L	30	11-2, 12-1, 18-1, 21-1
beaker, 2-L	1	For Disposal
buret, 50-mL	30	15-2
capillary tube	3 vials of 100	1-1, 4-1
chart of pH and indicator colors	1 or more	15-1
chemical scoop	30	9-1, 14-1
clamp, buret	30	15-2
clamp, pinch	30	10-2
clamp, thermometer	30	1-1, 4-1, 10-2
clamp, utility	60	10-2, 15-2, 18-1
connecting wire	1-lb spool	4-2, 17-1, 17-2
conducting tool	30	9-1
corks	100	2-2

*The quantities of chemicals and equipment listed are the minimum amounts commonly sold that would supply a class of 30 students working individually. Each teacher should adjust quantities as necessary for particular class and group sizes.

Copyright © Glencoe/McGraw-Hill, a division of The McGraw-Hill Companies, Inc.

Table T2 (Continued)

Class Equipment		
Equipment	**Quantity**	**Labs Used In**
crucible	90	2-2
dishpan, polyethylene	1	For Disposal
dropper	60	10-2, 18-1
evaporating dish	30	2-2, 20-3
file	30	6-1, 17-1
film can	30	21-1
filter paper	2 pkgs	2-2, 17-1, 17-2
flask, 125-mL Erlenmeyer	30	15-2
flask, 500-mL Erlenmeyer	30	10-2
foam balls, small	360	10-1
forceps	30	8-2, 13-1, 17-1, 18-1
funnel	30	2-2
galvanometer or voltmeter	30	17-1, 17-2
graduated cylinder, 10-mL	30	13-1, 15-1, 18-1, 18-2, 19-1, 20-2
graduated cylinder, 25-mL	30	12-3
graduated cylinder, 50-mL	30	2-2, 13-1
graduated cylinder, 100-mL	30	2-1, 12-1, 20-3
graduated cylinder, 500-mL	30	15-2
Handbook of Chemistry and Physics	1 or more	4-2
heat-resistant pad	30	10-2
hot plate	30	1-1, 4-1, 6-2, 10-2, 11-1, 11-2, 13-1, 19-1, 19-2, 20-1
ion model sheet	90	5-1
ketchup cups	120	13-2, 20-2
ketchup cup lids	60	13-2, 20-2
laboratory burner	30	1-1, 2-2, 4-1, 6-2, 7-1, 9-1, 10-2, 11-1, 13-1, 19-1, 19-2, 20-1
LED	30	4-2
marking pencil	30	13-1, 15-1, 18-1, 19-1
mass, 1 or 2 kg	30	20-1
metric ruler	30	2-1, 10-2, 11-2, 12-3, 13-1, 20-1, 20-2, 21-2
microplate, 24-well	30	3-2, 12-2, 14-1, 16-1, 16-2, 17-1, 20-1
microplate, 96-well	90	4-2, 6-1, 6-2, 7-2, 8-1, 8-2, 14-2, 18-2, 19-2
microscope or hand lens	30	10-1
microscope slides	90	10-1
mortar and pestle	30	18-2, 20-3
nail polish or model paint	2 bottles	16-2
nichrome wire	4-oz spool	2-2
paint-can lid	30	9-1
paper, black	30 sheets	8-1
paper, brown grocery	60 sheets	19-1, 19-2
paper punch	30	13-2, 20-2, 20-3
paper towels	3 rolls	4-1, 9-1, 11-1, 12-1, 12-2, 15-1, 15-2, 17-1, 20-2
paper, white	210 sheets	4-1, 15-2, 19-2, 21-1
pipets, microtip	300	1-2, 3-2, 6-1, 6-2, 7-2, 8-2, 10-1, 14-1, 14-2, 16-1, 18-2, 19-2
pipets, thin-stem	400	1-1, 2-2, 4-2, 8-1, 11-1, 12-2, 14-1, 15-1, 16-1, 16-2, 17-1, 17-2, 19-1, 20-1
plastic-foam cups	60	20-3
plastic-foam cup lids	30	20-3
plastic wrap	1 roll	1-2

Table T2 (Continued)

Class Equipment		
Equipment	**Quantity**	**Labs Used In**
pneumatic trough	30	6-2, 12-1
resistor, 1-K ohm	30	4-2
ring, iron	30	1-1, 2-2, 6-2, 9-1, 10-2, 11-1, 13-1, 19-1, 19-2, 20-1
ring, small iron	30	11-2
ring stand	60	1-1, 2-2, 4-1, 6-2, 9-1, 10-2, 11-1, 13-1, 15-2, 18-1, 19-1, 19-2, 20-1
rubber bands	120	1-1, 4-1, 13-2, 17-2, 20-2
rubber bands, thick	30	20-1
sand	10 lb	18-1
sandpaper or emery cloth	1 pkg	17-1
scissors	30	2-1, 3-1, 5-1, 5-2, 13-1, 14-1, 16-1, 17-1, 17-2, 20-1, 20-2
spatula	30	15-1
spectra chart, standard	1 or more	7-1
spectroscope, hand	1 or more	7-1
split peas, green	5 1-pound bags	21-1
split peas, yellow	1-pound bag	21-1
stirring rod, glass	30	1-1, 4-1, 9-1, 13-1, 15-2, 18-2, 20-3
stopper, 1-hole, to fit large test tube	30	10-2
stopper, solid, to fit:		
large test tube	240	18-1
small test tube	180	15-1
stopper, 2-hole, to fit:		
flask	30	10-2
large test tube	30	18-1
straw, plastic drinking	60	4-2, 6-1
string	1 roll	2-1, 9-2, 11-2
tape, clear	2 rolls	6-2, 9-2, 14-1
test tube, 13 × 100-mm	180	1-1, 10-1, 15-1
test tube, 18 × 150-mm	750	2-2, 10-2, 13-1, 18-1, 19-1
test-tube holder	30	10-2
test-tube rack	60	2-2, 15-1, 18-1, 19-1
thermal mitt	30	6-2, 10-2, 19-2
thermometer, -10°C to 100°C	60	1-1, 4-1, 6-2, 10-2, 11-1, 11-2, 12-1, 12-3, 13-1, 13-2, 19-2, 20-1, 20-2, 20-3
tongs, beaker	30	10-2
tongs, test-tube	30	11-2, 19-1
toothpicks	1 large box	1-2, 6-2, 7-2, 8-1, 10-1, 14-1, 18-2, 19-2, 20-1
tubing, glass	37.5 meters	10-2
tubing, rubber	16.5 meters	10-2, 18-1
watch glass	30	9-1
wire gauze	30	1-1, 2-2, 6-2, 10-2, 11-1, 13-1, 9-1, 19-2
wire stripper	30	4-2
wood splints	1 pkg	18-1

Table T3

Chemical and Equipment Suppliers

Aldrich Chemical Co., Inc.
1001 W. St. Paul Avenue
Milwaukee, WI 53233
(414) 273-3850

Arbor Scientiflc
P.O. Box 2750
Ann Arbor, MI 48106-2750
(313) 913-6200

Carolina Biological Supply Co.
2700 York Road
Burlington, NC 27215
(910) 584-0381

Central Scientific Co.
3300 Cenco Parkway
Franklin Park, IL 60131
(847) 451-0150

Edmund Scientific Co.
101 E. Gloucester Pike
Barrington, NJ 08007
(609) 547-3488

Fisher Scientific Co.
1600 W. Glenlake
Itasca, IL 60143
(630) 773-3050

Flinn Scientific Inc.
P.O. Box 219
Batavia, IL 60510
(630) 879-6900

Frey Scientific
900 Paragon Parkway
Mansfield, OH 44903
(419) 589-1900

Kemtec Educational Corp.
9889 Crescent Park Drive
West Chester, OH 45069
(513) 777-3535

LaPine Scientific Co.
13636 Western Avenue
P.O. Box 780
Blue Island, IL 60406
(708) 388-4030

McKilligan Supply Corp.
435 Main Street
Johnson City, NY 13790
(607) 798-9335

Nasco
901 Janesville Avenue
Fort Atkinson, WI 53538
(414) 563-2446

Pasco Scientific
10101 Foothills Blvd.
Roseville, CA 95678
(916) 786-3800

Sargent-Welch Scientific Co.
911 Commerce Court
Buffalo Grove, IL 60089
(847) 459-6625

Science Kit and Boreal Labs
777 E. Park Drive
Tonawanda, NY 14150
(716) 874-6020

Ward's Natural Science Establishment, Inc.
P.O. Box 92912
Rochester, NY 14692
(716) 359-2502

Table T4

Chemical List

Chemicals	Quantity	Labs Used In
acetone	500 mL	1-1, 4-2
albumin	120 g	19-1, 19-2
aluminum nitrate nonahydrate	500 g	4-2, 6-1, 8-2, 17-1
aluminum sulfate	500 g	15-1
aluminum sheet (22-30 gauge)	1 lb	9-1
aluminum wire	1-lb spool	4-1, 8-2, 17-1
ammonia, aqueous (ammonium hydroxide)	500 mL	7-2
ammonium chloride	500 g	4-2, 15-1, 20-1
ammonium vanadate	500 g	7-2, 16-1
aspirin	50	18-2

Chemical List		
Chemicals	**Quantity**	**Labs Used In**
barium nitrate	125 g	8-1
Benedict's solution	4 L	19-1, 19-2
benzoic acid	500 g	9-1
biuret reagent	1 L	19-1, 19-2
bromine water	100 mL	3-2
bromothymol blue	1 g	14-2
butane lighter	30	12-1
calcium carbide	500 g	18-1
calcium carbonate	500 g	12-2
calcium chloride	500 g	13-1
calcium nitrate tetrahydrate	25 g	7-2, 8-1
calcium oxide	500 g	14-1
chromium(III) nitrate	100 g	7-2
cobalt(II) nitrate hexahydrate	100 g	6-2, 7-2
copper sheet	500 g	9-1
copper wire	500 g	8-2, 16-2, 17-1
copper(II) carbonate, anhydrous	500 g	7-1
copper(II) nitrate	100 g	1-2, 6-1, 7-2, 8-2, 17-1
copper(II) sulfate, anhydrous	125 g	20-3
copper(II) sulfate pentahydrate	500 g	2-2, 20-3
dextrin	500 g	19-1
1, 4-dichlorobenzene (*p*-dichlorobenzene)	1 lb	4-1
dimes	30	17-2
ethyl alcohol	4 L	1-1, 4-2, 10-1, 14-2, 18-2
fructose	100 g	19-1
gelatin	500 g	19-1, 19-2
glucose	500 g	19-2
glycerol	500 mL	4-2, 10-2, 18-1
hydrochloric acid, concentrated	1 L	2-2, 3-2, 4-2, 6-1, 6-2, 7-2, 12-2, 12-3, 14-1, 14-2, 18-1, 19-1, 20-1, 20-2
hydrogen peroxide, 3%	500 g	6-1
iodine	30 mL	19-1, 19-2
iodine/potassium iodide	120 mL	3-2
iron nails, 3cm	360	16-2
iron wire	25 g	8-2, 17-1
iron(III) nitrate nonahydrate	500 g	1-2, 6-1, 7-2, 8-2, 17-1, 18-2
isopropyl alcohol	500 mL	1-1
lead foil	500 g	8-2, 17-1
lead(II) nitrate	125 g	1-2, 6-1, 8-2, 15-1, 17-1
lithium carbonate, anhydrous	500 g	7-1
magnesium bromide	100 g	4-2
magnesium nitrate hexahydrate	500 g	8-1, 8-2, 17-1
magnesium ribbon	30 g	6-1, 8-2, 12-3, 16-2, 17-1, 20-2
maltose	100 g	19-1
manganese(IV) dioxide	500 g	6-1
manganese(II) nitrate	500 mL	7-2
manganese(II) sulfate	500 g	15-1
marble chips	500 g	14-1
methyl alcohol	500 mL	4-2
napthalene	500 g	4-1, 9-1, 10-1
nickel wire	2 m	8-2, 17-1
nickel(II) nitrate hexahydrate	125 g	6-1, 7-2, 8-2, 17-1
nickels	120	17-2
nitric acid, concentrated	500 mL	1-2, 6-1, 17-1

Copyright © Glencoe/McGraw-Hill, a division of The McGraw-Hill Companies, Inc.

Table T4 (Continued)

Chemical List		
Chemicals	**Quantity**	**Labs Used In**
oxalic acid dihydrate	500 g	15-2
pennies	120	17-2
phenolphthalein	25 g	14-2, 15-2
potassium bromide	100 g	15-1
potassium chloride	500 g	9-1, 13-1
potassium hydrogen carbonate	500 g	12-2
potassium hydrogen tartrate (cream of tartar)	500 g	15-2
potassium iodide	125 g	9-1
potassium nitrate	500 g	7-2, 17-1
potassium permanganate	125 g	18-1
potassium thiocyanate	125 g	7-2
propyl alcohol	500 mL	1-1
quarters	30	17-2
salicylic acid	125 g	10-1, 18-2
silver foil	10 g	8-2, 17-1
silver nitrate	125 g	1-2, 6-1, 6-2, 8-2, 17-1
sodium acetate trihydrate	500 g	15-1
sodium bromide	500 g	3-2
sodium carbonate	500 g	1-2, 2-2, 6-1, 8-1, 15-1
sodium carbonate, anhydrous	500 g	7-1
sodium chloride	500 g	3-2, 4-2, 6-1, 10-1, 13-1, 13-2, 15-1, 16-2, 17-2
sodium chromate	100 g	6-1
sodium dichromate, dihydrate	100 g	6-1
sodium dihydrogen phosphate	100 g	15-1
sodium fluoride	125 g	3-2
sodium hydrogen carbonate	500 g	12-2, 15-1, 20-1
sodium hydroxide	500 g	1-2, 6-1, 7-2, 14-2, 15-2, 16-1
sodium hypochlorite (bleach)	500 mL	3-2, 6-1, 16-1, 18-1
sodium iodide	100 g	1-2, 3-2, 6-1
sodium oxalate	100 g	8-1
sodium phosphate	500 g	15-1
sodium sulfate decahydrate	500 g	6-1, 8-1
sodium thiocyanate	500 g	6-1
starch	500 g	19-1, 19-2
stearic acid	500 g	4-1
strontium nitrate	125 g	8-1
sulfuric acid, concentrated	500 mL	4-2, 7-2, 16-1
thymol blue	1 g	14-2
tin foil	25 g	17-1
tin(IV) sulfate dihydrate	5 g	17-1
trichlorotrifluorethane (TTE)	1 L	3-2, 9-1
universal indicator	120 mL	14-1, 14-2, 15-1
vegetable oil	473 mL (16 oz)	19-1, 19-2
water (deionized)	4 L	6-1, 9-1
zinc foil	1 roll	8-2, 17-1
zinc, granular	500 g	6-1, 16-1
zinc nitrate hexahydrate	125 g	7-2, 8-2, 17-1
zinc sulfate	100 g	15-1

Name Date Class

Determining Boiling Point 1-1

Laboratory Quiz

In the space at the left, write the letter of the answer that best completes each statement.

_______ **1.** The temperature at which a liquid boils _______.
a. never varies
b. is always 100°C
c. depends on atmospheric pressure
d. both a and b are correct

_______ **2.** Standard boiling points of liquids are usually determined at _______.
a. 101.3 kPa pressure
b. standard temperature
c. 100°C
d. varying temperatures

_______ **3.** When a substance _______, its vapor pressure equals atmospheric pressure.
a. freezes
b. sublimes
c. melts
d. boils

_______ **4.** _______ is not standard atmospheric pressure.
a. 1 atm
b. 760 mm Hg
c. 101.3 kPa
d. 273 K

_______ **5.** The boiling point of a substance was found in one laboratory to be 60°C. The standard boiling point listed for the substance in a reference book is 61.4°C. The percentage error of the laboratory work is _______.
a. 2.3% **b.** 22.8% **c.** 1.4% **d.** 14.0%

Name Date Class

The Eight-Solution Problem 1-2

Laboratory Quiz

In the space at the left, write the letter of the answer that best completes each statement.

_______ **1.** The eight-solution problem showed that compounds containing _______ are often colored.
a. silver **b.** copper **c.** tin **d.** gold

_______ **2.** In a chemical reaction the _______ always changes.
a. color of a solution
b. temperature of a solution
c. identity of a chemical
d. solubility of a chemical

_______ **3.** _______ is produced in a chemical reaction of lead nitrate with sodium hydroxide.
a. A white solid
b. A yellow solution
c. Gas
d. A green solid

_______ **4.** When a solution of sodium carbonate is mixed with a solution of _______, there is no observable evidence of a chemical reaction.
a. silver nitrate
b. sodium hydroxide
c. lead nitrate
d. copper nitrate

_______ **5.** _______ solution can burn the skin.
a. Sodium chloride
b. Sodium carbonate
c. Sodium nitrate
d. Nitric acid

Name Date Class

The Thickness of Aluminum Foil 2-1

Laboratory Quiz

In the space at the left, write the letter of the answer that best completes each statement.

_______ **1.** The density of aluminum is _______.
a. 2.7 g/mL **c.** 2.7 kg/mL
b. 2.7 g/L **d.** 2.7 mL

_______ **2.** _______ is a collection of only one type of atom.
a. Mass **c.** A compound
b. Volume **d.** An element

_______ **3.** _______ is **not** directly observable data necessary to compute the thickness of foil.
a. Mass **c.** Area
b. Volume **d.** Time

_______ **4.** The diameter of an aluminum atom is approximately _______.
a. 2.8×10^{8} cm **c.** 2.8×10^{8} m
b. 2.8×10^{-8} cm **d.** 2.8×10^{8} km

_______ **5.** Compared to the aluminum atoms in the block, the aluminum atoms in the foil are _________.
a. thicker **c.** more rectangular
b. flatter **d.** identical

Name Date Class

Identifying Elements by Flame Tests 2-2

Laboratory Quiz

In the space at the left, write the letter of the answer that best completes each statement.

_______ **1.** The flame color of an excited atom is produced by the ________ spectrum of the element.
a. ultraviolet **c.** absorption
b. emission **d.** color

_______ **2.** When ________ is excited, the electrons move from low energy positions to high energy positions.
a. a proton **c.** an atom
b. a flame **d.** a wave

_______ **3.** In this experiment, the hydrochloric acid was used to ________.
a. react with the wire **c.** clean the wire
b. test the flame **d.** flameproof the wire

_______ **4.** ________ should give a yellow flame.
a. Copper (II) bromide **c.** Scandium fluoride
b. Selenium chloride **d.** Sodium iodide

_______ **5.** After a chemical reaction, the elements that were present in the substances that reacted are ________.
a. no longer present **c.** changed to photons
b. different elements **d.** present in the products

Name Date Class

An Alien Periodic Table 3-1

Laboratory Quiz

In the space at the left, write the letter of the answer that best completes each statement.

_______ **1.** A horizontal row of the periodic table is called a _______.
a. group **c.** family
b. column **d.** period

_______ **2.** The lightest element is found in the _______ of the periodic table.
a. upper left **c.** lower left
b. upper right **d.** lower right

_______ **3.** Elements with similar properties are found in _______.
a. periods **c.** groups
b. series **d.** rows

_______ **4.** Most elements occupying the upper right corner of the periodic table are _______.
a. solids **c.** liquids
b. crystals **d.** gases

_______ **5.** As the atomic number of metals in a family increases, the melting point generally _______.
a. decreases **b.** increases **c.** remains the same

Name Date Class

Periodicity of Halogen Properties 3-2

Laboratory Quiz

In the space at the left, write the letter of the answer that best completes each statement.

_______ **1.** Elements within a group have _______.
a. identical nuclear charges **c.** similar properties
b. identical neutron counts **d.** similar isotopes

_______ **2.** The least active halogen is _______.
a. fluorine **c.** chlorine
b. bromine **d.** iodine

_______ **3.** The halogen with a reddish-brown color is _______.
a. fluorine **c.** chlorine
b. bromine **d.** iodine

_______ **4.** A common property of the halogens is they are all _______.
a. the same color **c.** liquids
b. poisonous **d.** salts

_______ **5.** As the atomic number of the halogens increases, their relative reactivity _______.
a. decreases **b.** increases **c.** remains the same

Name Date Class

Determining Melting Points 4-1

Laboratory Quiz

In the space at the left, write the letter of the answer that best completes each statement.

______ **1.** Melting is an example of a ________.
a. chemical change **c.** chemical reaction
b. change of state **d.** change of composition

______ **2.** Pure substances usually have ________.
a. distinct melting points **c.** melting points below 100°C
b. variable melting points **d.** melting points above 100°C

______ **3.** The melting point of a substance is the same as its ________.
a. point of decomposition **c.** freezing point
b. boiling point **d.** none of the above

______ **4.** If you recorded a melting point of 81°C for the unknown, the substance was probably ________.
a. stearic acid **c.** 1,4-dichlorobenzene
b. naphthalene **d.** contaminated with water

______ **5.** Errors in the experiment could be the result of ________.
a. water in the sample **c.** an inaccurate thermometer
b. an impure sample **d.** all of the above

Name Date Class

Distinguishing Ionic and Covalent Compounds 4-2

Laboratory Quiz

In the space at the left, write the letter of the answer that best completes each statement.

______ **1.** Ionic compounds, at room temperature, usually are ________.
a. solids **c.** gases
b. liquids **d.** solutions

______ **2.** The smallest unit of a covalent compound is a(n) ________.
a. ion **c.** molecule
b. crystal **d.** salt

______ **3.** A good conductor of electricity is a solution of ________.
a. HOH **c.** CH_3OH
b. NH_3 **d.** NaCl

______ **4.** Covalent compounds usually have ________.
a. low melting points and weak intermolecular forces
b. low melting points and strong intermolecular forces
c. high melting points and weak intermolecular forces
d. high melting points and strong intermolecular forces

______ **5.** Solutions that contain ________ conduct electricity.
a. water **c.** compounds
b. wires **d.** ions

Name Date Class

Making Models of Compounds 5-1

Laboratory Quiz

In the space at the left, write the letter of the answer that best completes each statement.

_______ **1.** The formula for aluminum ion is Al^{3+} and for sulfate ion is SO_4^{2-} The correct formula for aluminum sulfate is _______.
a. $AlSO_4$ **c.** $Al_3(SO_4)_2$
b. Al_2SO_4 **d.** $Al_2(SO_4)_3$

_______ **2.** The formulas of the two compounds formed by Fe^{2+} and Fe^{3+} with the sulfate ion are _______.
a. $FeSO_4$ and $Fe_2(SO_4)_3$ **c.** $FeSO_4$ and Fe_3SO_4
b. $FeSO_4$ and $FeSO_3$ **d.** Fe_2SO_4 and $Fe_2(SO_4)_3$

_______ **3.** The name of the compound formed by NH_4^+ and PO_4^{3-} is _______.
a. ammonia permanganate **c.** ammonium phosphate
b. ammonia phosphate **d.** ammonium phosphite

_______ **4.** The name and formula of the compound formed by the silver ion and the dichromate ion, $Cr_2O_7^{2-}$, are _______.
a. silver dichromate, $Ag(Cr_2O_7)_2$ **c.** silver dichromate, $Ag_2Cr_2O_7$
b. silver chromite, $AgCrO_4$ **d.** silver chromate, $AgCr_2O_7$

_______ **5.** The compound containing the copper (II) and acetate ions has the formula _______.
a. $CuC_2H_3O_2$ **c.** $C_2H_3O_2Cu$
b. $Cu_2C_2H_4O_2$ **d.** $Cu(C_2H_3O_2)_2$

Name Date Class

Formulas and Oxidation Numbers 5-2

Laboratory Quiz

In the space at the left, write the letter of the answer that best completes each statement.

_______ **1.** _______ is an example of a binary compound.
a. $NaC_2H_3O_2$ **c.** $NaNO_2$
b. ZnO **d.** $CuSO_3$

_______ **2.** _______ is the formula for nickel(II) chlorite.
a. $Ni(ClO_2)_2$ **c.** NiClO
b. Ni_2ClO **d.** $Ni_2(ClO_2)_2$

_______ **3.** The formula Cu_2S indicates that the formula unit is made of _______.
a. one copper ion and two sulfide ions
b. one copper ion and one sulfide ion
c. two copper ions and one sulfide ion
d. two copper ions and two sulfide ions

_______ **4.** In a neutral compound, the sum of the ion charges is _______.
a. 1+ **c.** 8
b. 1– **d.** 0

_______ **5.** The correct formula for tin(II) sulfite is _______.
a. Sn_2SO_3 **c.** $SnSO_2$
b. $SnHSO_3$ **d.** $SnSO_3$

Name Date Class

Types of Chemical Reactions 6-1

Laboratory Quiz

In the space at the left, write the letter of the answer that best completes each statement.

______ **1.** When lead nitrate and sodium sulfate are combined, a precipitate of ______ is formed.
a. lead sulfide **c.** lead sulfate
b. sodium nitride **d.** sodium nitrate

______ **2.** If nitrates of hydrogen, silver, aluminum, and iron are used in the rows of a microplate and $NaNO_3$ is used in column 8, precipitates would be expected ______.
a. in rows A–D **c.** nowhere
b. in rows A and B **d.** in rows B and D

______ **3.** The pattern of a double displacement reaction can be represented by ______.
a. $AB \rightarrow A + B$ **c.** $AB + C_2 \rightarrow AC + BC$
b. $AB + AB \rightarrow 2AB$ **d.** $AB + CD \rightarrow AD + CB$

______ **4.** One of the reactants in a single displacement reaction is ______.
a. a gas **c.** an acid
b. an element **d.** a solid

______ **5.** In the experiment, HCl and NaOCl were required for the reaction in well A1 so that ______.
a. the reaction was safe **c.** chemicals could be conserved
b. Cl_2 could be generated **d.** Mg would stay dry

Name Date Class

Effects of Concentration on Chemical Equilibrium 6-2

Laboratory Quiz

In the space at the left, write the letter of the answer that best completes each statement.

______ **1.** LeChatelier's principle makes it possible to predict ______.
a. the heat of a reaction **c.** the effect of changes on equilibrium
b. how fast a reaction goes **d.** the result of mixing reactants

______ **2.** The colored reactant and product in the reaction studied in the experiment are both ______.
a. complex ions **c.** acidic
b. cations **d.** nonmetals

______ **3.** The colors of the initial mixtures of $Co(H_2O)_6^{2+}$ and Cl^- show ______.
a. that a reaction occurred **c.** that cobalt is hydrated
b. the Cl^- concentration at equilibrium **d.** that charges were neutralized

______ **4.** A change of ______ will shift an equilibrium system.
a. catalyst **c.** time
b. color **d.** concentration

______ **5.** More SO_2 could be produced in the reaction $2SO_2(g) + O_2(g) \rightleftarrows 2SO_3(g)$ by ______.
a. increasing the pressure **c.** removing O_2
b. decreasing the concentration of SO_3 **d.** decreasing the temperature

Name Date Class

Atomic Spectra 7-1

Laboratory Quiz

In the space at the left, write the letter of the answer that best completes each statement.

_______ **1.** When an atom absorbs energy, _______.
a. electrons move from high energy levels to low energy levels
b. electrons move from low energy levels to high energy levels
c. protons move from high energy levels to low energy levels
d. protons move from low energy levels to high energy levels

_______ **2.** The flame color produced by sodium is _______.
a. red **c.** yellow
b. green **d.** blue

_______ **3.** If two different spectra are observed, it must mean _______.
a. one sample was more concentrated than the other
b. the temperature of the Bunsen flame changed
c. two identical samples are being analyzed
d. two different samples are being analyzed

_______ **4.** The observed spectra of lithium and sodium are different because _______.
a. both are Group 1 elements **c.** sodium is a metal
b. their electron configurations differ **d.** their masses are different

_______ **5.** The spectra of a mixture of two elements will show _______.
a. only one spectral line **c.** only two spectral lines
b. the spectral lines of the more concentrated substance **d.** the spectral lines of both substances

Name Date Class

Transition Metals 7-2

Laboratory Quiz

In the space at the left, write the letter of the answer that best completes each statement.

_______ **1.** Elements in Groups 3 to 12 are known as _______.
a. rare earths **c.** transition elements
b. alkali metals **d.** alkaline earth metals

_______ **2.** A blue solution is characteristic of an ion of _______.
a. calcium **c.** potassium
b. copper **d.** zinc

_______ **3.** In this activity, the ion that causes a precipitate to form is _______.
a. OH^- **c.** NO_3^-
b. Cl^- **d.** SCN^-

_______ **4.** The transition metal with a full *d* sublevel is _______.
a. cobalt **c.** nickel
b. iron **d.** zinc

_______ **5.** The element that did not form a colored precipitate is _______.
a. iron **c.** nickel
b. potassium **d.** chromium

Name Date Class

Periodicity and Chemical Reactivity 8-1

Laboratory Quiz

In the space at the left, write the letter of the answer that best completes each statement.

_______ **1.** The elements of a vertical column of the periodic table are known as _______.
- **a.** a group or family
- **b.** a period
- **c.** metals
- **d.** nonmetals

_______ 2. The periodic table can be used to predict _______.
- **a.** physical properties
- **b.** chemical properties
- **c.** both a and b
- **d.** neither a nor b

_______ **3.** The elements of Group 2 include all the following except _______.
- **a.** calcium
- **b.** beryllium
- **c.** sodium
- **d.** barium

_______ **4.** The precipitation you observed in the experiment was an indication of the _______.
- **a.** speed of the reactions
- **b.** solubility of the salts formed in the reactions
- **c.** temperature of the reactions
- **d.** concentration of the solutions

_______ **5.** As the atomic mass of the metal ions tested in the experiment increased, the solubility of their salts _______.
- **a.** increased
- **b.** stayed the same
- **c.** changed without a pattern
- **d.** decreased

Name Date Class

Comparing Activities of Selected Metals 8-2

Laboratory Quiz

In the space at the left, write the letter of the answer that best completes each statement.

_______ **1.** When metals combine with nonmetals, _______ are usually formed.
- **a.** radicals
- **b.** nitrates
- **c.** covalent bonds
- **d.** ionic bonds

_______ **2.** The activity of a metal depends on its _______.
- **a.** temperature
- **b.** mass number
- **c.** tendency to lose electrons
- **d.** tendency to gain electrons

_______ **3.** In this experiment, _______ was the most active metal.
- **a.** manganese
- **b.** silver
- **c.** aluminum
- **d.** magnesium

_______ **4.** Based on this experiment, you could predict that _______.
- **a.** silver would probably not react with water
- **b.** magnesium and water probably would not react
- **c.** silver is the least active of all metals
- **d.** gold is more active than silver

_______ **5.** Metals are elements that tend to _______.
- **a.** lose electrons and form positive ions
- **b.** lose electrons and form negative ions
- **c.** gain electrons and form positive ions
- **d.** gain electrons and form negative ions

Name Date Class

Diagnostic Properties of Bonds 9-1

Laboratory Quiz

In the space at the left, write the letter of the answer that best completes each statement.

_______ **1.** Two solids that do NOT dissolve in water or in paint thinner are _______.
a. KCl and Al **c.** KI and Al
b. Al and Cu **d.** Cu and KI

_______ **2.** Low melting point is a characteristic of _______.
a. metals **c.** ionic compounds
b. salts **d.** covalent compounds

_______ **3.** Salts generally have a _______.
a. high melting point and are soluble in paint thinner
b. low melting point and are soluble in paint thinner
c. high melting point and are insoluble in paint thinner
d. low melting point and are insoluble in paint thinner

_______ **4.** A solid that does not melt is _______.
a. benzoic acid **b.** napthalène **c.** potassium iodide **d.** copper

_______ **5.** Which material has the odor of moth flakes?
a. napthalene **b.** copper **c.** aluminum **d.** benzoic acid

---

Name Date Class

Electron Clouds 9-2

Laboratory Quiz

In the space at the left, write the letter of the answer that best completes each statement.

_______ **1.** The space around the nucleus is occupied by _______.
a. protons **c.** neutrons
b. electrons **d.** alpha particles

_______ **2.** The probable location of an electron is called a _______.
a. pear **c.** cloud
b. bond **d.** sphere

_______ **3.** The bond angle defined by a linear molecule is _______.
a. 90° **c.** 270°
b. 180° **d.** 360°

_______ **4.** A tetrahedron will be the typical shape of _______.
a. CH_3Cl **c.** CO_2
b. HCl **d.** H_2

_______ **5.** Two molecules with the same shape are _______.
a. CCl_4 and CO_2 **c.** HCl and BF_3
b. BF_3 and $CHCl_3$ **d.** HCl and Cl_2

Name ______________________ Date ______________ Class ______________

Crystal Shapes 10-1

Laboratory Quiz

In the space at the left, write the letter of the answer that best completes each statement.

_______ **1.** The crystal structure at the right is _______.
a. tetragonal **c.** orthorhombic
b. triclinic **d.** rhombohedral

_______ **2.** In the laboratory, you observed that growth of a crystal in a drop of saturated solution began _______.
a. at the edge of the drop
b. at the center of the drop
c. in scattered parts of the drop
d. either a or b depending on the crystal structure

_______ **3.** A crystal system in which the unit cell axes are all equal in length, and none of the angles between the unit cell axes are equal to 90° is a(n) _______ crystal system.
a. cubic **b.** orthorhombic **c.** rhombohedral **d.** triclinic

_______ **4.** Compared with crystals formed quickly from a saturated solution, crystals formed more slowly are typically _______.
a. smaller **c.** either smaller or larger
b. larger **d.** the same size

_______ **5.** The three corner angles of the crystal shown at the right are _______.
a. 90° 90° 90° **b.** 90° 90° >90° **c.** 90° <90° >90° **d.** 90° 60° 60°

Name ______________________ Date ______________ Class ______________

Relating Gas Temperature and Pressure 10-2

Laboratory Quiz

In the space at the left, write the letter of the answer that best completes each statement.

_______ **1.** In the experiment, as the gas molecules in the test tube were heated, _______.
a. their molecular motion decreased **c.** they transmitted higher pressure
b. their kinetic energy decreased **d.** their pressure decreased

_______ **2.** The shape of the graph line could have changed markedly if _______.
a. the room temperature had been different **c.** the test tube had been larger
b. a leak had occurred in the tubing **d.** the flask had been smaller

_______ **3.** In the air in the tires of a car, as _______.
a. temperature decreases, pressure increases
b. temperature increases, pressure increases
c. pressure increases, temperature decreases
d. pressure decreases, temperature increases

_______ **4.** As the temperature of a gas increases, the pressure will increase only _______.
a. on a warm day **c.** in an open container
b. in a small space **d.** in a confined area

_______ **5.** In the experiment, as the test tube cooled, the pressure exerted by the gas molecules changed in the _______.
a. test tube only **c.** flask only
b. tubing only **d.** test tube, tubing, and flask

Name Date Class

Determining Absolute Zero 11-1

Laboratory Quiz

In the space at the left, write the letter of the answer that best completes each statement.

_______ **1.** Charles's law describes the relationship between _______ for an ideal gas.
- **a.** temperature and pressure
- **b.** temperature and volume
- **c.** pressure and volume
- **d.** moles and volume

_______ **2.** If the temperature of a gas increases from 0° to 1°C, the volume will _______.
- **a.** increase by 1/273 of the original
- **b.** decrease by 1/273 of the original
- **c.** be half the original
- **d.** be twice the original

_______ **3.** At _______ the volume of a sample of ideal gas should be zero.
- **a.** 373°C
- **b.** 273 K
- **c.** 0°C
- **d.** 0 K

_______ **4.** The value of absolute zero is _______.
- **a.** 0°C
- **b.** –273°C
- **c.** –100 K
- **d.** –100°C

_______ **5.** In the experiment, the change in the volume of the sample of air _______.
- **a.** was equal to the volume of water drawn into the pipet
- **b.** was equal to 100 drops of water
- **c.** was equal to 100 drops of water minus the added drops
- **d.** could not be measured

Name Date Class

Charles's Law 11-2

Laboratory Quiz

In the space at the left, write the letter of the answer that best completes each statement.

_______ **1.** As the temperature of a gas decreases, the volume of the gas will _______.
- **a.** decrease
- **b.** increase
- **c.** remain the same
- **d.** decrease, then increase

_______ **2.** _______ would result in a gas turning to a liquid.
- **a.** Low temperature and high pressure
- **b.** Low temperature and low pressure
- **c.** High temperature and high pressure
- **d.** High temperature and low pressure

_______ **3.** To double the volume of a gas, it would be necessary to double the _______.
- **a.** pressure on the gas
- **b.** Fahrenheit temperature of the gas
- **c.** Celsius temperature of the gas
- **d.** Kelvin temperature of the gas

_______ **4.** At _______, gases lose electrons and eventually become plasma.
- **a.** room temperature
- **b.** very high temperatures
- **c.** very low temperatures
- **d.** boiling temperatures

_______ **5.** Which gas is most likely to behave as an ideal gas?
- **a.** nitrogen
- **b.** helium
- **c.** carbon dioxide
- **d.** oxygen

Name Date Class

Determining the Avogadro Constant 12-1

Laboratory Quiz

In the space at the left, write the letter of the answer that best completes each statement.

_______ **1.** The volume occupied by one mole of a gas varies with _______.
a. the identity of the gas **c.** temperature only
b. pressure only **d.** both temperature and pressure

_______ **2.** To calculate the volume of butane at standard conditions, it was necessary to use _______.
a. the Avogadro constant **c.** the combined gas law
b. the molar volume **d.** Dalton's law

_______ **3.** The number of molecules in 0.25 mol of butane is _______.
a. 1.5×10^{23} **c.** 3×10^{12}
b. 6.02×10^{23} **d.** 1.5×10^{6}

_______ **4.** The _______ is the same for all gases at a given temperature and pressure.
a. mean free path **c.** average kinetic energy
b. diameter of an atom **d.** chemical formula

_______ **5.** An industrial process generated _______ mol of methane gas, which had a volume of 448 L at standard temperature and pressure.
a. 40 **c.** 20
b. 10 **d.** 11.2

Name Date Class

Stoichiometry of a Chemical Reaction 12-2

Laboratory Quiz

In the space at the left, write the letter of the answer that best completes each statement.

_______ **1.** The coefficients in a balanced equation show that the masses of elements are conserved and tell the number of _______ of each element or compound.
a. grams **b.** molecules **c.** moles **d.** atoms

_______ **2.** The balanced equation representing a reaction is $LiAlH_4 + 4H_2O$ - $LiOH + Al(OH)_3 + 4H_2$. If 2 moles of $LiAlH_4$ and 8 moles of H_2O are reacted, _______ of $Al(OH)_3$ will be produced.
a. 2 grams **b.** 2 moles **c.** 0.05 moles **d.** 0.5 grams

_______ **3.** If 2 moles of $KHCO_3$ reacted with 2 moles of HCl, you would expect that _______ moles of CO_2 would be produced.
a. 2 **b.** 1 **c.** 44 **d.** 88

Use the information below to answer questions 4 and 5.

Element:	Calcium	Carbon	Hydrogen	Oxygen	Potassium
Atomic Mass:	40	12	1	16	39

_______ **4.** In a sample of $CaCO_3$, the percentage of carbon would be _______.
a. 0.12% **b.** 12% **c.** 0.20% **d.** 20%

_______ **5.** If 100 grams of $KHCO_3$ reacted with an excess of HCl, you would expect that _______ of CO_2 would be produced.
a. 100 moles **b.** 100 grams **c.** 44 grams **d.** 44 moles

Name Date Class

Molar Volume of a Gas 12-3

Laboratory Quiz

In the space at the left, write the letter of the answer that best completes each statement.

______ **1.** Avogadro's principle explains how the ______ of a gas are related.
a. volume and mass **c.** volume, mass, and number of particles
b. volume and pressure **d.** volume and temperature

______ **2.** The molar volume of any gas at STP is ______.
a. 22.4 liters **c.** 101 liters
b. 22.4 kPa **d.** 760 mm

______ **3.** ______ liters of a gas at STP equals a molar volume of the gas.
a. 22.4 **b.** 1.0 **c.** 2.0 **d.** 0.02

______ **4.** The ______ determines the number of moles of hydrogen produced in the reaction below.
$Mg (s) + 2HCl (aq) \rightarrow MgCl_2 (aq) + H_2 (g)$
a. volume of the water **c.** number of moles of Mg
b. temperature of the water **d.** temperature of the Mg

______ **5.** In the ideal gas law, *n* represents the ______.
a. number of moles of a gas **c.** pressure of a gas
b. number of moles of Mg **d.** volume of a gas

Name Date Class

Relating Solubility and Temperature 13-1

Laboratory Quiz

In the space at the left, write the letter of the answer that best completes each statement.

______ **1.** In general, increasing the temperature of a solvent allows ______.
a. more gas to dissolve **c.** less solid to dissolve
b. more solid to dissolve **d.** both a and b

______ **2.** The solvent in this experiment was ______.
a. KCl **c.** $CaCl_2$
b. $NaCl$ **d.** H_2O

______ **3.** In this experiment, the increase in temperature had the smallest effect on the solubility of ______.
a. KCl **c.** $NaCl$
b. $CaCl_2$ **d.** H_2O

______ **4.** If 1.92 grams of salt are recovered when 5 mL of solution are heated, the mass of salt in 100 mL of solution would be ______.
a. 100 grams **c.** 9.6 grams
b. 38.4 grams **d.** 1.92 grams

______ **5.** When the solutions in the foil dishes were heated, the water evaporated and the solute ______.
a. remained behind **c.** increased in mass
b. dissolved **d.** decreased in mass

Name Date Class

The Effect of a Solute on Freezing Point 13-2

Laboratory Quiz

In the space at the left, write the letter of the answer that best completes each statement.

_______ 1. A colligative property depends upon _______.
- a. size of the solute particles
- b. concentration of the solute
- c. nature of the solute
- d. surface area of the solute

_______ 2. _______ is not a colligative property.
- a. Freezing point lowering
- b. Boiling point elevation
- c. Partial pressure
- d. Osmotic pressure

_______ 3. In a 0.30 molal solution of calcium sulfate there are _______ moles of ions per kilogram of solvent.
- a. 0.15
- b. 0.30
- c. 0.45
- d. 0.60

_______ 4. In a ketchup cup calorimeter, _______ insulates the contents.
- a. the rubber band
- b. the two layers of plastic
- c. the air between the cups
- d. the lid

_______ 5. The addition of a(n) _______ substance to a liquid lowers the freezing point of the liquid.
- a. soluble volatile
- b. insoluble volatile
- c. soluble nonvolatile
- d. insoluble nonvolatile

Name Date Class

Acidic and Basic Anhydrides 14-1

Laboratory Quiz

In the space at the left, write the letter of the answer that best completes each statement.

_______ 1. When dissolved in water, the oxide of a metal forms _______.
- a. a basic solution with a pH less than 7
- b. a basic solution with a pH greater than 7
- c. an acidic solution with a pH less than 7
- d. an acidic solution with a pH greater than 7

_______ 2. Which compound will dissolve in water to form a basic solution?
- a. marble chips
- b. calcium oxide
- c. an indicator
- d. HCl

_______ 3. An acid and a base react to form a(n) _______.
- a. salt and water
- b. basic anhydride
- c. acid anhydride
- d. indicator

_______ 4. Which equation shows the formation of an acid?
- a. $CaO + H_2O \rightarrow Ca(OH)_2$
- b. $CO_2 + H_2O \rightarrow H_2CO_3$
- c. $CaCO_3 + 2HCl \rightarrow CO_2 + H_2O + CaCl_2$
- d. $Ca(OH)_2 + H_2CO_3 \rightarrow CaCO_3 + 2H_2O$

_______ 5. When carbon dioxide is bubbled into water, _______ is formed.
- a. calcium oxide
- b. calcium carbonate
- c. hydrochloric acid
- d. carbonic acid

Name Date Class

Using Indicators to Determine pH 14-2

Laboratory Quiz

In the space at the left, write the letter of the answer that best completes each statement.

______ **1.** A 0.1M HCl solution has a pH of approximately ______.
a. 4 **c.** 1
b. 5 **d.** 2

______ **2.** In a 0.1M HCl solution, phenolphthalein will be ______.
a. slightly pink **c.** colorless
b. red **d.** pink

______ **3.** Neither acid nor base was transferred to well 7 because ______.
a. distilled water has a pH of 7 **c.** the acid was too dilute
b. distilled water has a basic pH **d.** the base was too dilute

______ **4.** Of the indicators used in this experiment, the best choice for testing a solution that must not be highly acidic would be ______.
a. phenolphthalein **c.** thymol blue
b. bromothymol blue **d.** universal indicator or thymol blue

______ **5.** A perfect universal indicator will ______.
a. be a mixture of three indicators **c.** be colorless
b. have a different color at each pH **d.** not be poisonous

Name Date Class

Hydrolysis of Salts 15-1

Laboratory Quiz

In the space at the left, write the letter of the answer that best completes each statement.

______ **1.** If a salt solution gives a basic reaction, its pH could be ______.
a. 4.0 **b.** 6.8 **c.** 7.0 **d.** 9.2

______ **2.** In any hydrolysis reaction, water serves as a source of ______.
a. H_3O^+ ions only **c.** OH^- ions only
b. H^+ ions only **d.** H_3O^+ and OH^- ions

______ **3.** To ensure accurate results in determining the pH of each reaction, it is essential that you ______.
a. use only 10 mL of water **c.** write a balanced equation
b. use distilled water **d.** use very little salt

______ **4.** $(NH_4)_2SO_4$ dissolves in water to form H^+ and SO_4^{2-}, and NH_4OH. This reaction yields a pH ______.
a. < 7 **b.** ≈ 7 **c.** > 7 **d.** > 8.5

______ **5.** Each of the following equations represents a hydrolysis reaction that would occur *except* ______.
a. $KClO_3 + H_2O \rightarrow HClO_3 + KOH$
b. $CuSO_4 + 2H_2O \rightarrow Cu(OH)_2 + H_2SO_4$
c. $NaCl + H_2O \rightarrow NaOH + HCl$
d. $AlCl_3 + 3H_2O \rightarrow 3HCl + Al(OH)_3$

Name Date Class

Acid/Base Titration 15-2

Laboratory Quiz

In the space at the left, write the letter of the answer that best completes each statement.

_______ **1.** You know you have reached the endpoint of a titration when the _______.
a. acid becomes a base
b. base becomes an acid
c. buret becomes empty
d. indicator changes color

_______ **2.** To standardize KOH solution, you would probably use _______.
a. a standard solution of HCl
b. a standard solution of NaOH
c. HCl of unknown molarity
d. NaOH of unknown molarity

_______ **3.** If you want to measure 0.006 L in a graduated cylinder marked in mL, you will need _______.
a. 0.6 mL **b.** 6 mL **c.** 60 mL **d.** 600 mL

_______ **4.** When you prepare a standard solution of an acid, the *last* step is to determine the _______.
a. moles of acid reacted
b. moles of base reacted
c. molarity of the solution
d. formula of the solution

_______ **5.** In preparing a standard solution of NaOH, you _______.
a. must use a monoprotic acid
b. may use a monoprotic acid
c. cannot use a diprotic acid
d. must use a diprotic acid

Name Date Class

Oxidation/Reduction of Vanadium 16-1

Laboratory Quiz

In the space at the left, write the letter of the answer that best completes each statement.

_______ **1.** Multiple oxidation numbers are characteristic of _______.
a. alkali metals
b. noble gases
c. transition elements
d. alkaline earth metals

_______ **2.** Changes in the oxidation numbers of vanadium can be detected by observing _______.
a. atomic numbers
b. bonding patterns
c. color changes
d. molar mass

_______ **3.** The iron ion that has nutrient value is _______.
a. Fe^{3+} **b.** Fe^{2+} **c.** V^{3+} **d.** V^{2+}

_______ **4.** The control in this activity is the vanadium with an oxidation number of _______.
a. 5+ **b.** 2+ **c.** 3+ **d.** 4+

_______ **5.** If V^{2+} is allowed to stand for a period of time, it reverts to V^{5+}. This happens because the V^{5+} _______.
a. combines with V^{3+}
b. loses V^{3+}
c. is oxidized by the oxygen in air
d. is reduced by oxygen in air

Name Date Class

Corrosion as an Electrochemical Process 16-2

Laboratory Quiz

In the space at the left, write the letter of the answer that best completes each statement.

______ **1.** An oxidizing agent is a substance that ______

a. gains protons **c.** gains electrons
b. loses protons **d.** loses electrons

______ **2.** A piece of metal Y was placed in a microplate well along with a solution of a salt of metal X. One half-reaction that took place was $Y \rightarrow Y^{3+} + 3e^-$. From this, we know that ______

a. X is oxidized **c.** X gained three electrons
b. Y is oxidized **d.** both b and c

______ **3.** ______ is the reducing agent in the reaction

$$Mg + ZnCl_2 \rightarrow MgCl_2 + Zn$$

a. Mg **b.** $ZnCl_2$ **c.** $MgCl_2$ **d.** Zn

______ **4.** The equation ______ represents reduction.

a. $Na^+ + e^- \rightarrow Na^0$ **c.** $Cl^- + e^- \rightarrow Cl^0$
b. $Na^+ \rightarrow Na^0 + e^-$ **d.** $Cl^- \rightarrow Cl^0 + e^-$

______ **5.** Rusting of iron occurs when iron is ______.

a. reduced **c.** reduced and then oxidized
b. oxidized **d.** oxidized and then reduced

Name Date Class

Comparing the Abilities of Metals to Give Up Electrons 17-1

Laboratory Quiz

In the space at the left, write the letter of the answer that best completes each statement.

______ **1.** One metal is described as more active than another when it ______.

a. is a better conductor
b. gives up electrons more readily in a reaction
c. is a poorer conductor
d. is more positive

______ **2.** ______ is the most active of all metals.

a. Sodium **b.** Mercury **c.** Potassium **d.** Lithium

______ **3.** The least active metal used in this activity is ______.

a. magnesium **b.** copper **c.** silver **d.** nickel

______ **4.** In this experiment, current could be carried in the cells because ______.

a. the filter paper was soaked in potassium nitrate
b. electrodes dipped into solutions of their ions
c. metal activities differ
d. electrodes were connected to a galvanometer

______ **5.** In the reaction $Al + Fe^{2+} \rightarrow Fe + Al^{3+}$, the Fe^{2+} ion is ______.

a. losing electrons **c.** reduced
b. a reducing agent **d.** oxidized

Name Date Class

The Six-Cent Battery 17-2

Laboratory Quiz

In the space at the left, write the letter of the answer that best completes each statement.

______ **1.** A ________ consists of several cells connected together.
a. voltage **c.** metal
b. battery **d.** potential

______ **2.** Any two metals, when separated in a conducting solution produce a flow of ________ .
a. electrons **c.** conductors
b. paper **d.** electrolytes

______ **3.** An electrochemical ________ consists of two different metals separated by an electrolyte.
a. coin **c.** cell
b. solution **d.** electron

______ **4.** Connecting three identical cells in a series, each producing 0.2 volts, would result in a net voltage of ________ volts.
a. 0.2 **c.** 2.8
b. 0.6 **d.** 3.0

______ **5.** ________ is the electrolyte used in this experiment.
a. A coin **c.** Paper
b. A rubber band **d.** Sodium chloride

Name Date Class

Examples of Organic Reactions 18-1

Laboratory Quiz

In the space at the left, write the letter of the answer that best completes each statement.

______ **1.** An unsaturated hydrocarbon always has ________
a. only single bonds **c.** one or more multiple bonds
b. one double bond **d.** at least one ionic bond

______ **2.** When $KMnO_4$ is added to a saturated hydrocarbon, the violet color ________.
a. deepens **c.** changes to blue
b. disappears **d.** remains unchanged

______ **3.** A gas that burns with a smoky flame is ________
a. methane **b.** propane **c.** ethyne **d.** potassium permanganate

______ **4.** If household bleach, NaClO, and hydrochloric acid, HCl, are added to a test tube of ethyne gas, the ________.
a. chlorine produced becomes colorless
b. chlorine produced remains yellow-green
c. precipitate produced is colorless
d. precipitate produced is yellow-green

______ **5.** Which of the following is a hydrocarbon?
a. bleach **b.** propane **c.** chlorine **d.** carbon dioxide

Name Date Class

Analysis of Aspirin 18-2

Laboratory Quiz

In the space at the left, write the letter of the answer that best completes each statement.

______ **1.** Aspirin is formed by the reaction of ______.
- **a.** two molecules of acetic acid
- **b.** two molecules of salicylic acid
- **c.** acetic acid and salicylic acid
- **d.** salicylic acid and water

______ **2.** The part of the molecule that gives aspirin its medicinal value is its ______.
a. acetyl part **b.** salicylate part **c.** iron complex **d.** ester bonding

______ **3.** When iron (III) nitrate is added to solutions of salicylic acid, ______.
- **a.** aspirin precipitates
- **b.** gas is given off
- **c.** salicylic acid goes into solution
- **d.** free salicylic acid forms a complex with the iron

______ **4.** The darker the color that the solution becomes when iron (III) nitrate is added, the ______.
- **a.** higher the concentration of salicylic acid is
- **b.** lower the concentration of salicylic acid is
- **c.** more effective the aspirin is
- **d.** purer the aspirin is

______ **5.** Aspirin tablets that have been kept for a long time probably ______.
- **a.** are more effective
- **b.** are less effective
- **c.** are equally effective
- **d.** have become dangerous to use

Name Date Class

Biochemical Reactions 19-1

Laboratory Quiz

In the space at the left, write the letter of the answer that best completes each statement.

______ **1.** Proteins can be identified by using ______.
- **a.** Benedict's solution
- **b.** a biuret test
- **c.** unglazed brown paper
- **d.** an iodine solution

______ **2.** If a compound contains carbon, hydrogen, oxygen, nitrogen, and sulfur, it is a ______.
a. lipid **b.** starch **c.** protein **d.** sugar

______ **3.** A polysaccharide consists of a long chain of ______.
a. amino acids **b.** oils **c.** monosaccharides **d.** proteins

______ **4.** In the laboratory activity, you used Benedict's solution to detect the presence of ______.
a. protein **b.** lipid **c.** starch **d.** sugar

______ **5.** To a starch solution in a test tube in a warm water bath, you add the enzyme amylase, which digests starch. After about twenty minutes, the solution gives a positive test for ______.
- **a.** protein and starch
- **b.** sugar and starch
- **c.** lipid and starch
- **d.** starch only

Name Date Class

Qualitative Analysis of Food 19-2

Laboratory Quiz

In the space at the left, write the letter of the answer that best completes each statement.

_______ **1.** The basic nutrients in all foods are sugars, starches, proteins, _______.
a. and lipids
b. and vitamins
c. vitamins, minerals, water, and lipids
d. vitamins and minerals

_______ **2.** The use of an iodine solution is a specific test for _______.
a. sugars **b.** starches **c.** proteins **d.** water

_______ **3.** A positive test for lipid is a _______.
a. color change in Benedict's solution
b. color change in biuret solution
c. color change in iodine solution
d. translucent spot on brown paper

_______ **4.** The color change in Benedict's solution indicates the presence of _______.
a. sugar **b.** starch **c.** lipid **d.** protein

_______ **5.** A sample of ice cream would probably test positive for _______.
a. protein
b. lipid
c. sugar
d. all of the above

Name Date Class

Energy Changes in Physical and Chemical Processes 20-1

Laboratory Quiz

In the space at the left, write the letter of the answer that best completes each statement.

_______ **1.** As a system becomes more orderly, its entropy _______.
a. decreases
b. increases
c. decreases then increases
d. remains the same

_______ **2.** Chemical reactions have a tendency to move toward lower energy and _______.
a. greater entropy
b. increased temperature
c. decreased order
d. negative heat

_______ **3.** A sample of material has the least entropy if it is in _______.
a. aqueous solution
b. liquid state
c. solid state
d. gaseous state

_______ **4.** A reaction is always spontaneous if the energy change is _______.
a. endothermic and entropy decreases
b. endothermic and entropy increases
c. exothermic and entropy decreases
d. exothermic and entropy increases

_______ **5.** The heat content of a system _______ in an endothermic reaction.
a. decreases
b. increases
c. increases then decreases
d. remains the same

Name Date Class

Measuring the Heat of Reaction 20-2

Laboratory Quiz

In the space at the left, write the letter of the answer that best completes each statement.

__________ **1.** ______ is the amount of heat absorbed or given off when one mole of compound is formed from its elements.
a. Entropy **c.** Calorimetry
b. Heat of formation **d.** Exothermic

__________ **2.** If this experiment were done with an endothermic reaction, then the temperature change would ______
a. be hard to measure **c.** be small
b. be very large **d.** have a negative value

__________ **3.** In this experiment, the ______ of the reaction mixture could be taken as being equal to that of water.
a. heat **b.** boiling point **c.** specific heat **d.** caloric value

__________ **4.** The ______ provided the information needed to calculate an amount of heat from a temperature change.
a. specific heat **b.** thermometer **c.** heat of reaction **d.** calorimeter

__________ **5.** A reaction for which $\Delta H = +450$ kJ is ______.
a. exothermic **b.** slow **c.** impossible **d.** endothermic

Name Date Class

Heat of Hydration 20-3

Laboratory Quiz

In the space at the left, write the letter of the answer that best completes each statement.

__________ **1.** Chemical reactions form products that differ from reactants ______.
a. physically **c.** in energy content
b. chemically **d.** all of the above

__________ **2.** A reaction that absorbs energy is said to be ______.
a. endothermic **c.** hydrated
b. exothermic **d.** in solution

__________ **3.** The ______ form of a compound has one or more molecules of water attached.
a. anhydrous **c.** dissolved
b. hydrated **d.** dehydrated

__________ **4.** A reaction raises the temperature of 100 grams of water 2°C. The enthalpy of reaction is ______.
a. 200 J **b.** 50 J **c.** 8.36 J **d.** 836 J

__________ **5.** The specific heat of water is ______.
a. 100°C/mol **c.** 4.18 J/g•°C
b. 5.0 g **d.** 0.2 mol

Copyright © Glencoe/McGraw-Hill, a division of The McGraw-Hill Companies, Inc.

Name Date Class

Radioactive Dating—A Model 21-1

Laboratory Quiz

In the space at the left, write the letter of the answer that best completes each statement.

________ 1. Isotopes are elements with the same atomic number but different numbers of ______.
- a. protons
- b. neutrons
- c. electrons
- d. orbitals

________ 2. ______ is the isotope of hydrogen with an atomic mass of two.
- a. Protium
- b. Deuterium
- c. Tritium
- d. Hydronium

________ 3. The atomic mass of the element $^{14}_{6}C$ is ______.
- a. 6
- b. 8
- c. 14
- d. 20

________ 4. ______ is the time necessary for one-half of a sample of a radioactive element to decay into another element.
- a. Isotope
- b. Decay
- c. Radioactivity
- d. Half-life

________ 5. For any radioactive isotope, as the age of the isotope increases, the ______.
- a. half-life decreases
- b. amount of isotope increases
- c. half-life remains the same
- d. amount of isotope remains the same

Name Date Class

How Does Radiation Affect Seed Germination? 21-2

Laboratory Quiz

In the space at the left, write the letter of the answer that best completes each statement.

________ 1. A petri dish of seed that is not exposed to radiation is described as the ______ dish.
- a. timed
- b. microwaved
- c. control
- d. intensity

________ 2. ______ damage affects the living seed itself.
- a. Genetic
- b. Somatic
- c. Germination
- d. Extended

________ 3. Four groups of seeds were exposed to radiation for 0 sec, 10 sec, 20 sec, and 30 sec, respectively. Which germination percentages showed that the radiation had no effect on the seeds?
- a. 100%, 100%, 100%, 100%
- b. 25%, 50%, 75%, 100%
- c. 100%, 50%, 50%, 50%
- d. 100%, 50%, 50%, 100%

________ 4. ______ can cause chemical bonds to break and living cells to function eratically.
- a. Germination
- b. Genetics
- c. Organisms
- d. Energy

________ 5. As the intensity and duration of exposure of seeds to microwave radiation increases, the percent of germination ______.
- a. decreases
- b. increases
- c. remains the same

Copyright © Glencoe/McGraw-Hill, a division of The McGraw-Hill Companies, Inc.

Answers to Lab Quizzes

Lab 1-1
1. c
2. a
3. d
4. d
5. a

Lab 1-2
1. b
2. c
3. a
4. b
5. d

Lab 2-1
1. a
2. d
3. d
4. b
5. d

Lab 2-2
1. b
2. c
3. c
4. d
5. d

Lab 3-1
1. d
2. a
3. c
4. d
5. a

Lab 3-2
1. c
2. d
3. b
4. b
5. a

Lab 4-1
1. b
2. a
3. c
4. b
5. d

Lab 4-2
1. a
2. c
3. d
4. a
5. d

Lab 5-1
1. d
2. a
3. c
4. d
5. d

Lab 5-2
1. b
2. a
3. c
4. d
5. d

Lab 6-1
1. c
2. c
3. d
4. b
5. b

Lab 6-2
1. c
2. a
3. b
4. d
5. c

Lab 7-1
1. b
2. c
3. d
4. b
5. d

Lab 7-2
1. c
2. b
3. a
4. d
5. b

Lab 8-1
1. a
2. c
3. c
4. b
5. d

Lab 8-2
1. d
2. c
3. d
4. a
5. a

Lab 9-1
1. c
2. d
3. c
4. d
5. a

Lab 9-2
1. b
2. c
3. b
4. a
5. d

Lab 10-1
1. d
2. a
3. c
4. b
5. a

Lab 10-2
1. c
2. b
3. b
4. d
5. d

Lab 11-1
1. b
2. a
3. d
4. b
5. a

Lab 11-2
1. a
2. a
3. d
4. b
5. b

Lab 12-1
1. d
2. c
3. a
4. c
5. c

Lab 12-2
1. c
2. b
3. a
4. b
5. c

Lab 12-3
1. c
2. a
3. a
4. c
5. a

Lab 13-1
1. b
2. d
3. c
4. b
5. a

Lab 13-2
1. b
2. c
3. d
4. c
5. c

Lab 14-1
1. b
2. b
3. a
4. a
5. d

Copyright © Glencoe/McGraw-Hill, a division of The McGraw-Hill Companies, Inc.

Answers to Lab Quizzes

Lab 14-2
1. c
2. c
3. a
4. d
5. b

Lab 15-1
1. d
2. d
3. b
4. a
5. c

Lab 15-2
1. d
2. a
3. b
4. c
5. b

Lab 16-1
1. c
2. c
3. b
4. a
5. c

Lab 16-2
1. c
2. d
3. a
4. a
5. b

Lab 17-1
1. b
2. d
3. c
4. a
5. c

Lab 17-2
1. b
2. a
3. c
4. b
5. d

Lab 18-1
1. c
2. d
3. c
4. a
5. b

Lab 18-2
1. c
2. b
3. d
4. a
5. b

Lab 19-1
1. b
2. c
3. c
4. d
5. b

Lab 19-2
1. c
2. b
3. d
4. a
5. d

Lab 20-1
1. a
2. a
3. c
4. d
5. b

Lab 20-2
1. b
2. d
3. c
4. a
5. d

Lab 20-3
1. d
2. a
3. b
4. d
5. c

Lab 21-1
1. b
2. b
3. c
4. d
5. c

Lab 21-2
1. c
2. b
3. a
4. d
5. a

Copyright © Glencoe/McGraw-Hill, a division of The McGraw-Hill Companies, Inc.

Chemistry

CONCEPTS AND APPLICATIONS

LABORATORY MANUAL
STUDENT EDITION

Author: Tom Russo

McGraw-Hill
New York, New York Columbus, Ohio Mission Hills, California Peoria, Illinois

Chemistry

Concepts and Applications

Student Edition

Teacher Wraparound Edition

Teacher Classroom Resources:

- Laboratory Manual SE and TE
- Study Guide SE and TE
- Problems and Solutions Manual
- Supplemental Practice Problems
- Chapter Review and Assessment
- Section Focus Transparency Package
- Basic Concepts Transparency Package
- Problem Solving Transparency Package
- ChemLab and MiniLab Worksheets
- Critical Thinking/Problem Solving
- Chemistry and Industry
- Consumer Chemistry
- Tech Prep Applications
- Applying Scientific Methods in Chemistry
- Spanish Resources
- Lesson Plans
- Microcomputer-based Labs

Technology

- Computer Test Bank: Dos and Macintosh
- CD-ROM Multimedia System: Windows and Macintosh Versions
- Videodisc Program
- Chapter Summaries, English and Spanish Audiocassettes
- MindJogger Videoquizzes
- Mastering Concepts in Chemistry - Software

The Glencoe Science Professional Development Series

- Cooperative Learning in the Science Classroo
- Alternate Assessment in the Science Classroon
- Lab and Safety Skills in the Science Classroom
- Performance Assessment in the Science Classr

Glencoe/McGraw-Hill

A Division of The McGraw-Hill Companies

Copyright © by Glencoe/McGraw-Hill. All rights reserved. Except as permitted under the United States Copyright Act, no part of this publication may be reproduced or distributed in any form or by any means, or stored in a database or retrieval system, without prior written permission of the publisher.

Send all inquiries to:
Glencoe/McGraw-Hill
936 Eastwind Drive
Westerville, Ohio 43081

ISBN 0-02-827459-8
Printed in the United States of America.

2 3 4 5 6 7 8 9 024 03 02 01 00 99 8

Contents

To the Student v

Safety in the Laboratory vi

Laboratory Safety Agreement vii

First Aid in the Laboratory vii

Safety Symbols viii

Laboratory Equipment ix

Equipment Checklist xii

Laboratory Techniques xiii

Treating Data in Chemistry xix

Tables xxi

Laboratory Activities

1-1 Determining Boiling Point 1

1-2 The Eight-Solution Problem 5

2-1 The Thickness of Aluminum Foil 11

2-2 Identifying Elements by Flame Tests 15

3-1 An Alien Periodic Table 21

3-2 Periodicity of Halogen Properties 27

4-1 Determining Melting Points 33

4-2 Distinguishing Ionic and Covalent Compounds 37

5-1 Making Models of Compounds 41

5-2 Formulas and Oxidation Numbers 45

6-1 Types of Chemical Reactions 49

6-2 Effects of Concentration on Chemical Equilibrium 55

7-1 Atomic Spectra 61

7-2 Transition Metals 65

8-1 Periodicity and Chemical Reactivity 69

8-2 Comparing Activities of Selected Metals 73

9-1 Diagnostic Properties of Bonds 77

9-2 Electron Clouds 83

10-1 Crystal Shapes 87

10-2 Relating Gas Temperature and Pressure 91

11-1 Determining Absolute Zero 97
11-2 Charles's Law 101

12-1 Determining the Avogadro Constant 105
12-2 Stoichiometry of a Chemical Reaction 111
12-3 Molar Volume of a Gas 115

13-1 Relating Solubility and Temperature 119
13-2 The Effect of a Solute on Freezing Point 123

14-1 Acidic and Basic Anhydrides 127
14-2 Using Indicators to Determine pH 131

15-1 Hydrolysis of Salts 135
15-2 Acid / Base Titration 141

16-1 Oxidation / Reduction of Vanadium 147
16-2 Corrosion as an Electrochemical Process 151

17-1 Comparing the Abilities of Metals to Give Up Electrons 155
17-2 The Six-Cent Battery 161

18-1 Examples of Organic Reactions 167
18-2 Analysis of Aspirin 173

19-1 Biochemical Reactions 177
19-2 Qualitative Analysis of Food 183

20-1 Energy Changes in Physical and Chemical Processes 187
20-2 Measuring the Heat of Reaction 191
20-3 Heat of Hydration 195

21-1 Radioactive Dating — A Model 201
21-2 How Does Microwave Radiation Affect Seeds? 205

To the Student

Chemistry is the science of matter, its properties, and changes. In your classroom work in chemistry, you will learn a great deal of the information that has been gathered by scientists about matter. But, chemistry is not just information. It is also a process for finding out more about matter and its changes. Laboratory experiments are the primary means that chemists use to learn more about matter. The experiments in the ***Chemistry: Concepts and Applications Laboratory Manual*** require that you form and test hypotheses, measure and record data and observations, analyze those data, and draw conclusions based on those data and your knowledge of chemistry. These processes are the same as those used by professional chemists and all other scientists.

Organization of Experiments

- **Introduction:** Following the title and number of each activity, an introduction provides a background discussion about the problem you will study in the laboratory.
- **Objectives:** The objectives are statements of what you should accomplish by doing the experiment. Recheck this list when you have finished the activity.
- **Materials:** The materials list shows the apparatus you need to have on hand for the experiment. Above the materials list, you may find a group of **safety symbols**. These symbols warn you of potential hazards in the laboratory. Before beginning any laboratory, refer to page viii to see what these symbols mean.
- **Procedure:** The numbered steps of the procedure tell you how to carry out the experiment and sometimes offer hints to help you be successful in the laboratory. Some experiments have **CAUTION** statements in the procedure to alert you to hazardous substances or techniques.
- **Data and Observations:** This section presents a suggested table or form for collecting your laboratory data. Always record data and observations in an organized way as you do the experiment.
- **Analysis:** The Analysis section shows you how to perform the calculations necessary for you to analyze you data and reach conclusions. It also provides questions to aid you in interpreting data and observations in order to reach an experimental result.
- **Conclusions:** Here, you are asked to form a scientific conclusion based on what you actually observed, not what "should have happened."
- **Extension and Application:** The questions and activities in this section may ask you to apply what you have learned to other situations, to make additional hypotheses or conclusions, or to research a question related to the experiment.

Use of Microchemistry in the Lab

Many experiments in ***Chemistry: Concepts and Applications Laboratory Manual*** use the latest development in laboratory techniques — microchemistry. In microchemistry, you use plastic pipets and microplates instead of large glass beakers, flasks, and test tubes. You also use small amounts of chemicals in reactions. Still, when working with microchemistry, you should use the same care in obtaining data and making observations that you would use in large-scale labs. Likewise, you must observe the same safety precautions as for any chemistry experiment.

Safety in the Laboratory

The chemistry laboratory is a place to experiment and learn. You must assume responsibility for your own personal safety and that of people working near you. Accidents are usually caused by carelessness, but you can help prevent them by closely following the instructions printed in this manual and those given to you by your teacher. The following are some safety rules to help guide you in protecting yourself and others from injury in a laboratory.

1. The chemistry laboratory is a place for serious work. Do not perform experiments without your teacher's permission. **Never** work alone in the laboratory. Work only when your teacher is present.
2. Study your lab assignment **before** you come to the lab. If you are in doubt about any procedure, ask your teacher for help.
3. Safety goggles and a laboratory apron must be worn whenever you work in the lab. Gloves should be worn whenever you use chemicals that cause irritations or can be absorbed through the skin.
4. Contact lenses should not be worn in the lab, even if goggles are worn. Lenses can absorb vapors and are difficult to remove in an emergency.
5. Long hair should be tied back to reduce the possibility of it catching fire.
6. Avoid wearing dangling jewelry or loose, draping clothing. The loose clothing may catch fire and either the clothing or jewelry could catch on chemical apparatus.
7. Wear shoes that cover the feet at all times. Bare feet or sandals are not permitted in the lab.
8. Know the location of the fire extinguisher, safety shower, eyewash, fire blanket, and first-aid kit. Know how to use the safety equipment provided for you.
9. Report any accident, injury, incorrect procedure, or damaged equipment immediately to your teacher.
10. Handle chemicals carefully. *Check the labels of all bottles* ***before*** *removing the contents.* Read the label three times: before you pick up the container, when the container is in your hand, and when you put the bottle back.
11. Do **not** return unused chemicals to reagent bottles.
12. Do not take reagent bottles to your work area unless specifically instructed to do so. Use test tubes, paper, or beakers to obtain your chemicals. Take only small amounts. It is easier to get more than to dispose of excess.
13. Do not insert droppers into reagent bottles. Pour a small amount of the chemical into a beaker.
14. **Never** taste any chemical substance. **Never** draw any chemicals into a pipet with your mouth. Eating, drinking, chewing gum, and smoking are prohibited in the laboratory.
15. If chemicals come in contact with your eyes or skin, flush the area immediately with large quantities of water. Immediately inform your teacher of the nature of the spill.
16. Keep combustible materials away from open flames. (Alcohol and acetone are combustible.)
17. Handle toxic and combustible gases only under the direction of your teacher. Use the fume hood when such materials are present.
18. When heating a substance in a test tube, be careful not to point the mouth of the tube at another person or yourself. Never look down the mouth of a test tube.
19. Use caution and the proper equipment when handling hot apparatus or glassware. Hot glass looks the same as cool glass.
20. Dispose of broken glass, unused chemicals, and products of reactions only as directed by your teacher.
21. Know the correct procedure for preparing acid solutions. **Always** *add the acid slowly to the water.*
22. Keep the balance area clean. Never weigh chemicals directly on the pan of the balance. Use weighing paper or a small container.
23. Do not heat graduated cylinders, burets, or pipets with a laboratory burner.
24. After completing an experiment, clean and put away your equipment. Clean your work area. Make sure the gas and water are turned off. Wash your hands with soap and water before you leave the lab.

Laboratory Safety Agreement

MATERIAL	**DATE**
Review of First Aid in the Laboratory.	________________
Review of Laboratory Techniques	________________
In-class review of laboratory regulations, procedures, safety symbols, and safety equipment	________________

I, ________________________________, have read and understand the material entitled First Aid in the Laboratory and Laboratory Techniques. I agree to abide by the regulations and procedures outlined in this material. Furthermore, I agree to abide by any additional printed or verbal instructions provided by my teacher or school district during the school year.

____________________________________ ________________

student's signature *date*

First Aid in the Laboratory

REPORT ALL ACCIDENTS, INJURIES, AND SPILLS TO YOUR TEACHER IMMEDIATELY.

YOU MUST KNOW: safe laboratory techniques
where and how to report an accident, injury, or spill
location of first-aid equipment, fire alarm, phone, school nurse's office
evacuation procedure

Injury	Safe Response
burns	Flush with water. Call your teacher immediately.
cuts and bruises	Follow the instructions on the first-aid kit. Report to the school nurse.
fainting or collapse	Provide the person with fresh air. Have the person recline so that their head is lower than their body. Call your teacher. A nurse or doctor may be needed to provide artificial respiration.
fire	Wrap the person in fire blanket. Extinguish all flames.
foreign matter in eye	Flush with plenty of water. Use eyewash bottle or fountain.
poisoning	Note the suspected poisoning agent and call your teacher.
severe bleeding	Apply pressure or a compress directly to the wound and get medical attention.
spills on skin	Flush with water or use safety shower.
acid spills	For acid spills, apply baking soda, $NaHCO_3$, and call your teacher.
base spills	For base spills, apply boric acid, H_2BO_3, and call your teacher.

Safety Symbols

Disposal Alert
This symbol appears when care must be taken to dispose of materials properly.

Biological Hazard
This symbol appears when there is danger involving bacteria, fungi, or protists.

Open Flame Alert
This symbol appears when use of an open flame could cause a fire or an explosion.

Thermal Safety
This symbol appears as a reminder to use caution when handling hot objects.

Sharp Object Safety
This symbol appears when a danger of cuts or punctures caused by the use of sharp objects exists.

Fume Safety
This symbol appears when chemicals or chemical reactions could cause dangerous fumes.

Electrical Safety
This symbol appears when care should be taken when using electrical equipment.

Plant Safety
This symbol appears when poisonous plants or plants with thorns are handled.

Laser Safety
This symbol appears when care must be taken to avoid staring directly into the laser beam or at bright reflections.

Radioactive Safety
This symbol appears when radioactive materials are used.

Clothing Protection Safety
This symbol appears when substances used could stain or burn clothing.

Fire Safety
This symbol appears when care should be taken around open flames.

Explosion Safety
This symbol appears when the misuse of chemicals could cause an explosion.

Eye Safety
This symbol appears when a danger to the eyes exists. Safety goggles should be worn when this symbol appears.

Poison Safety
This symbol appears when poisonous substances are used.

Chemical Safety
This symbol appears when chemicals used can cause burns or are poisonous if absorbed through the skin.

Laboratory Equipment

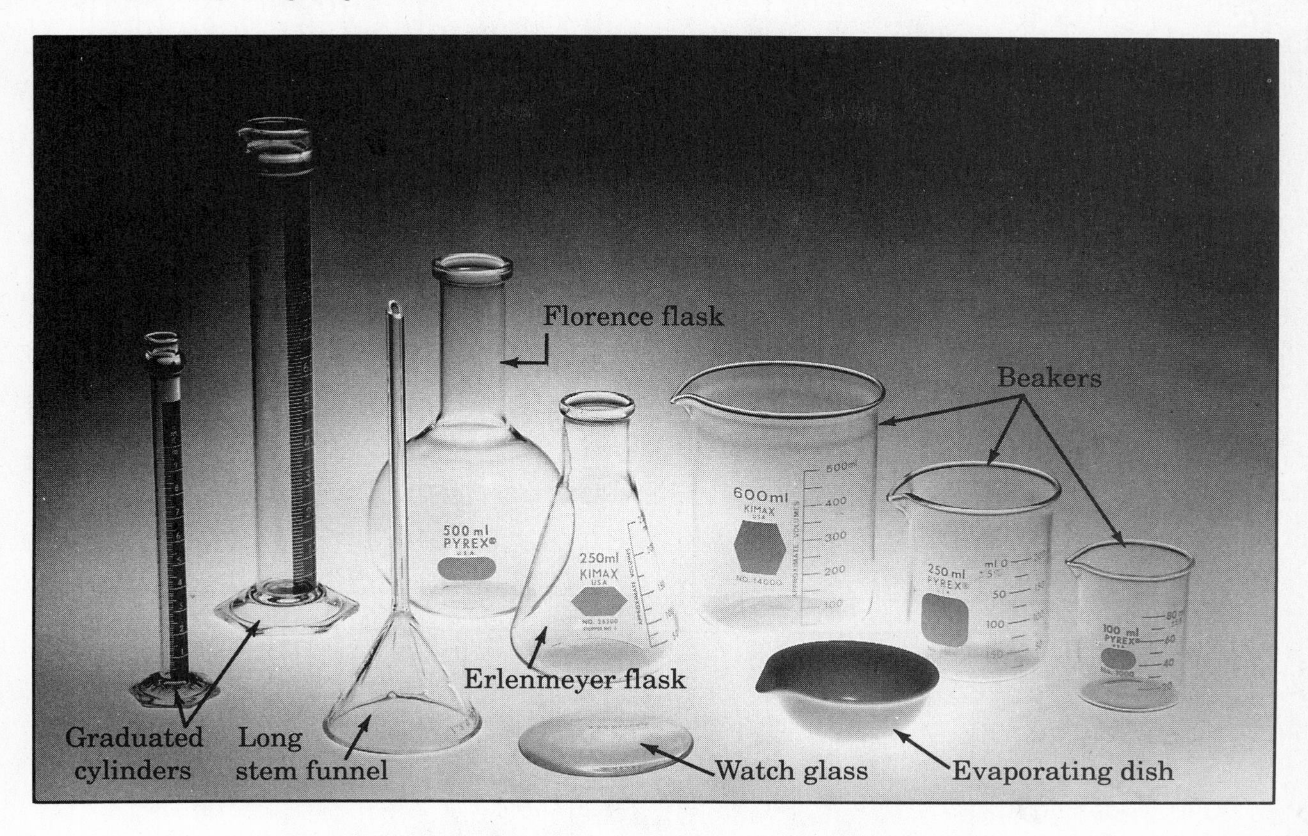

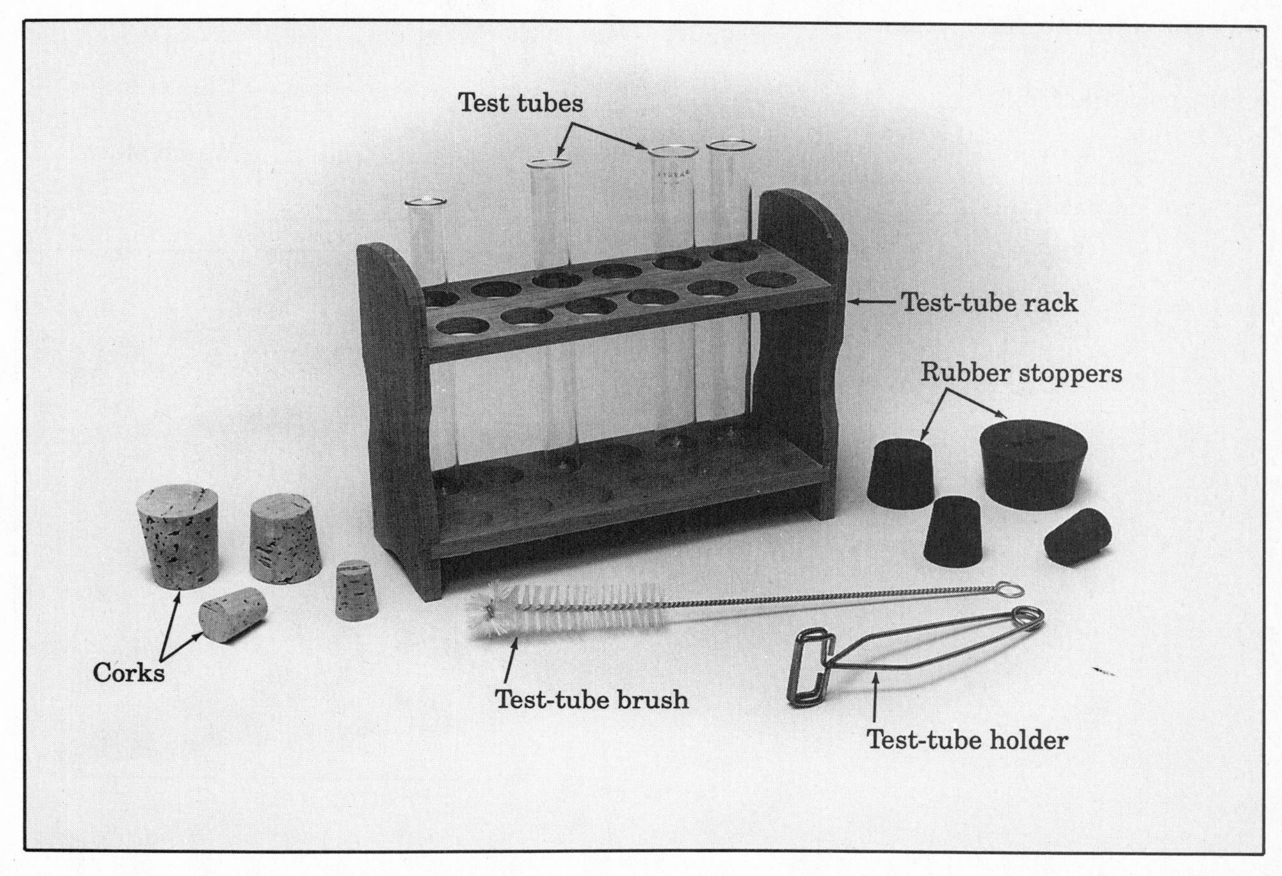

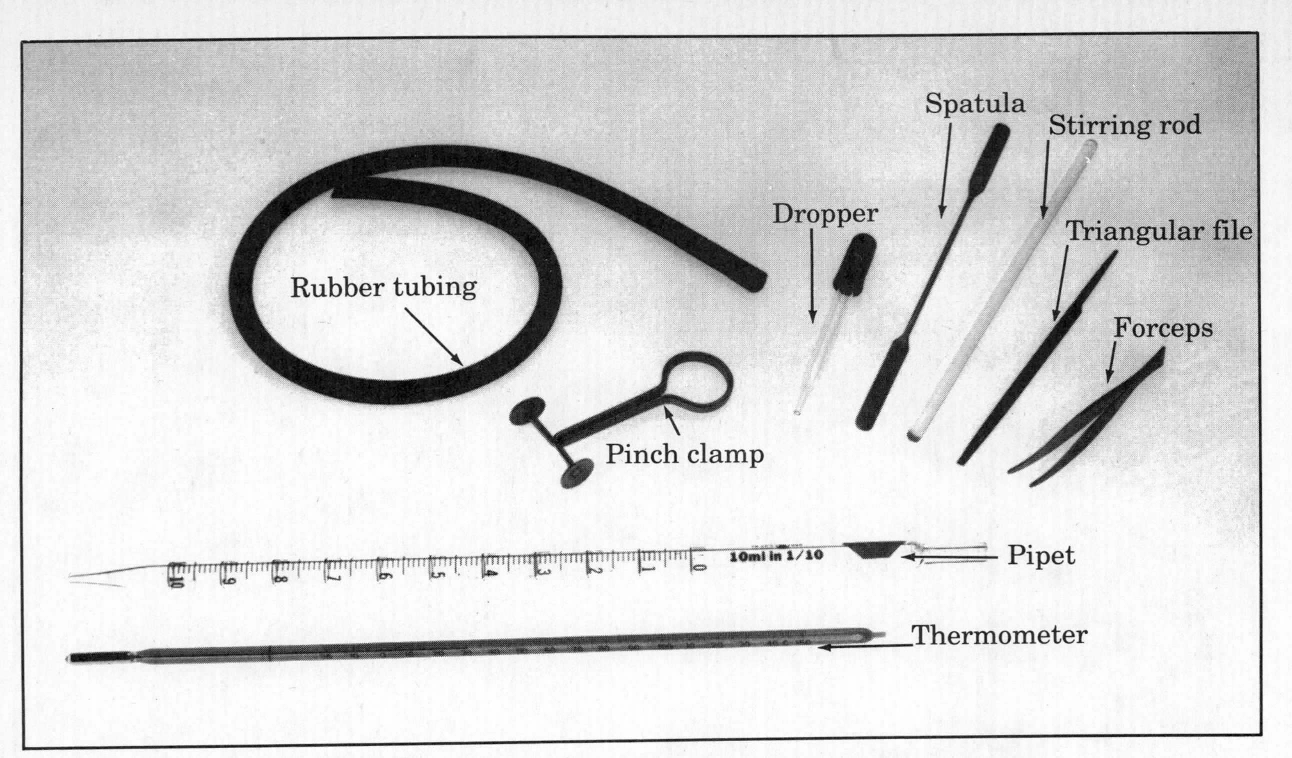
Spatula
Stirring rod
Dropper
Triangular file
Rubber tubing
Forceps
Pinch clamp
10ml in 1/10
Pipet
Thermometer

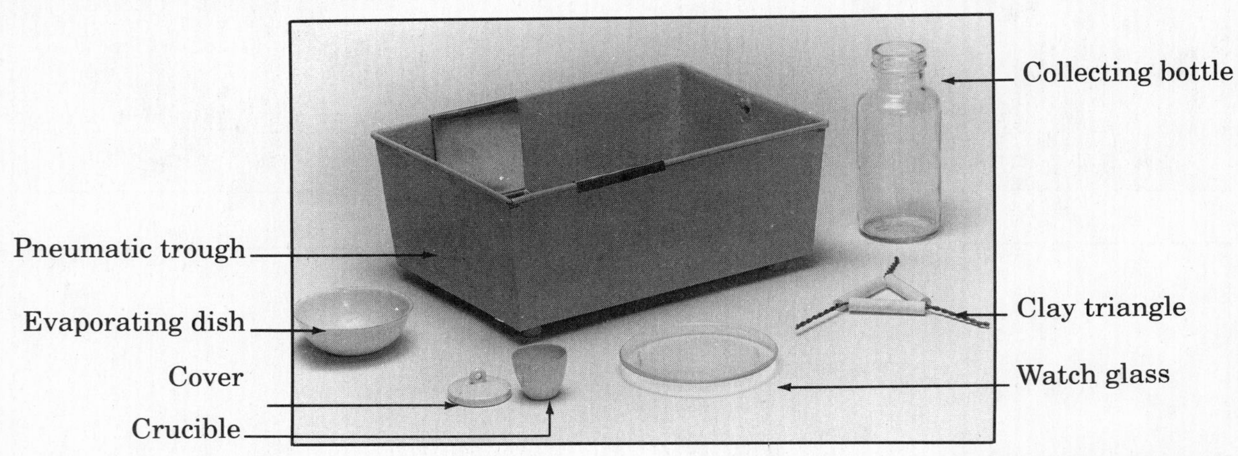
Collecting bottle
Pneumatic trough
Clay triangle
Evaporating dish
Watch glass
Cover
Crucible

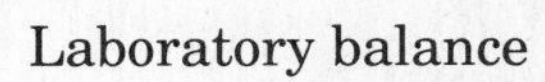
Laboratory balance

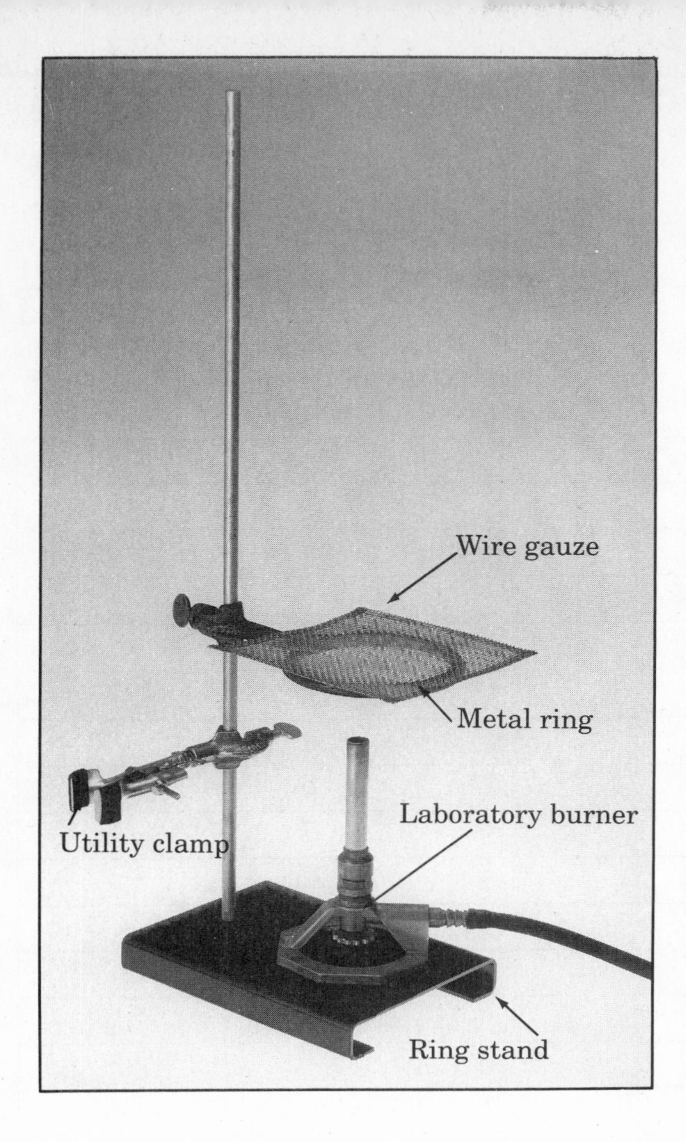
Wire gauze
Metal ring
Utility clamp
Laboratory burner
Ring stand

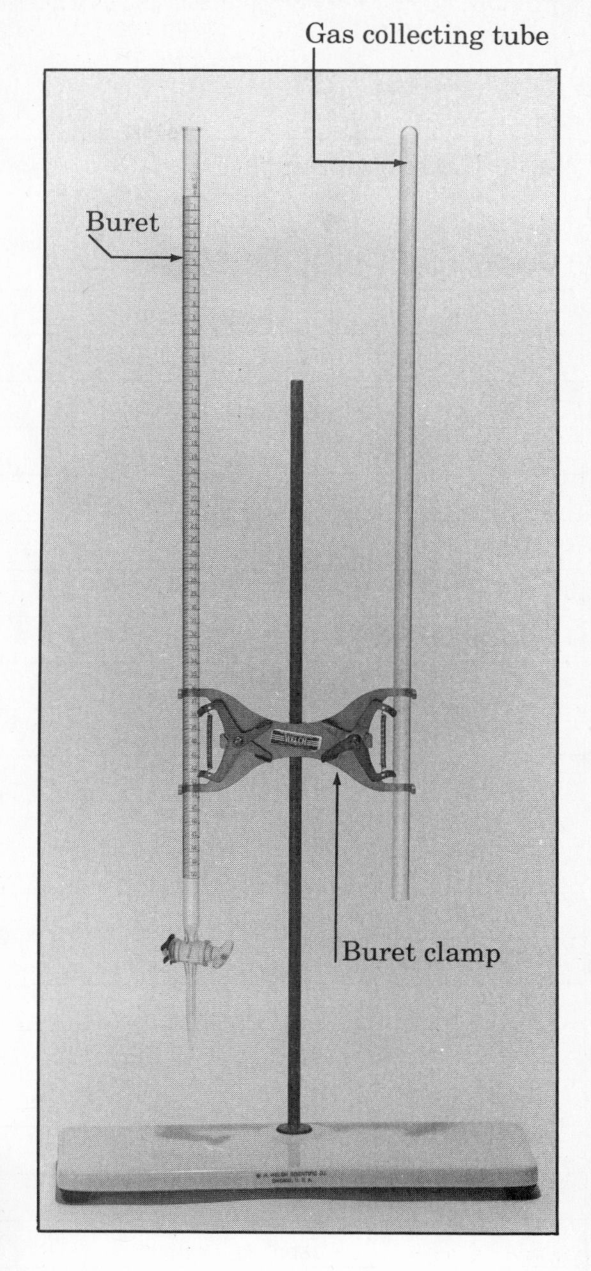
Gas collecting tube
Buret
Buret clamp

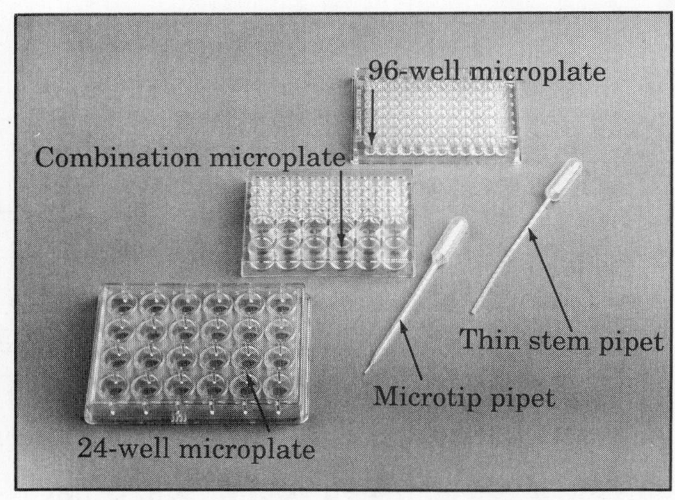
96-well microplate
Combination microplate
Thin stem pipet
Microtip pipet
24-well microplate

Equipment Checklist

Name (Last Name First) **Locker Number** **Class Period**

Quantity	Description	Check In	Check Out	Breakage
1	beaker (50-mL)			
1	beaker (100-mL)			
2	beakers (250-mL)			
4	beakers (400-mL)			
1	chemical scoop			
3	crucibles			
2	droppers			
1	evaporating dish			
1	flask, Erlenmeyer (125-mL)			
1	flask, Erlenmeyer (500-mL)			
1	funnel			
1	goggles			
1	graduated cylinder (10-mL)			
1	graduated cylinder (50-mL)			
1	laboratory apron			
1	24-well microplate			
3	96-well microplates			
10	pipets, microtip			
15	pipets, thin stem			
1	scissors			
1	spatula			
6	test tubes (small, 13 x 100-mm)			
25	test tubes (large, 18 x 150-mm)			
1	test-tube brush			
1	test-tube holder			
2	test-tube racks			

*The above list may be changed in some way by your teacher because of the available equipment. Be sure to note these changes on this list.

Return this paper to your teacher after you have completed your check. Other equipment used in the laboratory will be made available as the need arises.

Laboratory Techniques

1. Heat Sources

Lighting a Laboratory Burner and Adjusting the Flame

Connect the hose of the burner to a gas supply. Hold a lighted match or a striker to the edge of the top of the burner and then partly open the valve in the gas supply. See Figure A.

The size of the flame can be changed by opening and closing the valve on the gas supply. The color of the flame indicates the amount of air in the gas. The air supply is controlled by burning the tube of the burner. A yellow flame indicates more air is needed, and the burner tube can be turned to increase the amount of air. If the flame goes out, the air supply should be reduced by turning the burner tube in the opposite direction. The gas supply is controlled by the valve on the bottom of the burner. The hottest part of the flame is just above the tip of the inner cone of the flame. If the flame rises from the barrel of the burner, turn down the gas. If the flame is yellow, open the air regulator.

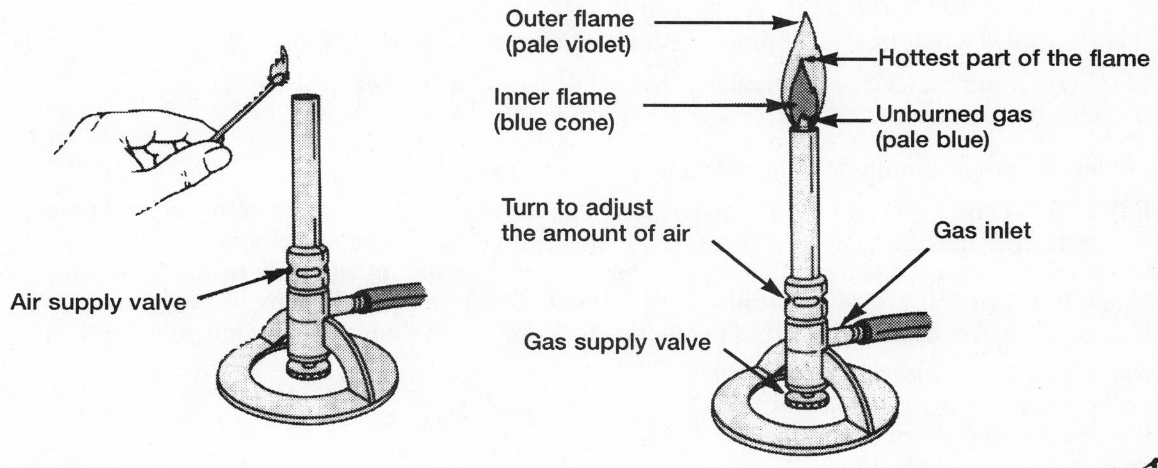

Figure A

Using Laboratory Burners

Never place a flame beneath a container and leave it.

Material in a beaker or flask. Place a square of wire gauze on a ring clamped to a ring stand. Set the container on the wire gauze. Adjust the amount of heat to be applied by altering the height of the flame.

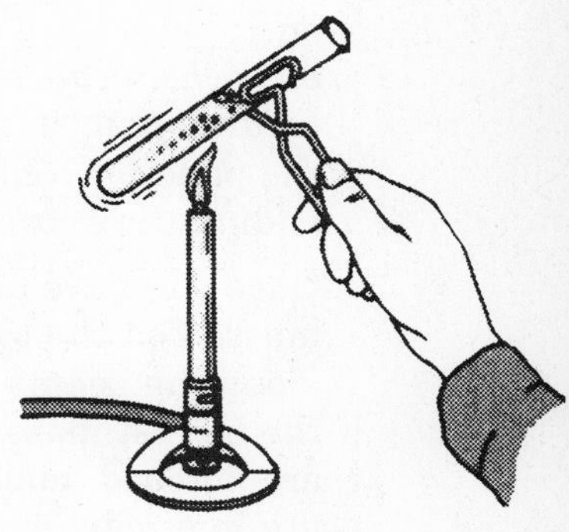

Figure B

Material in a test tube. Clamp the test tube at the top with a test-tube clamp. Hold the tube at about a 30° angle to the vertical. See Figure B. Pass the test tube back and forth through the laboratory burner flame until heating is complete. Do not point the open end of the test tube at any person, because sudden boiling or "bumping" may take place and the test tube contents may be violently ejected.

Using Hot Plates For safety reasons, use a hot plate instead of an open flame any time you can. To use a hot plate, it must first be plugged into the electric line. Some hot plates have an on-off switch that must be turned on for use. Also, some hot plates are adjustable and the control should be turned near to maximum to begin. After a material is brought to the correct temperature, the heating rate can be controlled by reducing the variable control setting or by alternately turning the hot plate on and off with the switch or by unplugging and plugging in the line cord.

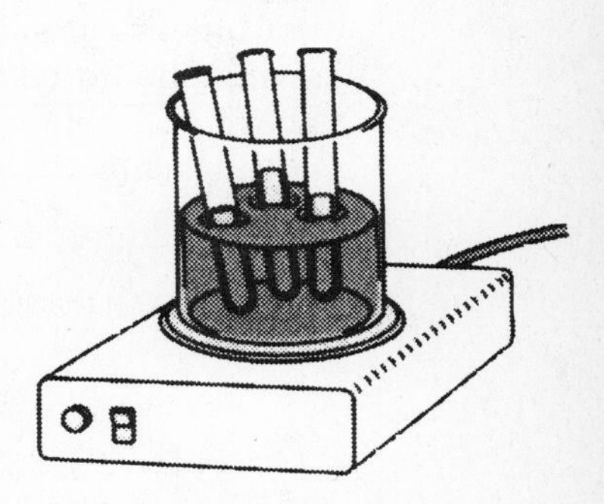

Figure C

If you know that during a laboratory period you are going to heat test tubes and other small containers, it is good practice to start a water bath at the beginning of the period. This bath may be set up by filling an appropriately-sized beaker half full of tap water and placing it on the hot plate. Bring the water almost to a boil and keep it there with proper heat adjustment. Small containers may then be heated by simply placing them in the water bath. See Figure C.

Never heat an empty container, even to dry it. The container will probably crack and you may burn out the hot plate heating element. Never run the hot plate without something on it. This will also lead to burn out.

2. Measuring Mass

Using the Balance

There are various tubes of laboratory balances. The beam balance you use may look somewhat different from the one in Figure D; however, all beam balances have some common features.

The following technique should be used to transport a balance from place to place.

1. Be sure all riders are back to the zero point.
2. If the balance has a lock mechanism to lock the pan(s), be sure it is on.
3. Place one hand under the balance and the other hand on the beams' support to carry the balance.

The following steps should be followed in using the balance.

1. Before determining the mass of any substance, slide all of the riders back to the zero point. Check to see that the pointer swings freely along the scale. You do not have to wait for the pointer to stop at the zero point. The beam should swing an equal distance above and below the zero point. Use the adjustment screw to obtain an equal swing of the beams, if necessary. You must repeat this procedure to "zero" the balance every time you use it.
2. *Never put a hot object directly on the balance pan.* Any dry chemical that is to be massed should be placed on paper or in a container. *Never pour chemicals directly on the balance pan.* Remember to mass the paper or container before adding the substance.
3. Once you have placed the object to be massed on the pan, move the riders along the beams beginning with the largest mass first. If the beams are notched, make sure all riders are in a notch before you take a reading. Remember, the pointer does not have to stop swinging, but the swing should be an equal distance above and below the zero point on the scale.

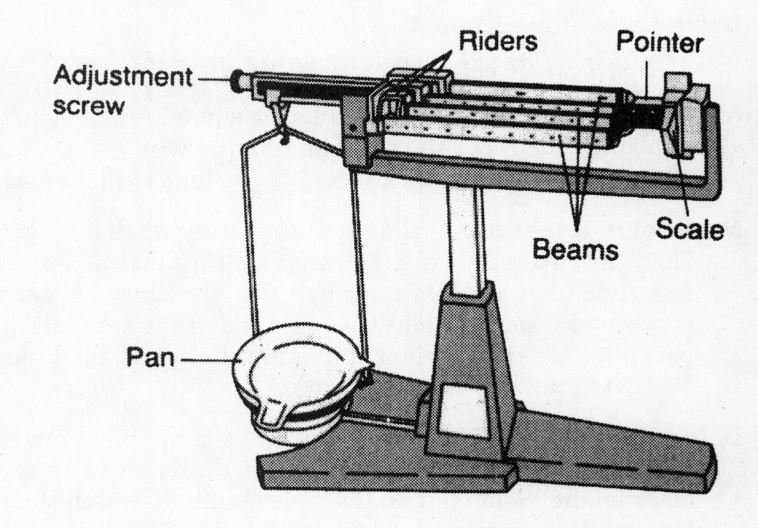

Figure D A triple beam balance

4. The mass of the object will be the sum of the masses indicated on the beams. For example:

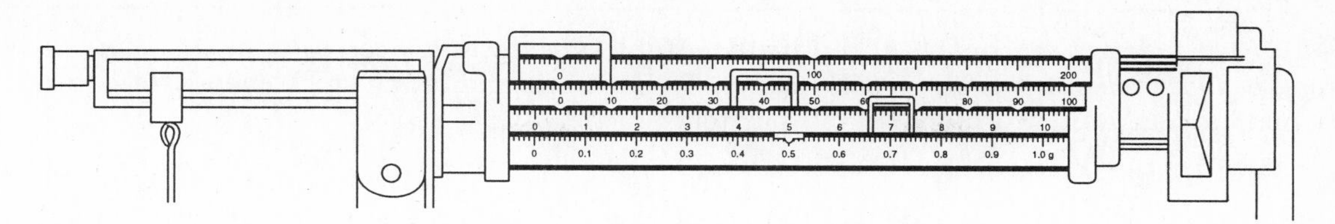

Figure E

The mass of this object would be read as 47.50 grams.

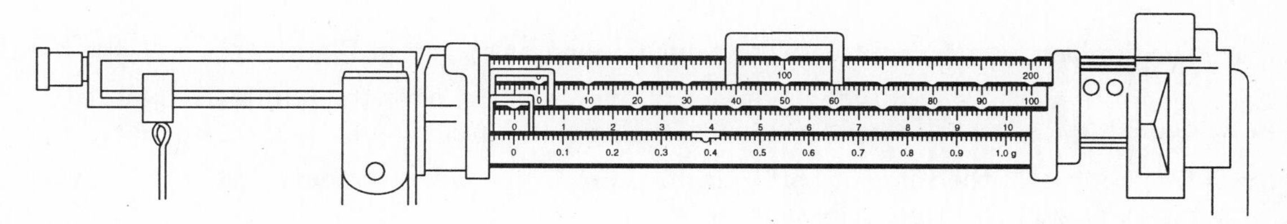

Figure F

The mass of this object would be read as 100.39 grams.

The electronic top-loading balance offers several advantages over the beam balance. The time requirement for mass measurement is much less and accuracy is increased with the direct digital readout.

When massing a sample in a container, the tare feature of the electronic balance can be used to determine the mass of just the sample. Follow these steps:

1. Place the container on the massing platform. As with the beam balance, materials should be cool before massing and materials should be massed in containers.
2. Press the tare button to obtain a reading of zero.
3. Add the material to be massed to the container and read the mass of the material.

3. Inserting Glass Tubing into a Rubber Stopper

1. Begin by lubricating the tip of the glass tubing and the rubber stopper with soapy water, glycerol, or some other suitable substance you teacher provides.
2. Never force the tubing into the stopper. Ease it in with a gentle twisting motion, as shown in Figure G. Protect your hands with a cloth or paper towels. Be sure that the ends of the tubing are directed away from the palms of your hands. **CAUTION:** *Excessive hand pressure on the tubing will cause it to break. Severe injury can occur.*
3. The end of the glass tubing should protrude from the stopper.

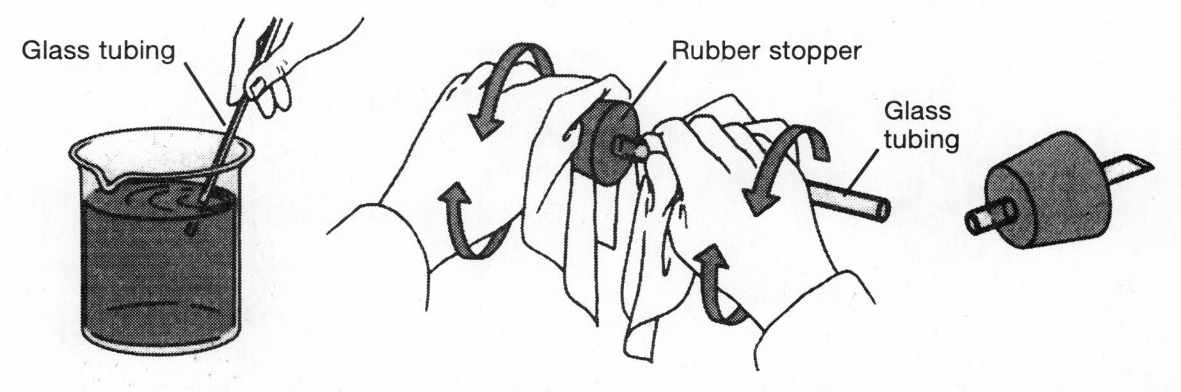

Figure G The proper method of inserting glass tubing into a rubber stopper

4. TRANSFERRING CHEMICALS

CAUTION: *Never touch chemicals with your hands. Many substances are corrosive or irritating to skin. Goggles must be worn when transferring substances.* To avoid contaminating stock chemicals, return unused portions to a container designated by your teacher.

Solids

1. Solids are generally kept in wide-mouth bottles. Use a clean spoon or spatula to remove solid material from its container, or rotate the bottle back and forth to shake out the solid.
2. Place the solid material on a piece of folded paper and slide the solid very carefully along the crease and into your container.

Liquids

1. Remove the stopper by holding it between your fingers, as shown in Figure H.
2. Wearing goggles, hold the graduated cylinder at eye level and pour the liquid slowly until the desired volume has been transferred. Read the volume from the bottom of the meniscus.
3. Replace the stopper in the reagent bottle. If any liquid runs down the outside of the bottle, rinse it with water before returning it to the shelf.

Diluting Acids

The acid is added to the water, and never the reverse. The acid should be poured slowly down the stirring rod, as shown in Figure H, and the solution continually stirred. *Diluting an acid produces heat.* Therefore, it is important to add the acid slowly and to stir the solution.

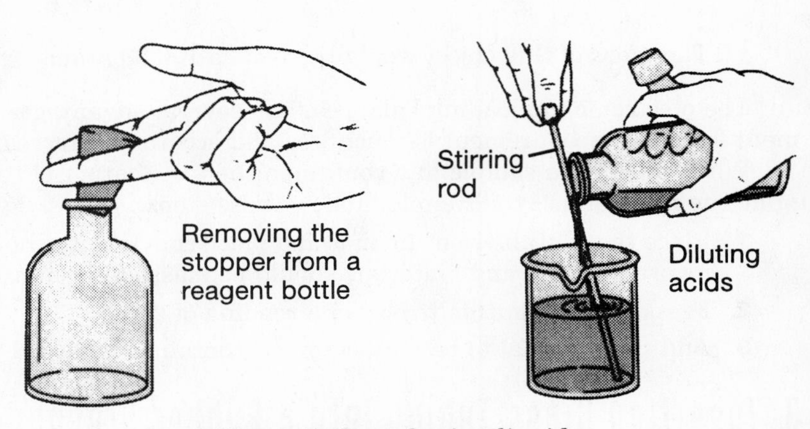

Figure H Transferring liquids

Decanting and Filtering

It is often necessary to separate a solid from a liquid. Filtration is a common process of separation used in most laboratories. The liquid is decanted, that is, the liquid is separated from the solid by carefully pouring off the liquid, leaving only the solid material. To avoid splashing and to maintain control, the liquid is poured down a stirring rod. See Figure I. The solution is usually filtered through filter paper to catch any solid that has not settled to the bottom of the beaker. See Figure J. The solid may be rinsed with distilled water to remove any solution.

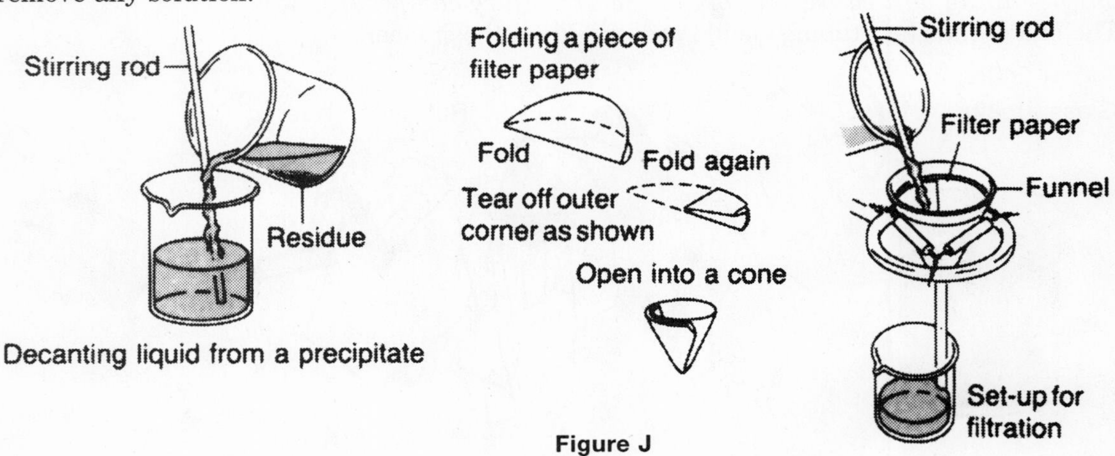

Figure I

Figure J

5. Titration

Titration is an operation performed using a long, graduated tube called a buret. Check the operation of the stopcock on a buret when you first get it. If it sticks or is stiff, have it lubricated by the instructor.

If you have a funnel to fill the buret, do not rest the funnel in the buret. Hold the funnel in place with your hand.

When filling a buret, do not spend time trying to fill to exactly the 0.0 mark. Begin anywhere between 0 and 5, but read the actual starting point accurately. Also, be sure you have filled the tip before getting the initial reading. Fill the tip by opening the stopcock slightly and letting the tip fill from the main tube. When turning the stopcock, always maintain a slight inward pressure to avoid solution leaking out around the stopcock plug.

Before attempting a titration, practice filling the buret with water and manipulating the stopcock until you can release one drop at a time.

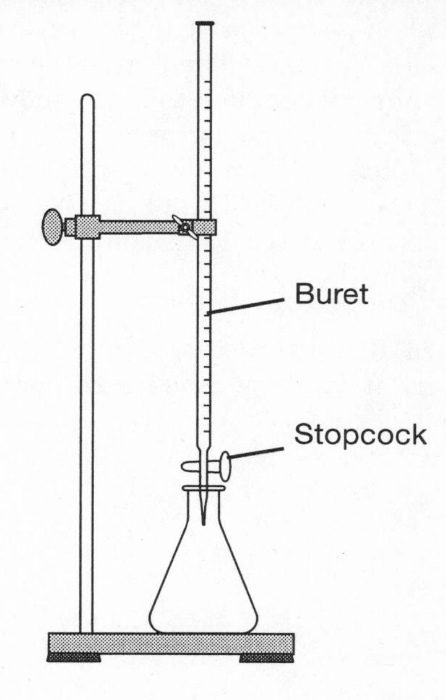

Figure K

6. Microchemistry

Microchemistry uses smaller amounts of chemicals than do other chemistry methods. The hazards of glass have been minimized by the use of plastic labware. If a chemical reaction must be heated, hot water will provide the needed heat. Open flames or burners are seldom used in microchemistry techniques. By using microchemistry, you will be able to do more experiments and have a safer environment in which to work.

Microchemistry uses two basic tools.

The Microplate

The first is a sturdy plastic tray called a microplate. The tray has shallow wells arranged in rows (running across) and columns (running up and down). These wells are used instead of test tubes, flasks, and beakers. Some microplates have 96 wells; other microplates have 24 larger wells.

The Plastic Pipet

Microchemistry uses a pipet made of a form of plastic that is soft and very flexible. The most useful property of the pipet is the fact that the stem can be stretched without heating into a thin tube. If the stem is stretched and then cut with scissors (Figure L), the small tip will deliver a tiny drop of reagent. You may also use a pipet called a microtip pipet, which has been pre-stretched at the factory. It is not necessary to stretch a microtip pipet.

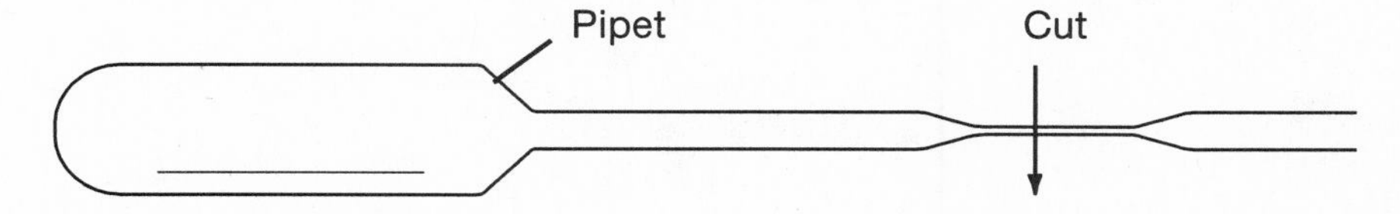

Figure L Cutting a stretched pipet

The pipet can be used over and over again simply by rinsing the stem and bulb between reagents. The plastic inside the pipet is non-wetting and does not hold water or solutions the way glass does.

The Microplate Template and Microplate Data Form

Your teacher can provide you with Microplate Templates and Microplate Data Forms whenever you carry out an activity that uses microchemistry techniques.

To help you with your observations, place the Microplate Template beneath your 24-well or 96-well microplate. The template is marked with the correct number of wells, and each row and column is labelled to help guide you with your placement of chemicals from the micropipets. The white paper background provided by the template allows you to observe color changes and precipitate formations with ease.

Use Microplate Data Forms to write down the chemicals used and to record your observations of the chemical reactions that occur in each well.

Waste Disposal

Discard all substances according to your teacher's instructions. All plastic microchemistry equipment can be washed with distilled water for reuse.

Treating Data in Chemistry

Data. Any qualitative or quantitative information that is collected under controlled conditions and used in determining the results of an experiment is classified as data. The data collected during an experiment must be properly identified, and correct significant digits and units must be used. When recording data, erasures are not permitted. If an error is made, a line should be drawn through the incorrect data, and then the correction should be entered. Equipment and instruments must be read to the correct number of significant digits.

Tables. In many experiments, the amount of information included in your data and observations will be extensive. Tables are an efficient means of organizing relevant information. If a table is organized correctly, it often shows patterns of data that can be used to develop conclusions.

Graphing of Experimental Data. In a controlled experiment, all conditions are held constant except for the condition being tested. This condition is called the independent variable. The independent variable may be temperature or time, for example. During the experiment, a single factor is measured and observed. This factor, called the dependent variable, changes with changes in the independent variable; that is, it is dependent upon the independent variable. An example of a dependent variable might be the rate of a reaction, which is affected by temperature.

Graphing the results of an experiment involving two variables helps to make the relationship between the variables more obvious. Consider the following graph for the direct relationship between the mass and the volume of an object.

Characteristic *A* Mass (g)	**Characteristic *B*** Volume (cm^3)
1	3
2	6
3	9
4	12

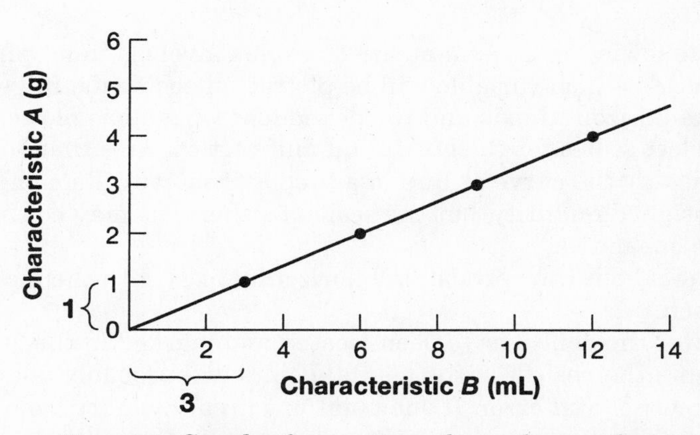

Figure A Graph of mass vs. volume for an object

Notice that a 1 g change in *A* causes a 3 mL change in *B*. We say that *A* varies directly as *B*. Mathematically,

$$A = kB$$

If the graph is constructed with enough care, related values can be read directly from the graph. For example, the volume of a 3.5 g object would be 10.5 mL.

The two variables in the graph in Figure A are represented by a straight line. The relationship between other variables, such as the pressure and volume of a gas, may turn out to be represented by a curve, as shown in Figure B.

Pressure (kPa)	Volume (L)
10	225.0
20	110.0
30	74.9
50	44.6
70	32.0
100	22.4

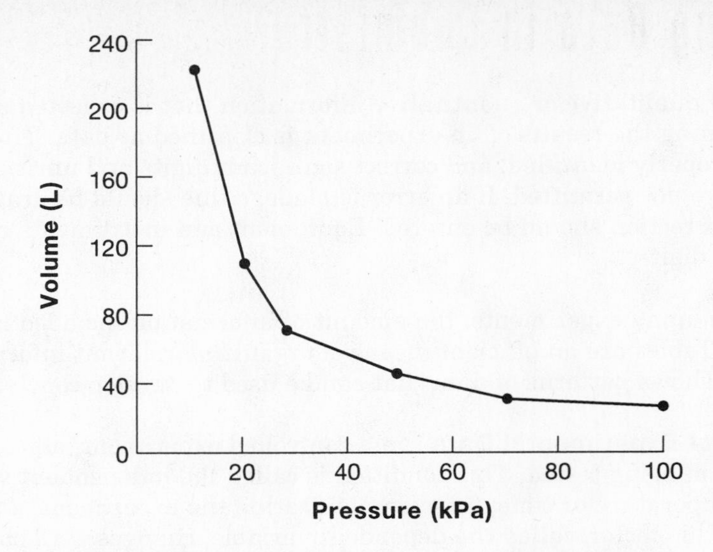

Figure B Graph of volume vs. pressure for a gas

Notice that as the pressure increases, the volume decreases. This type of variation is called an inverse relationship. Mathematically,

$$A = k/B$$

The following rules summarize the steps involved in graphing data.

1. Decide which variable will be plotted on each axis. In general, the independent variable is plotted on the horizontal axis and the dependent variable is plotted on the vertical axis.
2. Select scales for the horizontal and vertical axes that will reflect the precision of the measurements. Display the curve or line in a proportional way. Remember, each square on the graph is equal to an assigned quantity, but the scale of either axis may be changed if the graph is too compact or needs to be expanded.
3. Label both the vertical and horizontal axes with the factors being graphed. Indicate the units being used.
4. After the points have been located and marked in the graph, draw a smooth line or curve that represents the best fit of the points. Points will probably fall on both sides of the curve or line due to experimental error. If one point is a great distance from the curve, the data should be rechecked.

Calculations and Significant Digits. Methods used in making calculations must be clearly stated. The factor-label method, discussed in Appendix A of your textbook, is a convenient method for making calculations. All calculations should be properly labeled. The ratios should be arranged so that units can be divided out as factors. This procedure leaves the correct units in the answer and provides a check on the method used.

The rules for determining significant digits are described in Appendix A of your textbook. In recording data and making calculations, you must remember to consider the accuracy and precision of your measurements. When using a calculator for your work, remember to express your answer according to the rules for significant digits. Always double-check your answer against the data given.

Tables

Table 1 International System (SI) Standard Unit

International System (SI) Standard Unit

Measurement	Standard SI Unit	Common Lab Unit	Equivalency to SI Standard
mass	kilogram (kg)	gram (g)	1 g = 0.001 kg
length	meter (m)	centimeter (cm)	1 cm = 0.01 m
		millimeter (mm)	1 mm = 0.001 m
volume	cubic meter (m^3)	cubic decimeter (dm^3)	1 dm^3 = 0.001 m^3
		cubic centimeter (cm^3)	1 cm^3 = 0.001 dm^3
		milliliter (mL)	1 mL = 1cm^3
		liter (L)	1 L = 1000 mL = 1dm^3
time	second (s)		
temperature	kelvin (K)	degree Celsius (°C)	K = °C + 273

Table 2 SI Prefixes

Prefix	Symbol	Meaning	Multiplier: Numerical value	Multiplier: Expressed as scientific notation
Greater than 1				
giga-	G	billion	1 000 000 000	1×10^{9}
mega-	M	million	1 000 000	1×10^{6}
kilo-	k	thousand	1 000	1×10^{3}
Less than 1				
deci-	d	tenth	0.1	1×10^{-1}
centi-	c	hundredth	0.01	1×10^{-2}
milli-	m	thousandth	0.001	1×10^{-3}
micro-	µ	millionth	0.000 001	1×10^{-6}
nano-	n	billionth	0.000 000 001	1×10^{-9}
pico-	p	trillionth	0.000 000 000 001	1×10^{-12}

Table 3 Names and Charges of Polyatomic Ions

1−	2−	3−	4−
Acetate, CH_3COO^-	Carbonate, CO_3^{2-}	Arsenate, AsO_4^{3-}	Hexacyanoferrate(II), $Fe(CN)_6^{4-}$
Amide, NH_2^-	Chromate, CrO_4^{2-}	Arsenite, AsO_3^{3-}	Orthosilicate, SiO_4^{4-}
Astatate, AtO_3^-	Dichromate, $Cr_2O_7^{2-}$	Borate, BO_3^{3-}	Diphosphate, $P_2O_7^{4-}$
Azide, N_3^-	Hexachloroplatinate, $PtCl_6^{2-}$	Citrate, $C_6H_5O_7^{3-}$	
Benzoate, $C_6H_5COO^-$	Hexafluorosilicate, SiF_6^{2-}	Hexacyanoferrate(III), $Fe(CN)_6^{3-}$	
Bismuthate, BiO_3^-	Molybdate, MoO_4^{2-}	Phosphate, PO_4^{3-}	
Bromate, BrO_3^-	Oxalate, $C_2O_4^{2-}$	**1+**	**2+**
Chlorate, ClO_3^-	Peroxide, O_2^{2-}	Ammonium, NH_4^+	Mercury(I), Hg_2^{2+}
Chlorite, ClO_2^-	Peroxydisulfate, $S_2O_8^{2-}$	Neptunyl(V), NpO_2^+	Neptunyl(VI), NpO_2^{2+}
Cyanide, CN^-	Phosphite, HPO_3^{2-}	Plutonyl(V), PuO_2^+	Plutonyl(VI), PuO_2^{2+}
Formate, $HCOO^-$	Ruthenate, RuO_4^{2-}	Uranyl(V), UO_2^+	Uranyl(VI), UO_2^{2+}
Hydroxide, OH^-	Selenate, SeO_4^{2-}	Vanadyl(V), VO_2^+	Vanadyl(IV), VO^{2+}
Hypobromite, BrO^-	Selenite, SeO_3^{2-}		
Hypochlorite, ClO^-	Silicate, SiO_3^{2-}		
Hypophosphite, $H_2PO_2^-$	Sulfate, SO_4^{2-}		
Iodate, IO_3^-	Sulfite, SO_3^{2-}		
Nitrate, NO_3^-	Tartrate, $C_4H_4O_6^{2-}$		
Nitrite, NO_2^-			

Table D.4 Properties of Elements

Element	Symbol	Atomic Number (Z)	Atomic Mass* (u)	Melting Point (°C)	Boiling Point (°C)	Density (g/cm³) (gases measured at STP)	Atomic Radius (pm)	First Ionization Energy (kJ/mol)	Standard Reduction Potential (V) (for elements from or to oxidation state indicated)	Enthalpy of Fusion (kJ/mol)	Specific Heat (J/g · °C)	Enthalpy of Vaporization (kJ/mol)	Abundance in Earth's Crust (%)	Major Oxidation States
Actinium	Ac	89	[227.0278]	1050	3300	10.07	203	666	(3+)−2.13	14.3	0.120	293	trace	3+
Aluminum	Al	13	26.981539	660.37	2517.6	2.699	143	577.5	(3+)−1.67	10.71	0.9025	290.8	8.1	3+
Americium	Am	95	[243.0614]	994	2600	13.67	183	579	(3+)−2.07	10	—	238.5	—	2+, 3+, 4+
Antimony	Sb	51	121.757	630.7	1635	6.697	161	834	(3+)+0.15	19.5	0.2072	193	2×10^{-5}	3+, 5+
Argon	Ar	18	39.948	−189.37	−185.86	0.001784	191	1521	—	1.18	0.52033	6.52	4×10^{-6}	
Arsenic	As	33	74.92159	816 (2840 kPa)	615 (sublimes)	5.778	121	947	(3+)+0.24	27.7	0.3289	— (sublimes)	1.9×10^{-4}	3+, 5+
Astatine	At	85	[209.98037]	300	350	—	—	916	(1−)+0.2	23.8	—	90.3	trace	1−, 5+
Barium	Ba	56	137.327	726.9	1845	3.62	222	502.9	(2+)−2.92	8.012	0.2044	140	0.039	2+
Berkelium	Bk	97	[247.0703]	986	—	14.78	170	601	(3+)−2.01	—	—	—	—	3+, 4+
Beryllium	Be	4	9.012182	1287	2468	1.848	112	899.5	(2+)−1.97	7.895	1.824	297.6	2×10^{-4}	2+
Bismuth	Bi	83	208.98037	271.4	1564	9.808	151	703	(3+)+0.317	10.9	0.1221	179	8×10^{-7}	3+, 5+
Bohrium	Bh	107	[262]	—	—	—	—	—	—	—	—	—	—	
Boron	B	5	10.811	2080	3865	2.46	85	800.6	(3+)−0.89	50.2	1.026	504.5	9×10^{-4}	3+
Bromine	Br	35	79.904	−7.25	59.35	3.1028	119	1139.9	(1−)+1.065	10.571	0.47362	29.56	2.5×10^{-4}	1−, 1+, 3+, 5+
Cadmium	Cd	48	112.411	320.8	770	8.65	151	867.7	(2+)−0.4025	6.19	0.2311	100	1.6×10^{-5}	2+
Calcium	Ca	20	40.078	841.5	1500.5	1.55	197	589.8	(2+)−2.84	8.54	0.6315	155	4.66	2+
Californium	Cf	98	[251.0796]	900	—	—	186	608	(3+)−2	—	—	—	—	3+, 4+
Carbon	C	6	12.011	3620	4200	2.266	77	1086.5	(4−)+0.132	104.6	0.7099	711	0.018	4−, 2+, 4+
Cerium	Ce	58	140.115	804	3470	6.773	181.8	541	(3+)−2.34	5.2	0.1923	313	0.007	3+, 4+
Cesium	Cs	55	132.90543	28.4	674.8	1.9	262	375.7	(1+)−2.923	2.087	0.2421	67	2.6×10^{-4}	1+
Chlorine	Cl	17	35.4527	−101	−34	0.003214	91	1255.5	(1−)+1.3583	6.41	0.47820	20.41	0.013	1−, 1+, 3+, 5+
Chromium	Cr	24	51.9961	1860	2679	7.2	128	652.8	(3+)−0.74	20.5	0.4491	339	0.01	2+, 3+, 6+
Cobalt	Co	27	58.9332	1495	2912	8.9	125	758.8	(2+)−0.277	16.192	0.4210	382	0.0028	2+, 3+
Copper	Cu	29	63.546	1085	2570	8.92	128	745.5	(2+)+0.34	13.38	0.38452	304	0.0058	1+, 2+
Curium	Cm	96	[247.0703]	1340	3540	13.51	174	581	(3+)−2.06	—	—	—	—	3+, 4+
Dubnium	Db	105	[262]	—	—	—	—	—	—	—	—	—	—	—
Dysprosium	Dy	66	162.5	1407	2600	8.536	178.1	572	(3+)−2.29	10.4	0.1733	250	6×10^{-4}	2+, 3+
Einsteinium	Es	99	[252.0828]	860	—	—	186	619	(3+)−2	—	—	—	—	3+
Erbium	Er	68	167.26	1497	2900	9.045	176.1	589	(3+)−2.32	17.2	0.1681	293	3.5×10^{-4}	3+
Europium	Eu	63	151.965	826	1439	5.245	208.4	547	(3+)−1.99	10.5	0.1820	176	2.1×10^{-3}	2+, 3+
Fermium	Fm	100	[257.0951]	—	—	—	—	627	(3+)−1.96	—	—	—	—	2+, 3+
Fluorine	F	9	18.9984032	−219.7	−188.2	0.001696	69	1681	(1−)+2.87	0.51	0.8238	6.54	0.0544	1−
Francium	Fr	87	[223.0197]	24	650	—	280	375	—	2	—	63.6	trace	1+
Gadolinium	Gd	64	157.25	1312	3000	7.886	180.4	592	(3+)−2.29	15.5	0.2355	311.7	6.3×10^{-4}	3+
Gallium	Ga	31	69.723	29.77	2203	5.904	134	578.8	(3+)−0.529	5.59	0.3709	256	0.0018	1+, 3+
Germanium	Ge	32	72.61	945	2850	5.323	123	761.2	(4+)+0.124	31.8	0.3215	334.3	1.5×10^{-4}	2+, 4+

* [ ] indicates mass of longest-lived isotope.

Table D.4 Properties of Elements (continued)

Element	Symbol	Atomic Number (Z)	Atomic Mass* (u)	Melting Point (°C)	Boiling Point (°C)	Density (g/cm³) (gases measured at STP)	Atomic Radius (pm)	First Ionization Energy (kJ/mol)	Standard Reduction Potential (V) (for elements from or to oxidation state indicated)	Enthalpy of Fusion (kJ/mol)	Specific Heat (J/g · C°)	Enthalpy of Vaporization (kJ/mol)	Abundance in Earth's Crust (%)	Major Oxidation States
Gold	Au	79	196.96654	1064	2808	19.32	144	889.9	(3+)+1.52	12.4	0.12905	324.4	3×10^{-7}	1+, 3+
Hafnium	Hf	72	178.49	2227	4691	13.28	159	654.4	(4+)−1.56	29.288	0.1442	661	3×10^{-4}	4+
Hassium	Hs	108	[265]	—	—	—	—	—	—	—	—	—	—	—
Helium	He	2	4.002602	−269.7 (2536 kPa)	−268.93	0.00017847	122	2372	—	0.02	5.1931	0.084	—	
Holmium	Ho	67	164.9032	1461	2600	8.78	176.2	581	(3+)−2.33	17.1	0.1646	251	1.5×10^{-4}	3+
Hydrogen	H	1	1.00794	−259.19	−252.76	0.0000899	78	1312	(1+) 0.0000	0.117	14.298	0.904	—	1−, 1+
Indium	In	49	114.82	156.61	2080	7.29	167	558.2	(3+)−0.3382	3.26	0.2407	231.8	2×10^{-5}	1+, 3+
Iodine	I	53	126.90447	113.6	184.5	4.93	138	1008.4	(1−)+0.5355	15.517	0.21448	41.95	4.6×10^{-5}	1−, 1+, 5+, 7+
Iridium	Ir	77	192.22	2447	4550	22.65	135.5	880	(4+)+0.926	26.4	0.1306	563.6	1×10^{-7}	3+, 4+, 5+
Iron	Fe	26	55.847	1536	2860	7.874	126	759.4	(3+)−0.4	13.807	0.4494	350	5.8	2+, 3+
Krypton	Kr	36	83.8	−157.2	−153.35	0.0037493	201	1351	—	1.64	0.2480	9.03	—	
Lanthanum	La	57	138.9055	920	3420	6.17	187	538	(3+)−2.37	8.5	0.1952	402	0.0035	3+
Lawrencium	Lr	103	[260.1054]	—	—	—	—	—	(3+)−2.06	—	—	—	—	3+
Lead	Pb	82	207.2	327	1746	11.342	175	715.6	(2+)−0.1251	4.77	0.1276	178	0.0013	2+, 4+
Lithium	Li	3	6.941	180.5	1347	0.534	156	520.2	(1+)−3.045	3	3.569	148	0.002	1+
Lutetium	Lu	71	174.967	1652	3327	9.84	173.8	524	(3+)−2.3	11.9	0.1535	414	8×10^{-5}	3+
Magnesium	Mg	12	24.305	650	1105	1.738	160	737.8	(2+)−2.356	8.477	1.024	127.4	2.76	2+
Manganese	Mn	25	54.93805	1246	2061	7.43	127	717.5	(2+)−1.18	12.058	0.4791	219.7	0.1	2+, 3+, 4+, 6+, 7+
Meitnerium	Mt	109	[266]	—	—	—	—	—	—	—	—	—	—	—
Mendelevium	Md	101	[258.0986]	—	—	—	—	635	—	—	—	—	—	2+, 3+
Mercury	Hg	80	200.59	−38.9	357	13.534	151	1007	(2+)+0.8535	2.2953	0.13950	59.1	2×10^{-6}	1+, 2+
Molybdenum	Mo	42	95.94	2623	4679	10.28	139	685	(6+) 0.114	36	0.2508	590	1.2×10^{-4}	4+, 5+, 6+
Neodymium	Nd	60	144.24	1024	3111	7.003	181.4	530	(3+)−2.32	7.13	0.1903	283.7	0.004	2+, 3+
Neon	Ne	10	20.1797	−248.61	−246.05	0.0008999	131	2081	—	0.34	1.0301	1.77	—	
Neptunium	Np	93	237.0482	640	3900	20.45	155	597	(5+)−0.91	9.46	—	336	—	2+, 3+, 4+, 5+, 6+
Nickel	Ni	28	58.6934	1455	2883	8.908	124	736.7	(2+)−0.257	17.15	0.4442	375	0.0075	2+, 3+, 4+
Niobium	Nb	41	92.90638	2477	4858	8.57	146	664.1	(5+)−0.65	26.9	0.2648	690	0.002	4+, 5+
Nitrogen	N	7	14.00674	−210	−195.8	0.0012409	71	1402	(3−)−0.092	0.72	1.0397	5.58	0.002	3−, 2−, 1−, 1+, 2+, 3+, 4+, 5+
Nobelium	No	102	[259.1009]	—	—	—	—	642	(2+)−2.5	—	—	—	—	2+, 3+
Osmium	Os	76	190.2	3045	5025	22.57	135	840	(4+)+0.687	31.7	0.130	627.6	2×10^{-7}	4+, 6+, 8+
Oxygen	O	8	15.9994	−218.8	−183	0.001429	60	1313.9	(2−) 0.815	0.44	0.91738	6.82	45.5	2−, 1−
Palladium	Pd	46	106.42	1552	2940	11.99	137	805	(2+) 0.915	17.6	0.2441	362	3×10^{-7}	2+, 4+
Phosphorus	P	15	30.973762	44.2	280.5	1.823	109	1012	(3−)−0.063	0.659	0.76968	49.8	0.11	3−, 3+, 5+
Platinum	Pt	78	195.08	1769	3824	21.41	138.5	868	(4+)+1.15	19.7	0.1326	510.4	1×10^{-6}	2+, 4+
Plutonium	Pu	94	[244.0642]	640	3230	19.86	162	585	(4+)−1.25	2.8	0.138	343.5	—	3+, 4+, 5+, 6+

* [ ] indicates mass of longest-lived isotope.

Table 4 Properties of Elements (continued)

Element	Symbol	Atomic Number (Z)	Atomic Mass* (u)	Melting Point (°C)	Boiling Point (°C)	Density (g/cm³) (gases measured at STP)	Atomic Radius (pm)	First Ionization Energy (kJ/mol)	Standard Reduction Potential (V) (for elements from or to oxidation state indicated)	Enthalpy of Fusion (kJ/mol)	Specific Heat (J/g • C°)	Enthalpy of Vaporization (kJ/mol)	Abundance in Earth's Crust (%)	Major Oxidation States
Polonium	Po	84	[208.9824]	254	962	9.4	164	813	(4+)+0.73	3.81	0.125	103	—	2−, 2+, 4+, 6+
Potassium	K	19	39.0983	63.2	766.4	0.862	231	418.8	(1+)−2.925	2.334	0.7566	76.9	1.84	1+
Praseodymium	Pr	59	140.90765	935	3343	6.782	182.4	522	(3+)−2.35	11.3	0.1930	332.6	9.1×10^{-4}	3+, 4+
Promethium	Pm	61	[144.9128]	1168	2460	7.2	183.4	536	(3+)−2.29	8.17	—	293	trace	3+
Protactinium	Pa	91	231.03588	1552	4227	15.37	163	568	(5+)−1.19	14.6	—	481	trace	3+, 4+, 5+
Radium	Ra	88	226.0254	700	1630	5	228	509.1	(2+)−2.916	8.36	—	136.8	—	2+
Radon	Rn	86	[222.0176]	−71	−62	0.00973	232	1037	—	16.4	—	16.4	—	
Rhenium	Re	75	186.207	3180	5650	21.232	137	760	(7+)+0.34	33.4	0.1368	707	1×10^{-7}	3+, 4+, 6+, 7+
Rhodium	Rh	45	102.9055	1960	3727	12.39	134	720	(3+)+0.76	21.6	0.2427	494	1×10^{-7}	3+, 4+, 5+
Rubidium	Rb	37	85.4678	39.5	697	1.532	248	403	(1+)−2.925	2.19	0.36344	69.2	0.0078	1+
Ruthenium	Ru	44	101.07	2310	4119	12.41	134	711	(4+)+0.68	25.5	0.2381	567.8	—	2+, 3+, 4+, 5+
Rutherfordium	Rf	104	[261]	—	—	—	—	—	—	—	—	—	—	—
Samarium	Sm	62	150.36	1072	1800	7.536	180.4	542	(3+)−2.3	8.9	0.1965	191	7×10^{-4}	2+, 3+
Scandium	Sc	21	44.95591	1539	2831	3	162	631	(3+)−2.03	15.77	0.5677	304.8	0.0022	3+
Seaborgium	Sg	106	[263]	—	—	—	—	—	—	—	—	—	—	—
Selenium	Se	34	78.96	221	685	4.79	117	940.7	(2−)− 0.67	5.43	0.3212	26.3	5×10^{-6}	2−, 2+, 4+, 6+
Silicon	Si	14	28.0855	1411	3231	2.336	118	786.5	(4−)−0.143	50.2	0.7121	359	27.2	2+, 4+
Silver	Ag	47	107.8682	961	2195	10.49	144	730.8	(1+)+0.7991	11.65	0.23502	255	8×10^{-6}	1+
Sodium	Na	11	22.989768	97.83	897.4	0.968	186	495.9	(1+)−2.714	2.602	1.228	97.4	2.27	1+
Strontium	Sr	38	87.62	776.9	1412	2.6	215	549.5	(2+)−2.89	7.4308	0.301	137	0.0384	2+
Sulfur	S	16	32.066	115.2	444.7	2.08	103	999.6	(2−)−0.45	1.7272	0.7060	9.62	0.03	2−, 4+, 6+
Tantalum	Ta	73	180.9479	2980	5505	16.65	146	760.8	(5+)−0.81	36.57	0.1402	737	2×10^{-4}	4+, 5+
Technetium	Tc	43	97.9072	2200	4567	11.5	136	702	(6+)+0.83	23.0	—	577	—	2+, 4+, 6+, 7+
Tellurium	Te	52	127.6	450	990	6.25	138	869	(2−)−1.14	17.4	0.2016	50.6	2×10^{-7}	2−, 2+, 4+, 6+
Terbium	Tb	65	158.92534	1356	2800	8.272	177.3	564	(3+)−2.31	10.3	0.1819	293	1×10^{-4}	3+, 4+
Thallium	Tl	81	204.3833	303.5	1457	11.85	170	589.1	(1+)−0.3363	4.27	0.1288	162	7×10^{-5}	1+, 3+
Thorium	Th	90	232.0381	1750	4787	11.78	179	587	(4+)−1.83	16.11	0.1177	543.9	8.1×10^{-4}	4+
Thulium	Tm	69	168.93421	1545	1727	9.318	175.9	596	(3+)−2.32	18.4	0.1600	213	5×10^{-5}	
Tin	Sn	50	118.71	232	2623	7.265	141	708.4	(4+)+0.064	7.07	0.2274	296	2.1×10^{-4}	2+, 4+
Titanium	Ti	22	47.88	1666	3358	4.5	147	658.1	(4+)−0.86	14.146	0.5226	425	0.63	2+, 3+, 4+
Tungsten	W	74	183.85	3680	6000	19.3	139	770.4	(6+)−0.09	35.4	0.1320	806	1.2×10^{-4}	4+, 5+, 6+
Uranium	U	92	238.0289	1130	3930	19.05	156	584	(6+)−0.83	12.6	0.11618	423	2.3×10^{-4}	3+, 4+, 5+, 6+
Vanadium	V	23	50.9415	1917	3417	6.11	134	650.3	(4+)−0.54	22.84	0.4886	459.7	0.0136	2+, 3+, 4+, 5+
Xenon	Xe	54	131.29	−111.8	−108.09	0.0058971	218	1170	—	2.29	0.15832	12.64	—	
Ytterbium	Yb	70	173.04	824	1427	6.973	193.3	603	(3+)−2.22	7.66	0.1545	155	3.4×10^{-4}	2+, 3+
Yttrium	Y	39	88.90585	1530	3264	4.5	180	616	(3+)−2.37	17.15	0.2984	393	0.0035	3+
Zinc	Zn	30	65.39	419.6	907	7.14	134	906.4	(2+)−0.7626	7.322	0.3884	115	0.0076	2+
Zirconium	Zr	40	91.224	1852	4400	6.51	160	659.7	(4+)−1.7	20.92	0.2780	590.5	0.0162	4+

* [ ] indicates mass of longest-lived isotope.

Table 5 Barometric Conversions to Kilopascals

Pressure (mm Hg)	Pressure (kPa)	Pressure (mm Hg)	Pressure (kPa)	Pressure (mm Hg)	Pressure (kPa)
730	97.3	747	99.6	764	101.9
731	97.5	748	99.7	765	102.0
732	97.6	749	99.9	766	102.1
733	97.7	750	100.0	767	102.3
734	97.9	751	100.1	768	102.4
735	98.0	752	100.3	769	102.5
736	98.1	753	100.4	770	102.7
737	98.3	754	100.5	771	102.8
738	98.4	755	100.7	772	102.9
739	98.5	756	100.8	773	103.1
740	98.7	757	100.9	774	103.2
741	98.8	758	101.1	775	103.3
742	98.9	759	101.2	776	103.5
743	99.1	**760**	**101.325**	777	103.6
744	99.2	761	101.5	778	103.7
745	99.3	762	101.6	779	103.9
746	99.5	763	101.7	780	104.0

Table 6 Vapor Pressure of Water

Temperature	Pressure		Temperature	Pressure	
(°C)	(mm Hg)	(kPa)	(°C)	(mm Hg)	(kPa)
0	4.6	0.61	23	21.1	2.81
1	4.9	0.65	24	22.4	2.99
2	5.3	0.71	25	23.8	3.17
3	5.7	0.76	26	25.2	3.36
4	6.1	0.81	27	26.7	3.56
5	6.5	0.87	28	28.3	3.77
6	7.0	0.93	29	30.0	4.00
7	7.5	1.00	30	31.8	4.24
8	8.0	1.07	35	42.2	5.63
9	8.6	1.15	40	55.3	7.37
10	9.2	1.23	45	71.9	9.59
11	9.8	1.31	50	92.5	12.3
12	10.5	1.40	55	118.0	15.73
13	11.2	1.49	60	149.4	19.86
14	12.0	1.60	65	187.5	25.00
15	12.8	1.71	70	233.7	31.16
16	13.6	1.81	75	239.1	38.54
17	14.5	1.93	80	355.1	47.34
18	15.5	2.07	85	433.6	57.81
19	16.5	2.20	90	525.8	70.10
20	17.5	2.33	95	633.9	84.51
21	18.7	2.49	100	760.0	101.3
22	19.8	2.64	105	906.1	120.8

Table 7 Solubility Guidelines

A substance is considered soluble if more than 3 grams of the substance dissolve in 100 mL of water. The more common rules are listed below.

1. All common salts of the Group 1(IA) elements and ammonium ions are soluble.
2. All common acetates and nitrates are soluble.
3. All binary compounds of Group 17(VIIA) elements (other than F) with metals are soluble except those of silver, mercury(I), and lead.
4. All sulfates are soluble except those of barium, strontium, lead, calcium, silver, and mercury(I).
5. Except for those in Rule 1, carbonates, hydroxides, oxides, sulfides, and phosphates are insoluble.

Table 8 Some Common Acids and Their Conjugate Bases

(equilibrium constants and relative strengths)

Acid	Base	K_a	Acid Strength
HI	I^-	very large	very strong
HBr	Br^-	very large	very strong
$HClO_4$	ClO_4^-	very large	very strong
HCl	Cl^-	very large	very strong
H_2SO_4	HSO_4^-	large	strong
HNO_3	NO_3^-	large	strong
H_3O^+	H_2O	1	strong
HOOCCOOH	$HOOCCOO^-$	5.36×10^{-2}	moderate
H_2SO_3	HSO_3^-	1.29×10^{-2}	moderate
HSO_4^-	SO_4^{2-}	1.02×10^{-2}	moderate
H_3PO_4	$H_2PO_4^-$	7.08×10^{-3}	moderate
HF	F^-	6.61×10^{-4}	weak
HNO_2	NO_2^-	7.24×10^{-4}	weak
$Al(H_2O)_6^{3+}$	$Al(H_2O)_5OH^{2+}$	1.4×10^{-5}	weak
CH_3COOH	CH_3COO^-	1.75×10^{-5}	weak
H_2CO_3	HCO_3^-	4.37×10^{-7}	weak
H_2S	HS^-	1.07×10^{-7}	weak
$H_2PO_4^-$	HPO_4^{2-}	6.31×10^{-8}	weak
HSO_3^-	SO_3^{2-}	6.17×10^{-8}	weak
NH_4^+	NH_3	5.7×10^{-10}	weak
HCO_3^-	CO_3^{2-}	4.68×10^{-11}	weak
H_2O_2	HO_2	2.24×10^{-12}	very weak
HPO_4^{2-}	PO_4^{3-}	4.17×10^{-13}	very weak
HS^-	S^{2-}	1.26×10^{-13}	very weak
H_2O	OH^-	1.00×10^{-14}	very weak

Table 9 Acid-Base Indicators

Indicator	Lower Color	Range	Upper Color
Methyl violet	yellow-green	0.0–2.5	violet
Malachite green HCl	yellow	0.5–2.0	blue
Thymol blue	red	1.0–2.8	yellow
Naphthol yellow S	colorless	1.5–2.6	yellow
p-Phenylazoaniline	orange	2.1–2.8	yellow
Methyl orange	red	2.5–4.4	yellow
Bromophenol blue	orange-yellow	3.0–4.7	violet
Gallein	orange	3.5–6.3	red
2,5-Dinitrophenol	colorless	4.0–5.8	yellow
Ethyl orange	salmon	4.2–4.6	orange
Propyl red	pink	5.1–6.5	yellow
Bromocresol purple	green-yellow	5.4–6.8	violet
Bromoxylenol blue	orange-yellow	6.0–7.6	blue
Phenol red	yellow	6.4–8.2	red-violet
Cresol red	yellow	7.1–8.8	violet
m-Cresol purple	yellow	7.5–9.0	violet
Thymol blue	yellow	8.1–9.5	blue
Phenolphthalein	colorless	8.3–10.0	dark pink
o-Cresolphthalein	colorless	8.6–9.8	pink
Thymolphthalein	colorless	9.5–10.4	blue
Alizarin yellow R	yellow	9.9–11.8	dark orange
Methyl blue	blue	10.6–13.4	pale violet
Acid fuchsin	red	11.1–12.8	colorless
2,4,6-Trinitrotoluene	colorless	11.7–12.8	orange

Table 10 Solubilities in Water

(S = soluble; P = slightly soluble; I = insoluble; D = decomposes in water; — = compound does not exist or is unstable.)

Nonmetal anion → / Metal cation	Acetate, $C_2H_3O_2^-$	Bromide, Br^-	Carbonate, CO_3^{2-}	Chlorate, ClO_3^-	Chloride, Cl^-	Chromate, CrO_4^{2-}	Hydroxide, OH^-	Iodide, I^-	Nitrate, NO_3^-	Oxide, O^{2-}	Oxalate, $C_2O_4^{2-}$	Phosphate, PO_4^{3-}	Silicate, SiO_3^{2-}	Sulfate, SO_4^{2-}	Sulfide, S^{2-}	Sulfite, SO_3^{2-}
Aluminum, Al^{3+}	S	S	—	S	S	—	I	S	S	I	I	I	I	S	D	—
Ammonium, NH_4^+	S	S	S	S	S	S	S	S	S	—	P	S	—	S	S	S
Antimony, Sb^{3+}	—	D	—	—	S	—	—	D	—	P	I	—	—	D	D	—
Arsenic, As^{3+}	—	D	—	—	D	—	—	S	—	P	—	—	—	—	I	—
Barium, Ba^{2+}	S	S	I	S	S	I	S	S	S	S	I	I	—	I	D	I
Bismuth, Bi^{3+}	I	D	—	—	D	—	I	I	D	I	D	I	I	D	I	—
Cadmium, Cd^{2+}	S	S	I	S	S	I	I	S	S	I	I	I	I	S	I	P
Calcium, Ca^{2+}	S	S	I	S	S	S	I	S	S	I	I	I	I	I	P	I
Chromium, Cr^{3+}	S	S(I)*	—	—	I	—	I	S	S	I	S	P	—	S	D	I
Cobalt, Co^{2+}	S	S	I	S	S	I	I	S	S	I	I	I	I	S	I	I
Copper, Cu^{2+}	S	S	I	S	S	—	I	I	S	I	I	I	—	S	I	—
Iron (III), Fe^{3+}	S	S	—	S	S	I	I	S	S	I	S	P	—	S	D	—
Iron (II), Fe^{2+}	S	S	P	S	S	—	I	S	S	I	I	I	I	S	I	I
Lead, Pb^{2+}	S	S	I	S	S	I	I	I	S	I	I	I	I	I	I	I
Magnesium, Mg^{2+}	S	S	P	S	S	S	I	S	S	I	I	I	I	S	D	S
Mercury (II), Hg^{2+}	S	S	—	S	S	P	—	I	S	I	I	—	—	D	I	—
Mercury (I), Hg_2^{2+}	P	I	I	S	I	P	—	P	D	I	I	—	—	I	I	—
Nickel, Ni^{2+}	S	S	I	S	S	I	I	S	S	I	I	I	—	S	I	I
Potassium, K^+	S	S	S	S	S	S	S	S	S	S	S	S	S	S	S	S
Silver, Ag^+	P	I	I	S	I	I	—	I	S	I	I	I	—	I	I	I
Sodium, Na^+	S	S	S	S	S	S	S	S	S	S	S	S	S	S	S	S
Strontium, Sr^{2+}	S	S	I	S	S	I	I	S	S	I	I	I	I	I	I	I

* Certain salts occur in two modifications.

Table 11 Solubility Product Constants (at 25°C)

Substance	K_{sp}	Substance	K_{sp}	Substance	K_{sp}
AgBr	5.01×10^{-13}	$BaSO_4$	1.10×10^{-10}	Li_2CO_3	2.51×10^{-2}
$AgBrO_3$	5.25×10^{-5}	$CaCO_3$	2.88×10^{-9}	$MgCO_3$	3.47×10^{-8}
Ag_2CO_3	8.13×10^{-12}	$CaSO_4$	9.12×10^{-6}	$MnCO_3$	1.82×10^{-11}
AgCl	1.78×10^{-10}	CdS	7.94×10^{-27}	$NiCO_3$	6.61×10^{-9}
Ag_2CrO_4	1.12×10^{-12}	$Cu(IO_3)_2$	7.41×10^{-8}	$PbCl_2$	1.62×10^{-5}
$Ag_2Cr_2O_7$	2.00×10^{-7}	CuC_2O_4	2.29×10^{-8}	PbI_2	7.08×10^{-9}
AgI	8.32×10^{-17}	$Cu(OH)_2$	2.19×10^{-20}	$Pb(IO_3)_2$	3.24×10^{-13}
AgSCN	1.00×10^{-12}	CuS	6.31×10^{-36}	$SrCO_3$	1.10×10^{-10}
$Al(OH)_3$	1.26×10^{-33}	FeC_2O_4	3.16×10^{-7}	$SrSO_4$	3.24×10^{-7}
Al_2S_3	2.00×10^{-7}	$Fe(OH)_3$	3.98×10^{-38}	TlBr	3.39×10^{-6}
$BaCO_3$	5.13×10^{-9}	FeS	6.31×10^{-18}	$ZnCO_3$	1.45×10^{-11}
$BaCrO_4$	1.17×10^{-10}	Hg_2SO_4	7.41×10^{-7}	ZnS	1.58×10^{-24}

Table 12 Standard Reduction Potentials (at 25°C, 101.325 kPa, 1*M*)

↑ Weak Oxidizing Agents/Strong Reducing Agents

Half-Reaction	E° (Volts)	Half-Reaction	E° (Volts)
$Li^+ + e^- \rightarrow Li$	−3.040	$SO_4^{2-} + 4H^+ + 2e^- \rightarrow H_2SO_3 + H_2O$	0.158
$K^+ + e^- \rightarrow K$	−2.924	$Cu^{2+} + e^- \rightarrow Cu^+$	0.159
$Rb^+ + e^- \rightarrow Rb$	−2.924	$HAsO_2 + 3H^+ + 3e^- \rightarrow As + 2H_2O$	0.248
$Cs^+ + e^- \rightarrow Cs$	−2.923	$UO_2^{2+} + 4H^+ + 2e^- \rightarrow U^{4+} + 2H_2O$	0.27
$Ba^{2+} + 2e^- \rightarrow Ba$	−2.92	$Bi^{3+} + 3e^- \rightarrow Bi$	0.317 2
$Sr^{2+} + 2e^- \rightarrow Sr$	−2.89	$Cu^{2+} + 2e^- \rightarrow Cu$	0.340
$Ca^{2+} + 2e^- \rightarrow Ca$	−2.84	$O_2 + 2H_2O + 4e^- \rightarrow 4OH^-$	0.401
$Na^+ + e^- \rightarrow Na$	−2.713	$Cu^+ + e^- \rightarrow Cu$	0.520
$La^{3+} + 3e^- \rightarrow La$	−2.37	$I_2 + 2e^- \rightarrow 2I^-$	0.535 5
$Mg^{2+} + 2e^- \rightarrow Mg$	−2.356	$H_3AsO_4 + 2H^+ + 2e^- \rightarrow HAsO_2 + 2H_2O$	0.560
$Ce^{3+} + 3e^- \rightarrow Ce$	−2.34	$O_2 + 2H^+ + 2e^- \rightarrow H_2O_2$	0.695
$Nd^{3+} + 3e^- \rightarrow Nd$	−2.32	$Rh^{3+} + 3e^- \rightarrow Rh$	0.7
$H_2 + 2e^- \rightarrow 2H^-$	−2.25	$Tl^{3+} + 3e^- \rightarrow Tl$	0.72
$Sc^{3+} + 3e^- \rightarrow Sc$	−2.03	$Fe^{3+} + e^- \rightarrow Fe^{2+}$	0.771
$Be^{2+} + 2e^- \rightarrow Be$	−1.97	$NO_3^- + 2H^+ + e^- \rightarrow NO_2 + H_2O$	0.775
$Al^{3+} + 3e^- \rightarrow Al$	−1.676	$Hg_2^{2+} + 2e^- \rightarrow 2Hg$	0.796 0
$U^{3+} + 3e^- \rightarrow U$	−1.66	$Ag^+ + e^- \rightarrow Ag$	0.799 1
$Ti^{2+} + 2e^- \rightarrow Ti$	−1.63	$O_2 + 4H^+(10^{-7}M) + 4e^- \rightarrow 2H_2O$	0.815
$Hf^{4+} + 4e^- \rightarrow Hf$	−1.56	$AmO_2^+ + 4H^+ + e^- \rightarrow Am^{4+} + 2H_2O$	0.82
$No^{3+} + 3e^- \rightarrow No$	−1.2	$NO_3^- + 2H^+ + 2e^- \rightarrow NO_2^- + H_2O$	0.835
$Mn^{2+} + 2e^- \rightarrow Mn$	−1.18	$OsO_4 + 8H^+ + 8e^- \rightarrow Os + 4H_2O$	0.84
$Cr^{2+} + 2e^- \rightarrow Cr$	−0.90	$Hg^{2+} + 2e^- \rightarrow Hg$	0.853 5
$2H_2O + 2e^- \rightarrow H_2 + 2OH^-$	−0.828	$2Hg^{2+} + 2e^- \rightarrow Hg_2^{2+}$	0.911 0
$Zn^{2+} + 2e^- \rightarrow Zn$	−0.762 6	$Pd^{2+} + 2e^- \rightarrow Pd$	0.915
$Cr^{3+} + 3e^- \rightarrow Cr$	−0.74	$NO_3^- + 4H^+ + 3e^- \rightarrow NO(g) + 2H_2O$	0.957
$Ga^{3+} + 3e^- \rightarrow Ga$	−0.529	$Br_2 + 2e^- \rightarrow 2Br^-$	1.065 2
$U^{4+} + e^- \rightarrow U^{3+}$	−0.52	$SeO_4^{2-} + 4H^+ + 2e^- \rightarrow H_2SeO_3 + H_2O$	1.151
$2CO_2 + 2H^+ + 2e^- \rightarrow H_2C_2O_4$	−0.475	$Ir^{3+} + 3e^- \rightarrow Ir$	1.156
$S + 2e^- \rightarrow S^{2-}$	−0.447	$Pt^{2} + 2e^- \rightarrow Pt$	1.188
$Fe^{2+} + 2e^- \rightarrow Fe$	−0.44	$O_2 + 4H^+ + 4e^- \rightarrow 2H_2O$	1.229
$Cr^{3+} + e^- \rightarrow Cr^{2+}$	−0.424	$Tl^{3+} + 2e^- \rightarrow Tl^+$	1.25
$2H_2O + 2e^- \rightarrow H_2 + 2OH^-(10^{-7}M)$	−0.414	$Pd^{4+} + 2e^- \rightarrow Pd^{2+}$	1.263
$Cd^{2+} + 2e^- \rightarrow Cd$	−0.402 5	$Cl_2 + 2e^- \rightarrow 2Cl^-$	1.358 28
$Ti^{3+} + e^- \rightarrow Ti^{2+}$	−0.37	$Au^{3+} + 2e^- \rightarrow Au^+$	1.36
$PbI_2 + 2e^- \rightarrow Pb + 2I^-$	−0.365	$Cr_2O_7^{2-} + 14H^+ + 6e^- \rightarrow 2Cr^{3+} + 7H_2O$	1.36
$PbSO_4 + 2e^- \rightarrow Pb + SO_4^{2-}$	−0.350 5	$MnO_4^- + 8H^+ + 5e^- \rightarrow Mn^{2+} + 4H_2O$	1.51
$In^{3+} + 3e^- \rightarrow In$	−0.338 2	$Au^{3+} + 3e^- \rightarrow Au$	1.52
$Tl^+ + e^- \rightarrow Tl$	−0.336 3	$H_5IO_6 + H^+ + 2e^- \rightarrow IO_3^- + 3H_2O$	1.603
$Co^{2+} + 2e^- \rightarrow Co$	−0.277	$2HBrO + 2H^+ + 2e^- \rightarrow Br_2 + 2H_2O$	1.604
$H_3PO_4 + 2H^+ + 2e^- \rightarrow H_3PO_3 + H_2O$	−0.276	$PbO_2 + SO_4^{2-} + 4H^+ + 2e^- \rightarrow PbSO_4 + 2H_2O$	1.698
$Ni^{2+} + 2e^- \rightarrow Ni$	−0.257	$H_2O_2 + 2H^+ + 2e^- \rightarrow 2H_2O$	1.763
$Sn^{2+} + 2e^- \rightarrow Sn$	−0.136	$Au^+ + e^- \rightarrow Au$	1.83
$Pb^{2+} + 2e^- \rightarrow Pb$	−0.125 1	$Co^{3+} + e^- \rightarrow Co^{2+}$	1.92
$Hg_2I_2 + 2e^- \rightarrow 2Hg + 2I^-$	−0.040 5	$S_2O_8^{2-} + 2e^- \rightarrow 2SO_4^{2-}$	1.96
$Fe^{3+} + 3e^- \rightarrow Fe$	−0.04	$O_3 + 2H^+ + 2e^- \rightarrow O_2 + H_2O$	2.075
$2H^+ + 2e^- \rightarrow H_2$	0.000 0	$F_2 + 2e^- \rightarrow 2F^-$	2.87
$Sn^{4+} + 2e^- \rightarrow Sn^{2+}$	0.154	$F_2 + 2H^+ + 2e^- \rightarrow 2HF$	3.053

Strong Oxidizing Agents/Weak Reducing Agents ↓

Name

Date Class

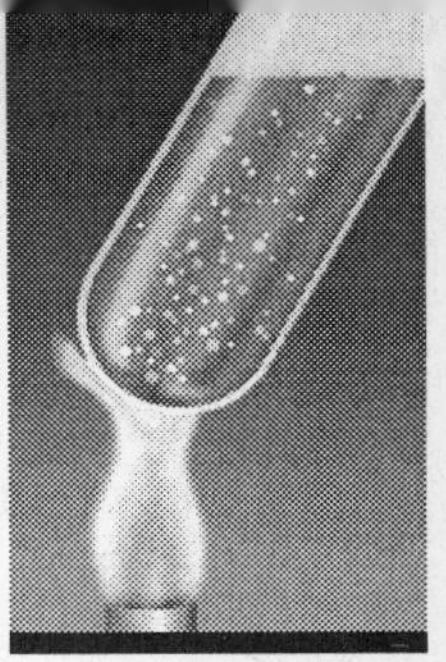

LAB MANUAL

1-1

Determining Boiling Point

Boiling point is an important physical property of chemical substances. The boiling point of a liquid is the temperature at which the vapor pressure of the liquid is equal to the atmospheric pressure above the liquid. When a liquid reaches the boiling point, characteristic bubbles usually appear in the liquid. Because the boiling point of any liquid varies with the pressure on the liquid, the standard boiling point is determined for all substances at standard atmospheric pressure, 101.3 kPa (760 mm of mercury or 1 atmosphere). In this experiment, you will determine the boiling points of several liquids.

OBJECTIVES

- **Measure** the boiling point of different liquids.
- **Identify** an unknown liquid by its boiling point.

MATERIALS

apron
goggles
thin-stem pipet
capillary tube
small test tube
100-mL beaker
rubber band
ring stand
thermometer (–10°C to 100°C) (2)
thermometer clamp
hot plate
glass stirring rod

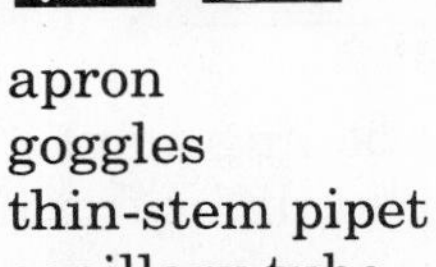

PROCEDURE

1. Obtain a sample liquid and a capillary tube that is sealed at one end.
2. Put the capillary tube with the sealed end up into a small test tube.
3. With a thin-stem pipet, put 3 mL of the sample liquid into the test tube.
4. Use a rubber band to attach the test tube to the thermometer. Position the bottom of the test tube at the same level as the bulb of the thermometer, as shown in Figure A.
5. Use the thermometer clamp to attach the test tube and-thermometer assembly to the ring stand.
6. Fill the small beaker about two-thirds full with tap water. Place the beaker on the hot plate beneath the test tube-and-thermometer assembly.
7. Begin heating the water bath. With the second thermometer, measure the temperature from time to time. Do not leave this thermometer in the water bath.
8. When the water bath has reached a temperature specified by your teacher, loosen the thermometer clamp and lower the test tube-and-thermometer assembly into the beaker of water, as shown in Figure B on the next page. Arrange the assembly so that the test tube and thermometer are in the water but the water level does not go above the top of the test tube. Tighten the clamp to hold the assembly in this position.

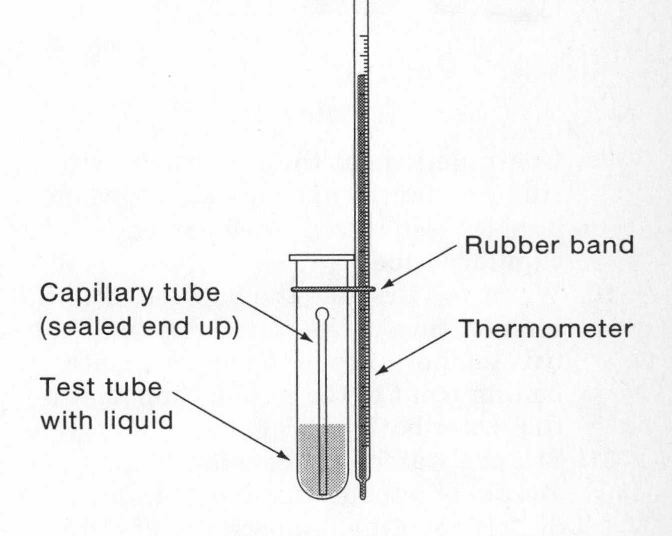

Figure A

Copyright © Glencoe/McGraw-Hill, a division of The McGraw-Hill Companies, Inc.

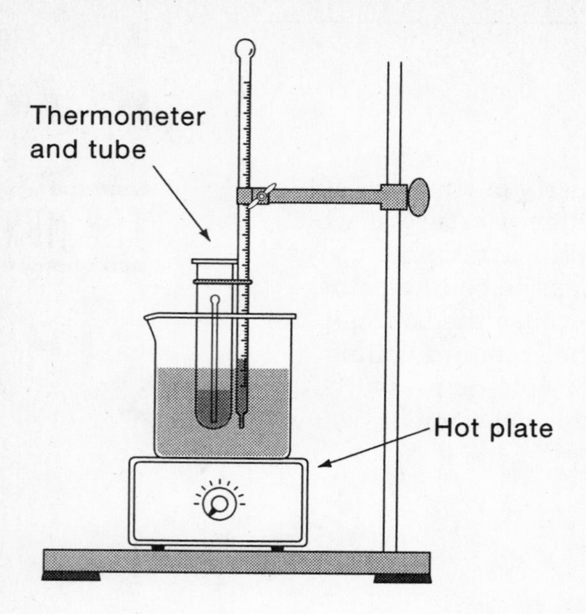

Figure B

9. Continue to heat the water bath, stirring constantly, until a steady stream of bubbles issues from the open end of the capillary tube.
10. When you first see the bubbles, note the temperature of the water. Assume that this temperature is the approximate boiling point of the liquid. Stop heating the water bath.
11. Stir the water bath constantly until the stream of bubbles just stops. Immediately read the temperature of the water and record it in the table under Data and Observations. This is the exact boiling point of the liquid.
12. Compare the boiling point you obtained with the standard boiling point for the liquid, which is listed in the *Handbook of Chemistry and Physics* or will be provided by your teacher.
13. Discard the remaining liquid according to your teacher's directions.
14. Repeat steps 1 through 13 with each of the other liquids provided.
15. Repeat steps 1 through 13 with the unknown liquid provided. Determine the possible identity of the unknown by comparing its boiling point with the boiling points of other liquids, as directed by your teacher.

DATA AND OBSERVATIONS

Liquid	Boiling Point		Percentage Error
	Measured (°C)	Listed (°C)	
Ethyl alcohol			
Acetone			
N-propanol			
Isopropyl alcohol			
Unknown			

Identity of unknown: ____________________

ANALYSIS

1. Why did you record the temperature when the bubbles *stopped* appearing as the boiling point of each liquid?

____

2. Determine the percentage error for each of your results and record it in the table. Use the equation below to determine the percentage error.

$$\% \text{ error} = \frac{\text{observed temp.} - \text{expected temp.}}{\text{expected temp.}} \times 100\%$$

3. Suggest reasons why the boiling point you found for a liquid might differ from the listed boiling point for the liquid.

____

____

CONCLUSIONS

Boiling point is a physical property. How can this property be used to identify an unknown liquid?

____

____

EXTENSION AND APPLICATION

What do you know about the boiling point for the liquids used in this experiment simply by examining the Procedure directions? How can the boiling point of a liquid with a boiling point greater than 100°C be determined?

Copyright © Glencoe/McGraw-Hill, a division of The McGraw-Hill Companies, Inc.

Teacher Guide

LAB MANUAL

1-1

Determining Boiling Point

Since boiling point is dependent on the absolute pressure on a liquid, a liquid may boil at any temperature. The higher the pressure on a liquid, the higher is the boiling point of the liquid. The molecules of a liquid under high pressure require much more thermal energy to go into the gaseous phase than do the molecules of a liquid under normal atmospheric pressure.

Misconception: Many students think that all liquids boil at 100°C. They usually do not know that the boiling point of a liquid changes with the pressure on the liquid.

PROCESS SKILLS

1. Students will determine the boiling points of several liquids.
2. Students will identify an unknown liquid by its boiling point.

PREPARATION HINTS

- If possible, use hot plates instead of burners. If burners must be used, eliminate acetone as a sample liquid. You might be able to avoid use of both these heat sources if the water from the tap is hot enough.
- **CAUTION:** *Do not use any liquid other than those suggested. Never, under any circumstances, use gasoline, ether, or other organic liquids that have a low flash point.*
- Distribute 3 mL-samples of the liquids in stoppered test tubes. Never allow students access to large quantities of sample liquids.
- Prepare capillary tubes sealed at one end for each sample liquid.
- Provide duplicates of one of the sample liquids for use as the unknown.

MATERIALS

ethyl alcohol
isopropyl alcohol
n-propanol
acetone
capillary tube
13 × 100 test tube
apron
goggles
thin-stem pipet
100-mL beaker
rubber band
ring stand
thermometer (–10°C to 100°C) (2)
thermometer clamp
hot plate (or burner, ring, and wire gauze)
glass stirring rod

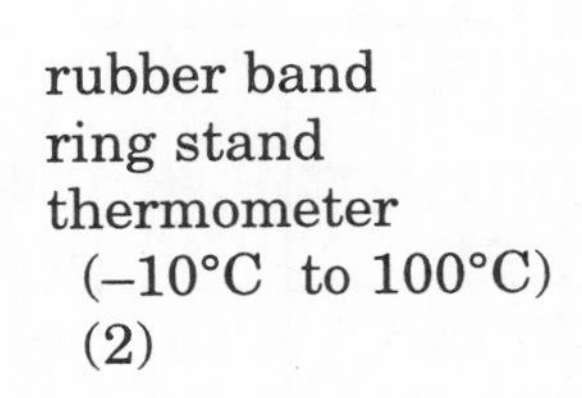

PROCEDURE HINTS

- Before proceeding, be sure students have read and understood the purpose, procedure, and safety precautions for this laboratory activity.
- Students should use a different dry test tube for each boiling point determination. One droplet of water in the bottom of a tube can cause inaccurate results.
- Before students lower the test tubes containing sample liquids into the water bath, have them heat the water to a temperature at least 5°C above the boiling point of the sample liquid.
- If hot water from the tap is at least 5°C higher than the boiling point of the liquid, it can be used in the water bath, eliminating the need for a burner or hot plate.
- Provide a list of the standard boiling points of the sample liquids for students to compare with their data or make a reference such as *Handbook of Chemistry and Physics* available for students' use.
- To identify the unknown liquid, students should compare its boiling point with the data they obtained for the sample liquids.
- Allow test tubes to dry thoroughly before they are used by another group or class.

- **Safety Precaution:** *If burners must be used, they must be turned off before the sample liquid is placed in the water bath. This will prevent the combustion of vapors from the sample.*
- **Troubleshooting:** The most common cause of error is failure to notice a steady stream of bubbles from the capillary tube. Emphasize that the stream indicates that the liquid is already boiling. When the stream stops as the liquid cools, the boiling temperature has been reached.
- **Disposal:** Sample liquids may be disposed of according to the instructions in **Disposal** J.

DATA AND OBSERVATIONS

Liquid	Boiling Point		Percentage Error
	Measured (°C)	Listed (°C)	
Ethyl alcohol	78.0	78.5	0.64%
Acetone	57.0	56.5	0.88%
N-propanol	96.5	97.2	0.72%
Isopropyl alcohol	81.5	82.5	1.2%

ANALYSIS

1. When a steady stream of bubbles can be seen in a liquid, the liquid is already boiling. Letting the liquid cool to the exact point the bubbles stop gives the exact boiling point temperature.
2. See the table for percentage errors of sample data.
3. Answers will vary, but should include the suggestion that the atmospheric pressure in the laboratory might be higher or lower than the standard atmospheric pressure.

CONCLUSIONS

Since boiling point is a physical property, the boiling point of an unknown liquid can be determined and then compared with a list of known boiling points for substances. This information alone may be enough to identify the unknown, or it may be used in combination with other properties to identify the substance.

EXTENSION AND APPLICATION

The boiling points for all liquids used in this experiment must be below 100°C because this temperature is the upper limit of the boiling water bath used to heat the sample.

To determine the boiling point of a liquid that has a boiling point higher than 100°C, the liquid must be heated in a material that has a boiling point higher than that of water. Instead of a water bath, an oil bath of vegetable or mineral oil, with a boiling point higher than 250°C, could be used.

Copyright © Glencoe/McGraw-Hill, a division of The McGraw-Hill Companies, Inc.

Name

Date Class

The Eight-Solution Problem

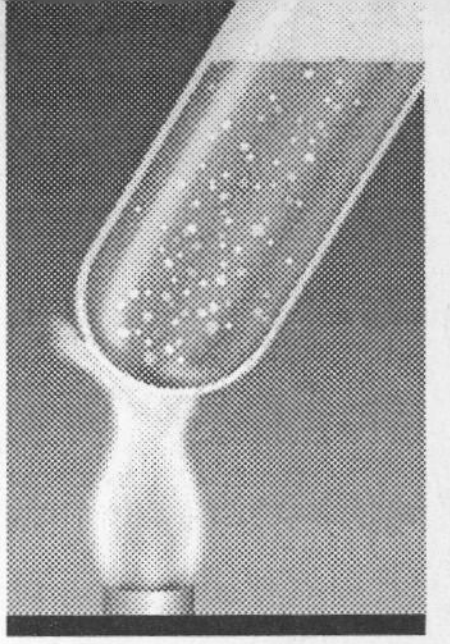

LAB MANUAL

1-2

Scientists are problem solvers, and one of the most important ways they gain the information they need to solve problems is by making detailed and accurate observations. Making observations is a key process of the experimental procedure; often, the observations made in one experiment can be applied to solving a different problem. This laboratory experiment is an example of problem solving by use of observations.

When substances are mixed together, one of two things can happen. First, no change may occur. The substances are stirred together but remain the same substances. Second, a change may occur. The substances interact in some way that results in the formation of different substances. This type of change is called a chemical change. The process or procedure that brings the change about is called a chemical reaction. A chemical reaction is often accompanied by a change that can be seen, such as a change in color, the formation of a solid, or the generation of a gas. In this experiment, you will observe the changes brought about when eight known substances are combined in different ways. You will then use the observations to identify the same eight substances when they are supplied as unknowns.

OBJECTIVES

- **Observe** the results of mixing water solutions of eight different known substances.
- **Record** observations.
- **Identify** the same substances in unknown solutions.

MATERIALS

96-well Microplate Data Form (3)
plastic wrap
microtip pipets
beaker
distilled water
toothpicks
goggles
apron

PROCEDURE

Part 1

1. In this activity you will observe what happens when solutions of eight substances are mixed two by two. Before you begin, form a **hypothesis** about how you could use these observations to identify unknown solutions of the same substances. Record your hypothesis under Data and Observations.
2. Place a Microplate Data Form on your lab bench so that the numbered columns are away from you and the lettered rows are at the left.
3. Cover the Microplate Data Form with a piece of plastic wrap.
4. Assign one of the eight known solutions to each of rows A–H. Nitric acid, for example, might be in row A, silver nitrate in row B, and so on. Label a second Microplate Data Form by writing the name of each solution next to the letter of its row. Part of a labeled data form is shown in Table 1 under Data and Observations.
5. *In the same order as in step 4,* assign a known solution to each of columns 1–8. Write the names above the columns on the labeled Microplate Data Form.
6. The labels on your Microplate Data Form show the solutions to be mixed in each square. For example, if your data form matched Table 1 in Data and Observations, you would mix silver nitrate and nitric acid in well B1. To avoid wasting chemicals, do not mix a chemical with itself and do not mix the

Copyright © Glencoe/McGraw-Hill, a division of The McGraw-Hill Companies, Inc.

same two substances more than once. Place X's in the squares of your labeled Microplate Data Form to show where you do not need to mix solutions.

7. Using a clean microtip pipet, place 3 drops of the solution you assigned to row B onto the plastic wrap over each square in row B that does not have an X. **CAUTION:** *Many of the chemicals you will use are toxic. Nitric acid and sodium hydroxide can burn the skin and make holes in clothing. Silver nitrate can stain the skin black. Follow proper chemical hygiene procedures. Wash your hands thoroughly after completing this laboratory activity*. Once you have placed solutions on the plastic wrap, do not move the data form and be careful not to touch the chemicals on it.
8. Fill the beaker with distilled water. Rinse the microtip pipet three times in the water. Then add 3 drops of the solution you assigned to row C to every square in row C that does not have an X. Repeat for rows D–H, *being sure to rinse the pipet three times each time you change chemicals.*
9. Now add 3 drops of the solution you assigned to column 1 to every square in column 1 that does not have an X. Rinsing the pipet each time you change substances, repeat this procedure for columns 2–8.
10. Stir the substances on each square with a toothpick, using a different toothpick for each square.
11. Examine each square for any observable changes in the substances. Record your observations in your labeled Microplate Data Form. Where you cannot observe a change, write *NR* for "no reaction."
12. Discard the solutions and plastic sheet as directed by your teacher.

Part 2

1. Set up your Microplate Data Form again with a clean piece of plastic wrap, repeating steps 2 and 3 of Part 1.
2. Each of the unknown solutions you will work with is identified by a letter. Using these letters, assign the unknowns to lettered rows and numbered columns of the Microplate Data Form as you did for the known solutions (steps 4 and 5 of Part 1). Label a new Microplate Data Form, as in Table 2, to show where you will place the unknown solutions. Mark each square where you will not mix solutions with an X (step 6 of Part 1).
3. Repeat steps 7–11 of Part 1. Make careful observations and record your observations in your Microplate Data Form.
4. Discard the solutions and plastic sheet as directed by your teacher.

DATA AND OBSERVATIONS

Hypothesis: ____________________

____

____

Your Microplate Data Forms should be set up following the method used in the samples below, each of which shows only a part of a Microplate Data Form.

Table 1. Known Solutions

		Nitric acid 1	Silver nitrate 2	Sodium iodide 3
Nitric acid	A	X	X	X
Silver nitrate	B		X	X
Sodium iodide	C			X

In Table 2, the unknown solutions are identified by the letters *S*–*U*. You should use the code your teacher provides for the unknowns.

Table 2. Unknown Solutions

		S 1	T 2	U 3
S	A	X	X	X
T	B		X	X
U	C			X

Copyright © Glencoe/McGraw-Hill, a division of The McGraw-Hill Companies, Inc.

ANALYSIS

1. Compare your observations of the known solutions with your observations of the unknown solutions. Identify one combination of known solutions and one combination of unknown solutions that clearly gave the same result. Complete the following sentences.

   Known solution of ____________________

   + known solution of ____________________

   produced ____________________

   Unknown solution ____________________

   + unknown solution ____________________
   produced the same result.

2. Using the information from step 1 in Analysis as a clue, continue to compare and match up your results for the knowns and unknowns until you have identified all the unknown solutions. Write the identity of each unknown in the table provided under Conclusions.

CONCLUSIONS

1. Complete the following table.

Unknown Solution	Name of Substance
S	
T	
U	
V	
W	
X	
Y	
Z	

2. What three different types of changes did you observe that can indicate a chemical reaction has occurred?

   ____

   ____

   ____

3. According to your observations, how many different chemical reactions occurred among the known solutions?

   ____

   ____

   ____

4. Was the method you used to identify the unknowns similar to what you hypothesized? Explain.

   ____

   ____

   ____

   ____

   ____

EXTENSION AND APPLICATION

Put about one-half teaspoon of table sugar on a plate and drop a few drops of vinegar onto the sugar. Repeat this experiment with baking soda (not baking powder) and vinegar. Do sugar and vinegar give a chemical reaction? Do baking soda and vinegar give a chemical reaction? Which two known solutions in the experiment reacted in a similar way to the reaction you observed here?

Copyright © Glencoe/McGraw-Hill, a division of The McGraw-Hill Companies, Inc.

Teacher Guide

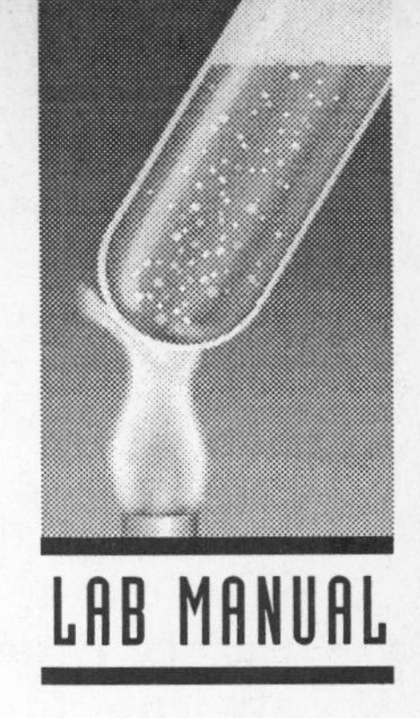

The Eight-Solution Problem

The main focus of this experiment is problem solving. It also provides students with their first experiences in using micropipets and in making and recording observations.

Students are given minimal information on how to use their observations of reactions between known solutions to identify unknown solutions. To make the experiment even more challenging, you can choose not to distribute the Microplate Data Forms but, instead, allow students to devise their own methods for solving the problem. After they have solved the problem, it is rewarding for them to see the differences in the methodology used by different lab groups. Students also like to compare their methods and results with the "book" method. There are many ways to solve this problem. The best way is the way that gives the best results.

Misconception: At first, students may perceive this problem as impossible to solve. Allow them time to think and plan what they will do.

PROCESS SKILLS

1. Students will mix solutions, observe reactions, and observe changes.
2. Students will identify eight unknowns by observing reactions the unknowns undergo and comparing these observations with those they made using known solutions.

PREPARATION HINTS

- All solutions except the silver nitrate solution can be prepared many months in advance. The silver nitrate solution must be freshly prepared and stored in a brown glass bottle. Old silver nitrate becomes black, and observant students may use the black color alone to identify the unknown.
- In preparing the salt solutions, be sure to check the hydration of your stock salt and adjust the masses needed, if necessary.
- Three copies of the Microplate Data Form for a 96-well microplate are needed for each lab group. Make and distribute additional copies as required.

MATERIALS

0.1 *M* solutions of the following compounds:

HNO_3 (6 mL of concentrated nitric acid in 1 L of solution)
$AgNO_3$ (17 g in 1 L of solution)
NaI (14.9 g in 1 L of solution)
$Pb(NO_3)_2$ (33.1 g in 1 L of solution)
Na_2CO_3 (10.6 g in 1 L of solution)
$Fe(NO_3)_3 \cdot 6H_2O$ (28.7 g in 1 L of solution)
NaOH (4.0 g in 1 L of solution)
$Cu(NO_3)_2 \cdot 3H_2O$ (24.1 g in 1 L solution)

Microplate Data Forms for 96-well microplate (3)
plastic wrap
microtip pipets (2)
beaker
distilled water
toothpicks

PROCEDURE HINTS

- Before proceeding, be sure students have read and understood the purpose, procedure, and safety precautions for this laboratory activity.
- You might want to precede this lab with a brief introduction to chemical reactions, a topic that is not fully covered until later in the course. To avoid misconceptions, you can note that not all chemical reactions are accompanied by changes visible to the naked eye nor do visible changes always indicate chemical reactions.
- The experiment itself does not take a great deal of time. The most time is spent in planning, observing, and analyzing the results. Encourage students to test their

Copyright © Glencoe/McGraw-Hill, a division of The McGraw-Hill Companies, Inc.

identifications of the unknowns by experiment before giving you their final list.

- Do not use the same letter code for the unknown solutions from class to class or from year to year.
- Remind students not to lean onto the plastic wrap with solutions on it.
- **Safety Precautions:** Be sure that students do not mix the solutions in a haphazard manner. Encourage only orderly and structured experimentation. Remind students to handle the solutions carefully.
- **Disposal:** Dispose of chemicals from this experiment as solutions of heavy metal ions. See **Disposal** J.

DATA AND OBSERVATIONS

Student's hypotheses will vary but should mention comparing known reactions with the reactions of unknown solutions.

Table 1: Students' Microplate Data Forms will differ depending on the order in which the students assigned solutions. A sample table is given below.

Table 2: Students' Microplate Data Forms will differ according to the letter code and the order in which the students assigned the solutions.

ANALYSIS

1. Students' answers can vary, but most commonly, the combination of lead(II) nitrate and sodium iodide leads to the first identification of an unknown. The combination of these solutions gives a distinctive yellow precipitate unlike any other in chemistry.

   Known solution of *lead(II) nitrate* + known solution of *sodium iodide* produced a *bright yellow solid.*

   The letters that students write in the blanks of the second sentence will vary according to the lettering code of the unknown solutions.
2. The combination that most commonly leads to the second identification of an unknown is sodium iodide and silver nitrate. Students' work may vary.

CONCLUSIONS

1. Table will vary according to the lettering code of the unknown solutions.
2. Observable changes that can indicate a chemical reaction has occurred include a change in color, the formation of a solid, and the formation of gas bubbles.
3. The number of chemical reactions observed is 13.
4. Students' answers will vary, depending on their hypotheses.

EXTENSION AND APPLICATION

Sugar and vinegar do not react. Baking soda and vinegar do react. Sodium carbonate and nitric acid react in the same way as baking soda and vinegar by producing bubbles of gas.

		Nitric acid 1	Silver nitrate 2	Sodium iodide 3	Lead(II) nitrate 4	Sodium carbonate 5	Iron(III) nitrate 6	Sodium hydroxide 7	Copper(II) nitrate 8
Nitric acid	A	X	X	X	X	X	X	X	X
Silver nitrate	B	NR	X	X	X	X	X	X	X
Sodium iodide	C	yellow solution	gray-white solid	X	X	X	X	X	X
Lead(II) nitrate	D	NR	NR	bright yellow solid	X	X	X	X	X
Sodium carbonate	E	gas bubbles	white solid	NR	white solid	X	X	X	X
Iron(III) nitrate	F	NR	NR	NR	NR	small amount of solid	X	X	X
Sodium hydroxide	G	NR	white solid	NR	white solid	NR	brown solid	X	X
Copper(II) nitrate	H	NR	NR	brown solid	NR	green solid	NR	blue-white solid	X

Copyright © Glencoe/McGraw-Hill, a division of The McGraw-Hill Companies, Inc.

Name ____________________

Date ____________ Class ____________

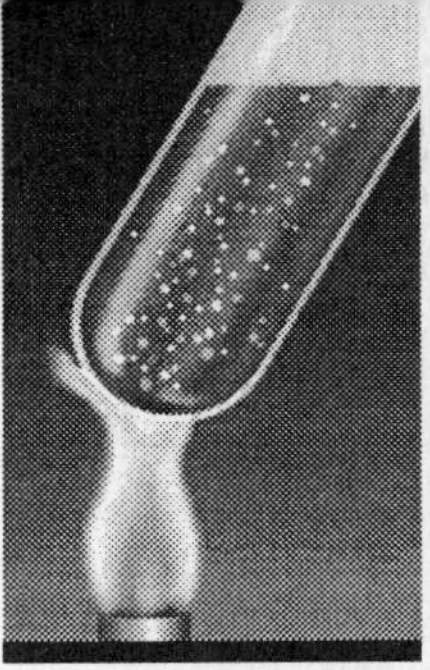

LAB MANUAL

2-1

The Thickness of Aluminum Foil

All matter is made of atoms. A collection of only one type of atom is an element. Aluminum is an element. One use of aluminum is to make foil, which is used to wrap food for storage. Large numbers of aluminum atoms are found in any sample of aluminum metal. Because each aluminum atom is extremely small, a sample of aluminum foil is many atoms thick.

In this activity, you will calculate the size of an aluminum atom by determining the thickness and mass of a piece of aluminum foil.

OBJECTIVES

- **Measure**, indirectly, the thickness of aluminum foil.
- **Estimate** the number of aluminum atoms that give the foil its thickness.

MATERIALS

balance
metric ruler
aluminum block
100-mL graduated cylinder
piece of aluminum foil
scissors
string

PROCEDURE

1. Before you begin, form a hypothesis regarding the directly observable data that are necessary to compute the thickness of aluminum foil. Write your hypothesis under Data and Observations.
2. Use a balance to determine the mass of a block of aluminum metal. Record the mass of the block in the data table.
3. Fill a 100-mL graduated cylinder with water to the 50 mL mark.
4. Tie a string to the block of aluminum and lower the block into the water until it is completely immersed. Record the new volume of water.
5. Calculate the volume of the aluminum block by subtracting the initial volume of water from the volume of water with the aluminum block in it. Record in the table the volume of the aluminum block.
6. Remove the block from the cylinder, dry the block and return it to your teacher.
7. Cut a piece of aluminum foil. The piece should be approximately 12 cm × 12 cm. Measure the exact length and width of the foil and calculate its area. Record the area in the table.
8. Use a balance to determine the mass of the aluminum foil. Record the mass.

DATA AND OBSERVATIONS

Hypothesis: ____________________

____

____

Mass of aluminum block	___ g
Volume of water in graduated cylinder	___ mL
Volume of water with aluminum block	___ mL
Volume of aluminum block	___ mL
Width of the piece of aluminum foil	___ cm
Length of the piece of aluminum foil	___ cm
Area of foil ($l \times w$)	___ cm^2
Mass of aluminum foil	___ g

ANALYSIS

1. What is the purpose of the block of aluminum?

   ___

   ___

2. What is the density of aluminum? (Density = Mass of aluminum block/volume of aluminum block)

   ___

   ___

3. What is the volume of the aluminum foil? (Volume of foil = Mass of foil/Density of aluminum)

   ___

   ___

4. What is the thickness of the foil? (Thickness of foil = Volume of foil/Area of foil)

   ___

   ___

CONCLUSIONS

If one aluminum atom is 2.5×10^{-8} cm in diameter, how many atoms thick is the aluminum foil?

___

___

EXTENSION AND APPLICATION

Compare the diameters of magnesium, calcium, strontium, and barium atoms. What is the relationship between the elements' atomic numbers and the diameters of their atoms? Why do you think this is so?

Teacher Guide

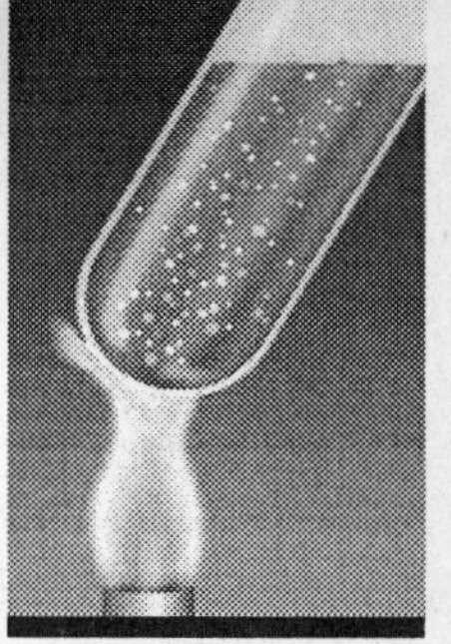

LAB MANUAL

2-1

The Thickness of Aluminum Foil

Aluminum is an extremely useful metal. One useful property of this metal is its malleability. Malleability means that the material can be hammered into thin sheets. A large number of atoms makes up a thin sheet of aluminum foil, and gives the foil its thickness. The thickness of foil in terms of number of atoms, can be approximated by making some measurements and using these measurements to calculate the thickness.

Misconception: Students may think that very thin materials, such as aluminum foil, are only a few atoms thick. Emphasize that a very large number of atoms give aluminum foil its thickness.

PROCESS SKILLS

1. Students will make measurements of a block of aluminum and a piece of aluminum foil.
2. Students will use the measurements to calculate the density of aluminum.
3. Students will conceptualize the size of an atom by calculating the number of aluminum atoms that give aluminum foil its thickness.

PREPARATION HINTS

Be aware that aluminum foil comes in various thicknesses. For example, broiler foil is very thick. For consistency of results, use foil from a single roll.

MATERIALS

balance	piece of aluminum foil
metric ruler	scissors
aluminum block	string
100-mL graduated cylinder	

PROCEDURE HINTS

- Before proceeding, be sure students have read and understood the purpose, procedure, and safety precautions for this laboratory activity.
- The physical, manipulative part of this activity is quite simple. The students' challenge will be in analyzing and interpreting the data.
- **Disposal** Aluminum blocks should be disposed of according to directions in **Disposal** F, and aluminum foil according to **Disposal** F or **Disposal** A.

DATA AND OBSERVATIONS

Students' hypotheses will vary but should indicate that the area and density of the foil are necessary to compute its thickness.

	Sample data:
Mass of aluminum block	17.63 g
Volume of water in graduated cylinder	50.0 mL
Volume of water with aluminum block	56.5 mL
Volume of aluminum block	6.5 mL
Width of aluminum foil	12.1 cm
Length of aluminum foil	12.3 cm
Area of foil ($l \times w$)	148.8 cm^2
Mass of aluminum foil	0.67 g

ANALYSIS

1. The block of aluminum provides a macroscopic amount of aluminum that can be used to calculate aluminum's density.
2. Sample answer: 17.53 g/6.5 mL = 2.7 g/mL = 2.7 g/cm^3, since 1 mL = 1 cm^3
3. Sample answer: 0.67 g/2.7 g/cm^3 = 0.25 mL = 0.25 cm^3

4. Sample answer: $0.25\ cm^3/48.8\ cm^2 = 0.0051\ cm$

CONCLUSIONS

Number of atoms thick = Height of aluminum foil / 2.5×10^{-8} cm/atom = 0.0051 cm / 2.5×10^{-8} cm/atom = 2.0×10^5 atoms

EXTENSION AND APPLICATION

For elements in the same group on the periodic table, atomic size increases with atomic number. The increase occurs because each successive element has one more energy level containing electrons.

Name

Date Class

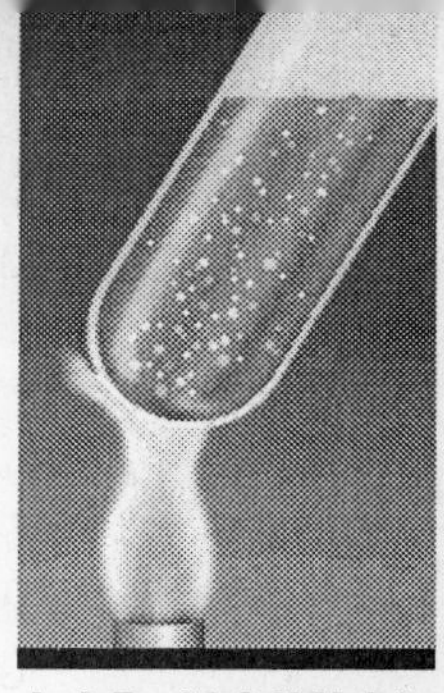

LAB MANUAL

2-2

Identifying Elements by Flame Tests

Have you ever wondered what gives fireworks their colors? Atoms in the fireworks are excited by being heated, causing their electrons to move into higher energy levels. Atoms do not stay excited for long, however; their electrons move back to their normal energy levels, releasing the extra energy as electromagnetic energy. If this energy is in the visible region of the spectrum, we see it as colored light. Atoms of each element have uniquely different electron arrangements. Therefore, as their electrons fall back to normal energy levels, they release energy in the distinctive patterns known as emission spectra. Since each element emits a unique set of wavelengths, emission spectra and their colors can be used to identify the elements.

In this experiment, you will observe the flame colors characteristic of two elements and will use the colors to trace the path of these elements in a chemical reaction.

OBJECTIVES

- **Perform** flame tests of two elements.
- **Isolate** the substances formed in a reaction.
- **Locate**, using flame tests, the elements copper and sodium in the products of the chemical reaction.

MATERIALS

test tubes (2)
test-tube rack
crucibles (3)
thin-stem pipet
laboratory burner
nichrome wire with loop at end
balance
50-mL graduated cylinder
corks (optional)
filter paper
funnel
ring stand
small iron ring
evaporating dish
250-mL beaker
wire gauze
goggles
apron
distilled water

PROCEDURE

1. Form a **hypothesis** to describe a method you could use to locate the metals copper and sodium in the products of a chemical reaction.
2. Label one test tube *copper(II) sulfate*. Label the other *sodium carbonate*.
3. Place small samples of solid copper(II) sulfate, solid sodium carbonate, and hydrochloric acid in three different crucibles. **CAUTION:** *Copper Compounds are toxic. Also, handle the acid carefully. It can cause burns and damage clothing.*
4. With a thin-stem pipet, moisten the copper(II) sulfate and sodium carbonate samples with a few drops of distilled water.
5. Light the laboratory burner.
6. The nichrome wire must be clean. Dip the loop at the end of the wire into the hydrochloric acid and heat it in the flame of the burner until the acid solution has evaporated.
7. Dip the wire loop into the moistened sample of copper(II) sulfate.
8. Place the loop into the flame of the laboratory burner. Record the color of the flame under Data and Observations.
9. Repeat steps 6–8 using sodium carbonate instead of copper(II) sulfate.
10. Weigh a 1.0 g sample of copper(II) sulfate and add it to the test tube with the appropriate label.
11. Weigh a 1.5 g sample of sodium carbonate and add it to the other labeled test tube.
12. Add 10 mL of distilled water to each test tube. Swirl each test tube or close each with a clean cork and shake until the salt is completely dissolved.

13. Pour the copper(II) sulfate solution into the sodium carbonate solution. Mix thoroughly. Record your observations.
14. Fold a piece of filter paper as shown in the figure. Open the folded paper into a cone and place it in a filter funnel. Set up the funnel and an evaporating dish as shown.

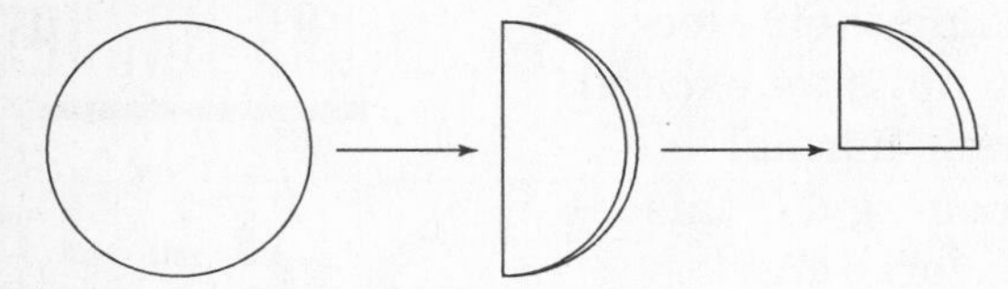

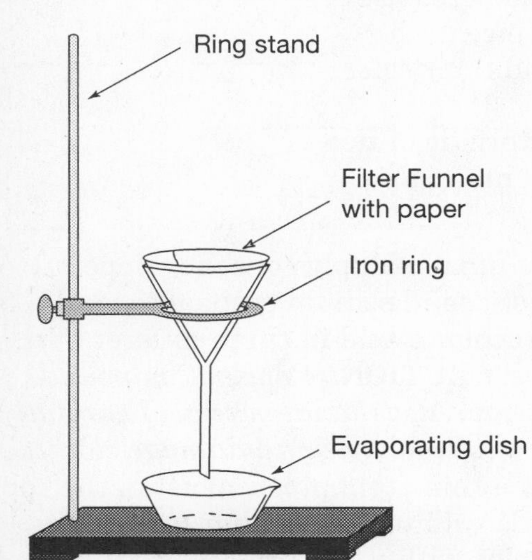

15. Pour the contents of the test tube into the filter paper cone. Catch the filtrate —the fluid that runs through the filter paper—in the evaporating dish. If necessary, dislodge any remaining solid in the test tube by adding a few mL of distilled water to the tube and swirling it. Then empty the test tube into the filter paper cone again.
16. Set aside the evaporating dish and the filtrate it contains. Put a beaker in its place. Rinse the solid on the filter paper cone three times with distilled water. Discard the rinse water that drops into the beaker.
17. Remove the funnel holding the filter paper and the solid from the iron ring and carefully set it aside in the test-tube rack.
18. Place the wire gauze on the iron ring and place the evaporating dish with the filtrate on the gauze.
19. Gently heat the filtrate in the dish until all the water has evaporated.
20. Repeat steps 6–8 using the solid in the evaporating dish instead of copper(II) sulfate. Record the color of the flame.
21. Repeat steps 6–8 using the solid on the filter paper. Record the color of the flame.
22. Discard all substances according to your teacher's instructions.

DATA AND OBSERVATIONS

Hypothesis: ______________________

____

____

____

1. Color of the flame with solid copper(II) sulfate:

____

2. Color of the flame with solid sodium carbonate:

____

3. Result of mixing copper(II) sulfate and sodium carbonate solutions:

____

____

4. Color of the flame with the solid in the evaporating dish:

____

5. Color of the flame with the solid from the test tube:

____

ANALYSIS

1. It is the metals in chemical compounds that give color to the flame when flame tests are done. What are the flame colors for copper and sodium?

Copper flame color: ______________

Sodium flame color: ______________

2. When the two solutions were mixed (step 13), did a chemical reaction occur? Give the evidence that supports your answer.

____

____

____

3. Were copper and sodium changed when you mixed the two solutions (step 13)? Give the evidence that supports your answer.

____

____

____

CONCLUSIONS

1. Describe what happened to the copper from the solid copper(II) sulfate when the copper and sodium salts were dissolved and the two solutions were mixed. What evidence supports your answer?

____

____

____

____

2. What happened to the sodium from the solid sodium carbonate when copper and sodium salts were dissolved and the two solutions were mixed? What evidence supports your answer?

____

____

____

____

3. Was the method you hypothesized similar to the procedure you followed to locate copper and sodium after the salt solutions of these metals were mixed?

____

____

____

____

____

____

____

____

EXTENSION AND APPLICATION

1. Flame tests are very sensitive. Look up in a college chemistry textbook or a reference book the kind of spectroscopy in which flame colors are used to detect small quantities of metals. What is the name of this kind of spectroscopy? Describe in a few sentences how it works.
2. Would it be possible for a space vehicle that landed on Mars to use flame colors to detect metals in the soil?
3. Find out about some of the metals used in fireworks and the colors that they give.

Identifying Elements by Flame Tests

LAB MANUAL

2-2

The characteristic flame colors of copper and sodium are demonstrated in this laboratory experience. The colors are then used to locate these metals in the reaction products of a simple double displacement reaction. The experiment introduces the idea that elements have specific properties and that these properties can be used to test for the presence of the elements. By detecting the metals after the reaction, the students can see that the elements themselves are not altered in a chemical reaction.

Misconception: The fact that the physical and chemical properties of products are different from those of reactants is often emphasized. This emphasis can cause students to think that the atoms of the elements making up the reactants also change. Remind students of the indivisibility of atoms and the constancy of their properties.

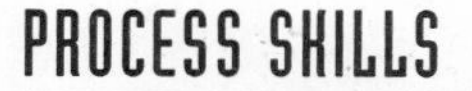

PROCESS SKILLS

1. Students will carry out flame tests.
2. Students will use flame tests to identify elements in the products of a reaction.

PREPARATION HINTS

- The nichrome wires must be clean for this experiment, since traces of substances from previous experiments would give false results. It is recommended that the nichrome wires be cleaned in 6*M* HCl prior to their use in the lab. Students will also clean the wires before use; it is important that they learn this essential step of a flame test.
- To prevent burning fingers, the nichrome wires can be secured in cork, if they are not in glass rods.

MATERIALS

6*M* HCl (Add 100 mL concentrated HCl to 100 mL distilled water. **CAUTION:** *Add acid to water. Do not add water to acid.*)

copper(II) sulfate (s)
sodium carbonate (s)
distilled water
18 × 150 mm test tubes (2)
test-tube rack
crucibles (3)
thin-stem pipet
laboratory burner
nichrome wire with loop at end
balance
50-mL graduated cylinder
corks (optional)
filter paper
funnel
ring stand
small iron ring
evaporating dish
250-mL beaker
wire gauze
goggles
apron

PROCEDURE HINTS

- Before proceeding, be sure students have read and understood the purpose, procedure, and safety precautions for this laboratory activity.
- The solid substances should be distributed in small quantities. Do not allow students to have access to stock bottles of either salt.
- Caution students that the flame test is only good as long as a substance remains on the wire loop. Once the solid has been vaporized, the flame test is no longer effective.
- **Troubleshooting:** The main problem in this lab comes from contaminated wire loops or chemical reactants. Be sure that the loops are cleaned as recommended in Preparation Hints.
- **Safety Precaution:** Adequate ventilation is required. If necessary, have a single laboratory burner in a fume hood for use in doing the flame tests.

- **Disposal:** Dispose of excess HCl according to the instructions in **Disposal** D. Dispose of copper(II) sulfate and sodium carbonate according to instructions in **Disposal** F.

DATA AND OBSERVATIONS

1. Students' hypotheses will vary but should include a description of a method similar to that of the experiment.
2. The color of the flame with solid copper(II) sulfate is green.
3. The color of the flame with solid sodium carbonate is yellow.
4. The result of mixing copper(II) sulfate and sodium carbonate solutions is the formation of a solid where there was none before.
5. The color of the flame with the solid in the evaporating dish is yellow.
6. The color of the flame with the solid from the test tube is green.

ANALYSIS

1. The copper flame color is green. The sodium flame color is yellow.
2. A chemical reaction occurred when the two solutions were mixed. The salts of copper and sodium were both soluble in water. When the solutions of the salts were mixed, however, a solid appeared, showing that something new that was not soluble in water had been produced.
3. The flame colors for copper and sodium were not changed; therefore, the elements could not have changed. If the copper or sodium atoms had been changed, the colors would have disappeared or been different.

CONCLUSIONS

1. The copper from the solid salt went into solution when the salt was dissolved, and then it became part of the solid that formed when the solutions of the salts were mixed. The green flame color of the solid showed that copper was present in it.
2. The sodium from the solid salt went into solution when the salt was dissolved, and it stayed in the water after the reaction. When the water was evaporated, the solid compound containing the sodium remained. The presence of sodium was shown by the yellow color of the flame when the solid was used in a flame test.
3. Answers will vary, depending on students' hypotheses.

EXTENSION AND APPLICATION

1. Metals are detected by atomic absorption spectroscopy. A solution that contains the element is sprayed into a flame. The color of light absorbed identifies the element, and the amount of light absorbed shows the amount of the element present (its concentration).
2. It should be possible to detect metals on Mars by a flame test because location has no effect on the emission or absorption of light. The space vehicle could carry fuel, oxygen, and a device to measure emission or absorption spectra. Data could be radioed to Earth for analysis.
3. In fireworks, barium gives a green color, strontium gives red, magnesium gives white, and sodium gives yellow.

Name

Date Class

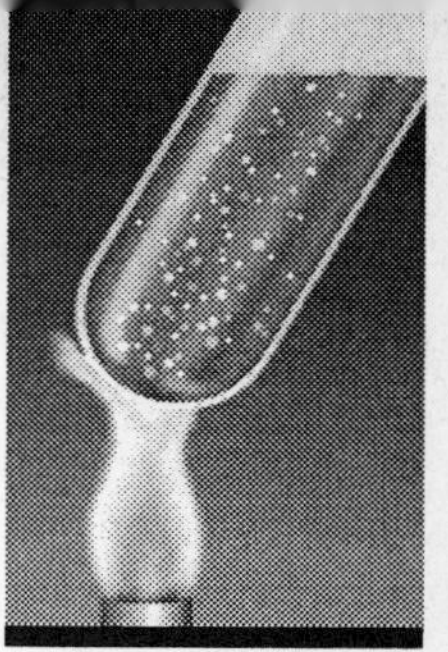

LAB MANUAL

3-1

An Alien Periodic Table

Your spaceship has arrived at a planet on the edge of the galaxy. You have discovered an ancient laboratory containing many artifacts of scientific investigations. Among the artifacts is information on what appears to be a periodic table of elements found on the planet.

Your ship's computer has been able to decipher the numerical values of each alien element's boiling point and melting point and a brief description of one of the elements that is labelled Σ. Fortunately, the compounds formed by this one element with the other elements have also been indicated in the table. Your mission is to arrange the elements in a table similar to your own planet's periodic table, keeping elements with similar properties in the same column.

OBJECTIVES

- **Use information** about unknown elements to arrange them into correct periods and groups.
- **Relate** a model periodic table to the actual periodic table.

MATERIALS

copy of table of information for alien elements
scissors

PROCEDURE

1. Study the information on your table of alien elements. Notice that the element Σ is a green-yellow gas. The number of atoms of element Σ that combine with each of the unknown elements is shown at the top left corner of each square. For example, $-\Sigma_2$ means that the unknown element combines with two atoms of Σ.
2. Cut out the box for each element and arrange the boxes into a periodic pattern like the one shown below. Take into account that in families of metals, melting points and boiling points generally decrease as elements get heavier, but the opposite is true for families of nonmetals. Also note that the way elements combine with other elements tends to vary with regularity across the periodic table. Another important factor to notice is the state of each element.

DATA AND OBSERVATIONS

Periodic Pattern

ANALYSIS

Explain how you decided the order of the families of elements.

CONCLUSIONS

1. Which is the lightest element in the alien periodic table? Which is the heaviest? Explain.

2. How are the methods you used similar to or different from the methods of Dmitri Mendeleev?

EXTENSION AND APPLICATION

Mendeleev was able to predict the existence and properties of ekasilicon from his knowledge of the periodic table. Chemists today use periodic relationships to predict the properties of elements heavier than element 103. What elements give clues to the properties of elements 104 and 105?

Copyright © Glencoe/McGraw-Hill, a division of The McGraw-Hill Companies, Inc.

Σ-Σ Σ					
œ (g) −157°C* −153°C**	Σ-Σ Σ green-yellow gas −101°C −34°C	__Σ_2 ‡ (g) 650°C 1105°C	¥ (g) −112°C −108°C	π (g) −248°C −246°C	__Σ_2 å (s) 776°C 1412°C
__Σ_2 ß (s) 841°C 1500°C	__Σ_4 ∂ (s) 3600°C 4200°C	*f* (g) −189°C −186°C	• (g) −270°C −268°C	__Σ_3, __Σ_5 Δ (s) 44°C 280°C	__Σ æ (s) 40°C 697°C
__Σ Ω (s) 63°C 766°C	__Σ ≈ (s) 113°C 184°C	__Σ ç (g) −219°C −188°C	__Σ_3, __Σ_5 √ (s) 630°C 1635°C	__Σ ∫ (s) 98°C 897°C	__Σ_3, __Σ_5 µ (g) -210°C -195°C
__Σ_3 ≤ (s) 2080°C 3865°C	__Σ_2, __Σ_4 ≥ (s) 232°C 2623°C	__Σ 8 (s) 180°C 1347°C	__Σ_2 * (g) -218°C -183°C	__Σ_2, __Σ_4 ● (s) 221°C 685°C	__Σ ((l) −7°C 59°C
__Σ_2, __Σ_4 ¡ (s) 945°C 2850°C	__Σ_3, __Σ_5 ¢ (s) 816°C 615°C	__Σ ∞ (g) −259°C −252°C	__Σ_3 § (s) 660°C 2517°C	__Σ_2, __Σ_4 ¶ (s) 450°C 990°C	__Σ, __Σ_3 ª (s) 30°C 2203°C
__Σ, __Σ_3 Å (s) 156°C 2080°C	__Σ_2 £ (s) 1287°C 2468°C	__Σ_2, __Σ_4 ± (s) 115°C 444°C	__Σ_4 ¿ (s) 1411°C 3231°C		

* The top number gives the melting point.

** The bottom number gives the boiling point.

Teacher Guide

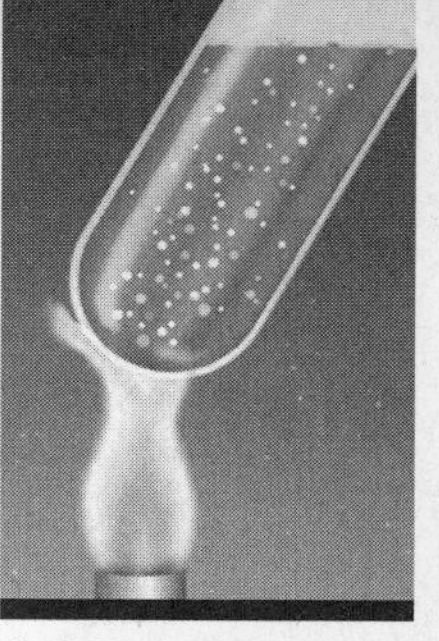

LAB MANUAL

3-1

An Alien Periodic Table

Students should see the periodic table as a source of information about the elements. They will probably not appreciate the insight of early chemists in organizing the known information into a table that implied other physical and chemical properties. This exercise helps students see the way in which this resource was developed. Some students will notice that the mythical alien elements are, in fact, the real elements of Periods 1 to 5, without the transition series.

PROCESS SKILLS

1. Students will position unknown elements in a periodic table on the basis of their physical and chemical properties.
2. Students will observe how certain properties vary within the periodic table.

PREPARATION HINTS

Prepare overhead transparencies of the "Periodic Pattern" and the "Alien Elements." Cut the alien element squares as per student directions. Have these materials on hand to illustrate construction of the periodic table should students experience difficulty with the concept.

MATERIALS

copy of table of information for alien elements
scissors

PROCEDURE HINTS

Before proceeding, be sure students have read and understood the purpose, procedure, and safety precautions for this laboratory activity.

- Be sure students understand how Σ is used in representing compounds formed by the alien elements. You may want to give them some examples of formulas of compounds formed by the halogens with other elements.
- Point out to students that in two of the elements they form compounds with the element Σ where Σ is written first, indicating that Σ has the lower electronegativity of the two elements. See the completed alien periodic table below.

DATA AND OBSERVATIONS

Periodic Pattern

∞							·
8	£	≤	∂	µ	*	ç	π
∫	‡	§	¿	∆	±	Σ	ƒ
Ω	ß	ª	¡	¢	•	(	œ
æ	å	Å	≥	√	¶	≈	¥

ANALYSIS

The number of Σ atoms in the compounds increases across each period up to the family with Σ in it and the final family that forms no compounds with Σ. Families at the right are all, or at least partly, nonmetals because there are gases in them. These elements should therefore be arranged with increasing melting and boiling points. The families at the left consist of only solids and are therefore, as in the modern periodic table, all metals, so here melting and boiling points should generally decrease.

CONCLUSIONS

1. The lightest element is ∞. It occupies the same position in the alien periodic table as hydrogen occupies in the modern periodic table. Hydrogen is the lightest element. The heaviest element is ¥. It occupies the position in the alien periodic table of the highest atomic number. Generally, as atomic number increases, atomic mass increases.
2. Methods here are similar in using physical properties to determine the arrangement. They differ because Mendeleev wanted to include elements with similar properties in the same columns, but also arrange all elements in order of their atomic masses.

EXTENSION AND APPLICATION

The properties of elements 104 and 105 should be related to the properties of sixth-period elements hafnium (72) and tantalum (73).

Name

Date Class

Periodicity of Halogen Properties

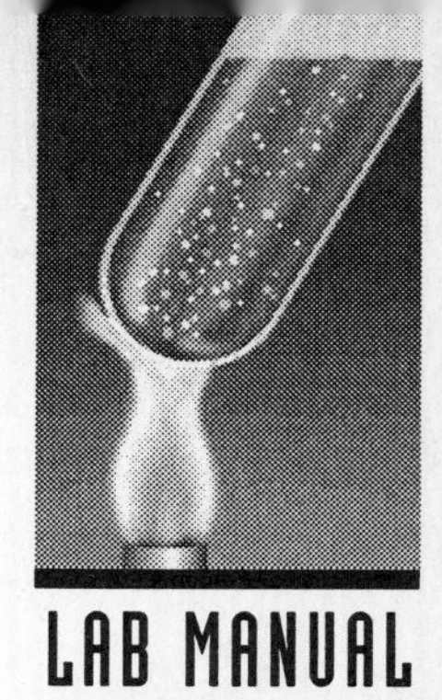

The elements in Group 17 of the periodic table are known as the halogens. The word *halogen* means "salt-former." The chemical property of forming salts, along with other chemical and physical characteristics, reflects the placement of the halogens in the same periodic table group. The halogens have common chemical reactions, but their chemical activity varies. A more active halogen will displace the ion of a less active halogen, for example,

$$F_2(g) + 2NaCl(aq) \rightarrow 2NaF(aq) + Cl_2(g)$$

In this activity, you will observe the difference in activity of the halogens by observing single displacement reactions.

OBJECTIVES

- **Observe** some properties of chlorine, bromine, iodine, and their ions in solution.
- **Determine** the order of activity of the halogen elements.

MATERIALS

24-well microplate
24-well template
microtip pipets

PROCEDURE

1. Consult the periodic table and **hypothesize** in what order the activity of the halogen elements will increase.Write your hypothesis under Data and Observations.
2. Label a 24-well microplate template as shown at the right. Place a 24-well microplate on your template with the numbered columns away from you and the lettered rows to the left.
3. Add approximately ½ pipetful of NaF solution to each of the wells in column 1, rows A through D.
4. Repeat step 3 by adding approximately ½ pipetful of NaCl solution to the wells in column 2, NaBr to the wells in column 3, and NaI to the wells in column 4.
5. Add approximately ½ pipetful of water to each well in row A, columns 1 through 4.
6. Add 10 drops of HCl solution to each well in row B, columns 1 through 4.
7. Add approximately ½ pipetful of NaOCl solution to each well in row B, columns 1 through 4.
8. Add approximately ½ pipetful of Br_2 water to the wells in row C, columns 1 through 4. **CAUTION:** *Do not inhale the bromine fumes from the solution.*
9. Finally, add approximately ½ pipetful of I_2/KI solution to the wells in row D, columns 1 through 4.

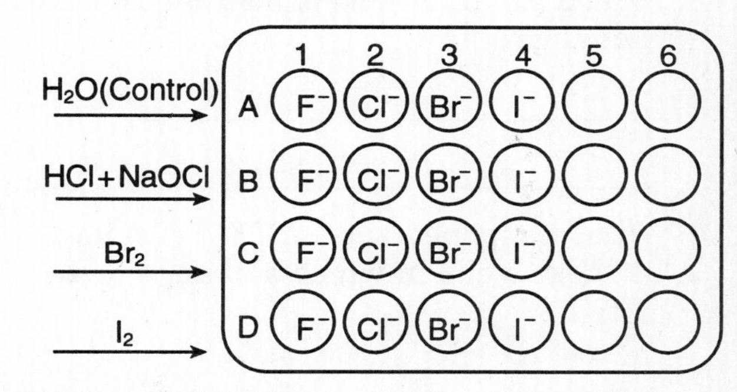

10. Label the rows and columns of your Microplate Data Form the same way as on your template. Compare all column 1 solutions to well A1; compare all column 2 solutions to well A2; and so on. Record your observations in a Microplate Data Form.
11. Add ¼ pipetful of TTE to any well that showed a change in appearance. **CAUTION:** *TTE is an irritant. Avoid contact with skin, nose, and eyes.*
12. One by one, draw up into separate, clean plastic pipets the contents of each well to which you added TTE. Invert the pipet and be sure no drop of liquid remains in the stem. Shake the pipet to ensure complete mixing. Record your observations in your Microplate Data Form.

DATA AND OBSERVATIONS

Hypothesis: ______________________

1. What were the physical characteristics of the column solutions before the row solutions were added?

2. What were the physical characteristics of the row solutions before adding them to the column solutions?

3. Which combinations of solutions showed a change?

4. What happened when TTE was added to those wells in which a change was noted?

ANALYSIS

1. Which column of solutions showed the greatest number of changes?

2. Which column of solutions showed the least number of changes?

CONCLUSIONS

1. If a physical change is a sign of a chemical reaction, which halogen is the most active, that is, which displaces other halogen ions most easily? Which is least active?

2. How is reactivity of the halogens related to position in the periodic table? Was your hypothesis from Procedure step 1 correct?

EXTENSION AND APPLICATION

1. Fluorine is the most active nonmetal. Francium is the most active metal. Predict whether cesium or potassium will be a more active element. Explain.
2. Do you think a reaction will occur between fluorine and sodium iodide? Why?
3. All of the halogens are harmful to living things. Describe how this property has been useful to humans.

Name Date Class

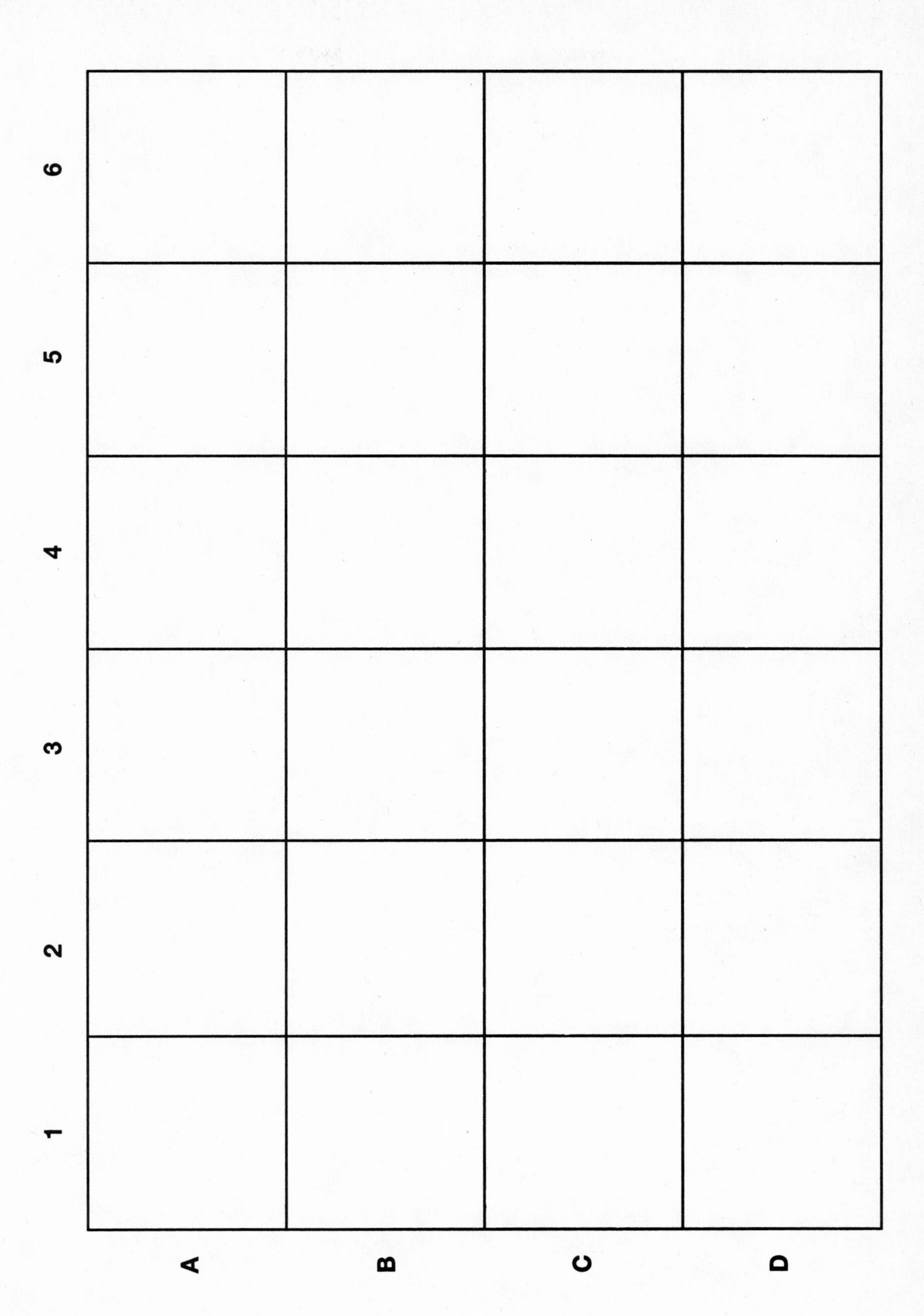

Copyright © Glencoe/McGraw-Hill, a division of The McGraw-Hill Companies, Inc.

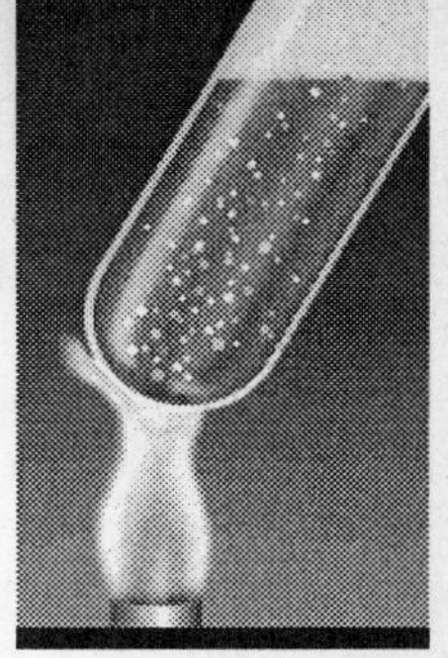

Periodicity of Halogen Properties

The elements in the halogen family are often referred to when describing periodic properties. The physical as well as the chemical properties of the halogens follow a predictable progression for a number of properties. This activity focuses on the chemical reactivity of the halogens as shown in halogen displacement reactions.

PROCESS SKILLS

1. Students will observe single displacement reactions of nonmetals.
2. Students will deduce the relative reactivity of the halogens.

PREPARATION HINTS

- 2*M* salt solutions may be prepared by dissolving the following masses of individual salts in 100 mL of water:
  NaF 8.40 g   NaCl 11.7 g
  NaBr 20.6 g   NaI 30.0 g
- Prepare a solution of $NaHCO_3$ (baking soda) to neutralize any acid spills that may occur.
- For each group, make a copy of the 24-well template on page T6.

MATERIALS

TTE (trichlorotrifluorethane)(commercial preparation)
Row solutions
 6*M* HCl (Add 500 mL concentrated HCl to 500 mL distilled water)
 5% sodium hypochlorite (NaOCl; bleach) (commercial preparation)
 bromine (Br_2) water (commercial preparation)
 iodine/potassium iodide (I_2/KI) (commercial preparation)
Column solutions
 2*M* solutions of
 sodium fluoride (NaF)
 sodium chloride (NaCl)
 sodium bromide (NaBr)
 sodium iodide (NaI)
24-well microplate
24-well template
microtip pipets

PROCEDURE HINTS

- Before proceeding, be sure students have read and understood the purpose, procedure, and safety precautions for this laboratory activity.
- **Safety Precaution:** Always add acid to water. Bromine is toxic by inhalation and by ingestion. TTE is a body tissue irritant.
- Slightly acidified household bleach functions as chlorine water. Acid should not be added to the bleach until called for in the directions.
- **Disposal:** Dispose of the solutions according to the directions in **Disposal J.**

DATA AND OBSERVATIONS

Students' hypotheses will vary, but a correct hypothesis is that halogen activity increases from the bottom to the top of the halogen family in the periodic table.

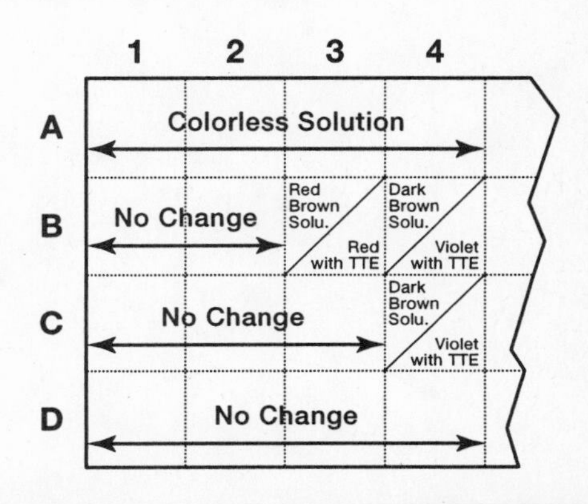

1. The solutions in the columns were all clear and colorless.
2. The chlorine solution has a faint yellow tint. Bromine water has a red-brown tint. The iodine solution has a deeper red-brown tint.
3. Row B, wells B3 and B4:
   $Cl_2 + Br^-$, turns red-brown
   $Cl_2 + I^-$, turns dark red-brown
   Row C, well C4:
   $Br_2 + I^-$, turns dark red-brown
4. When TTE was added to the changed solutions, the TTE developed a color. The TTE with bromine (well B3) has a reddish tint, while the TTE/iodine mixture (wells B4 and C4) is violet.

ANALYSIS

1. The chlorine column had the greatest number of changes.
2. The iodine column showed no change.

CONCLUSIONS

1. Chlorine is most active. Iodine is least active.
2. The order of reactivity observed was $Cl_2 > Br_2 > I_2$. The halogens decrease in activity down the family.

EXTENSION AND APPLICATION

1. Cesium will be more active because it is below potassium in the periodic table.
2. Yes, fluorine should be more reactive than iodine because it is the first element in the group.
3. Chlorine and iodine are valued for their antiseptic properties.

Name

Date Class

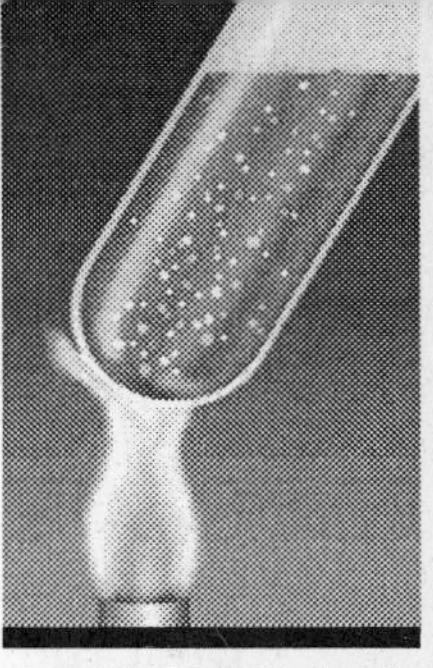

Determining Melting Points

LAB MANUAL

4-1

All pure substances are made up of atoms, molecules, or ions. These particles arrange themselves in order to minimize their energy at the temperature and pressure of the environment. The physical form of matter under these conditions is referred to as the state of a substance. The state of a substance at a specific temperature and pressure is a physical property of that substance.

In this activity, you will observe melting, a change in state from solid to liquid. You will determine the melting points of several pure substances and identify an unknown substance from its melting point alone.

OBJECTIVES

- **Determine** melting points of several substances, and compare with accepted values.
- **Identify** a substance from its melting point.

MATERIALS

apron
goggles
capillary tubes (4)
glass stirring rod
hot plate
ring stand
100-mL beaker
fine wire
thermometers (2)
thermometer clamp
small rubber bands
pieces of paper (4)
laboratory burner
paper towel

PROCEDURE

1. Place a small sample of a solid on a piece of paper.
2. Seal one end of a capillary tube (about 6 cm long) in the flame of a burner. **CAUTION:** *Do not touch the hot end of the tube.*
3. When the capillary tube has cooled, push the open end into the solid.
4. When you have a small amount of solid in the tube, invert the tube and gently tap the closed end on the table. If the solid does not drop to the bottom, push the sample down the tube with a piece of fine wire.
5. Wipe off the open end of the tube, then seal it in the burner flame.
6. Use a rubber band to attach the capillary tube to a thermometer. The sample should be at the same level as the bulb of the thermometer.
7. Set up a water bath. Put the hot plate on the base of the ring stand, and place the beaker one-half full of water on the hot plate.
8. Put the thermometer and attached capillary tube into the water bath. Clamp the thermometer to the ring stand so that it does not touch the bottom or sides of the beaker. Your setup should look like the one in the diagram.

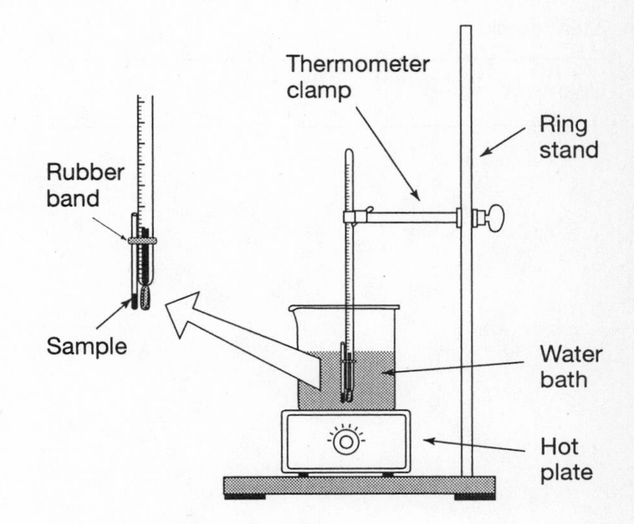

Copyright © Glencoe/McGraw-Hill, a division of The McGraw-Hill Companies, Inc.

9. Heat the water bath, stirring constantly with a glass stirring rod.
10. Observe the crystals of the sample. Quickly note the temperature when the crystals melt.
11. Stop heating the water bath and remove the thermometer and capillary tube. Cool the tube by running it under tap water.
12. Add cool tap water to the water bath to decrease the temperature noted in step 10 by about 10°C. Check the temperature with the second thermometer.
13. Put the thermometer with the attached capillary tube and sample back into the water bath.
14. Slowly heat the water bath again. As the temperature of the water nears the solid's melting point, reduce the rate of heating and watch the sample carefully. When the sample melts, have a partner read the temperature to the nearest 0.1°C. Record this temperature in the table under Data and Observations.
15. Repeat steps 1–14 with the other solids. Record each melting point in the table.

DATA AND OBSERVATIONS

Sample	Melting Point (°C)		Error (%)
	Observed	Accepted	
1,4-dichlorobenzene			
Naphthalene			
Stearic acid			
Unknown			

ANALYSIS

1. Your teacher will provide you with the accepted values for the melting points of the known solids. Record these temperatures in the table.
2. Calculate the percentage error of each melting point you determined and fill in the last column of the table.

CONCLUSIONS

1. What is the unknown substance?

____

____

2. How did you arrive at its identification?

____

____

____

____

EXTENSION AND APPLICATION

Suppose that naphthalene were mixed in with the 1,4-dichlorobenzene. Do you think the impurity would affect the melting point? How? Make a prediction. Then, with your teacher's permission, test your prediction by repeating the activity with a sample of 1,4-dichlorobenzene contaminated with naphthalene.

Teacher Guide

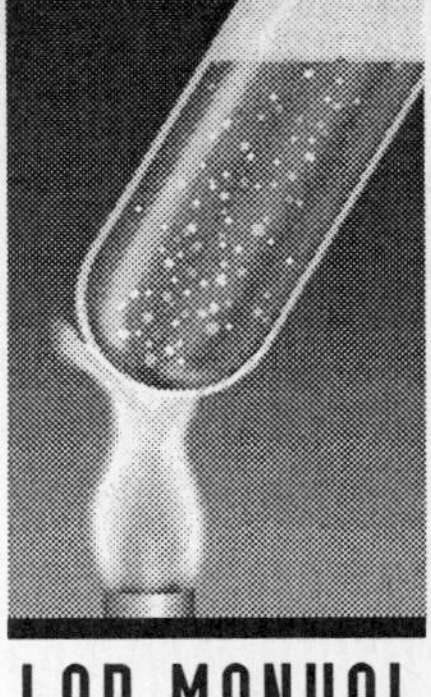

Determining Melting Points

If the temperature or the pressure of the surroundings of a pure substance changes, the state of the substance may also change. Most changes in state on Earth's surface are due to changes in temperature at constant pressure.

Misconception: Students may not realize that pure substances have distinct melting points. Remind them that for a substance to be classified as a solid, the substance must have a distinct melting point. Glass, for example, is considered amorphous because, instead of melting as it is heated, it only gets softer and softer.

PROCESS SKILLS

1. Students will observe the temperatures at which several solids melt.
2. Students will compare experimental results with accepted values.
3. Students will identify an unknown based on its melting point.

PREPARATION HINTS

- If commercial grades of naphthalene or 1,4-dichlorobenzene are used, grind any large lumps in a mortar and pestle to form a powder.
- 1,4-dichlorobenzene is also called para-dichlorobenzene or *p*-dichlorobenzene.
- Use only the materials listed. Do not use compounds with melting points above 100°C. Do not use an oil bath instead of a water bath.
- **Safety Precaution:** There is some concern that 1,4-dichlorobenzene and naphthalene may be toxic or carcinogenic. To reduce students' contact with these chemicals, you can prepare the sealed capillary tubes of samples ahead of the lab. This will also eliminate the need for using lab burners. Work in a fume hood to reduce the amount of vapor in the room. These tubes can be used many times, since the compounds will solidify when they are removed from the water bath.

MATERIALS

unknown (one of the known substances, unlabeled)
1,4-dichlorobenzene
naphthalene
stearic acid
apron
goggles
capillary tubes (4)
hot plate
ring stand
100-mL beaker
thermometers, –10°C to 110°C (2)
thermometer clamp
glass stirring rod
fine wire
small rubber bands
pieces of paper, about 10 cm × 10 cm (4)
laboratory burner
paper towel

PROCEDURE HINTS

- Before proceeding, be sure students have read and understood the purpose, procedure, and safety precautions for this laboratory activity.
- If thermometer clamps are not available, substitute utility clamps or buret clamps, using pieces of heavy vacuum tubing or folded cardboard as shims. Or, insert thermometers through one-hole stoppers and clamp the stoppers to the ring stands.
- You may want to give different unknowns chosen from among the substances on the list to different groups of students.
- **Troubleshooting:** If capillary tubes are not completely sealed, water may leak in and make the results unreliable. If liquid or bubbles appear inside a tube before the

Copyright © Glencoe/McGraw-Hill, a division of The McGraw-Hill Companies, Inc.

melting point of the substance has been reached, discard the tube with its contents.

- **Safety Precaution:** Be sure that students observe safety procedures when using the burner.

DATA AND OBSERVATIONS

Sample data are given.

Sample	Melting Point (°C)		Error (%)
	Observed	Accepted	
1,4-dichlorobenzene	54°	53.1°	2
Naphthalene	79.5°	80.2°	<1
Stearic acid	70°	69.6°	<1
Unknown	78.5°	80.2°	2

ANALYSIS

1. See the table for values for melting points.
2. The table shows percentage errors calculated for sample data. If necessary, help students recall how to calculate percentage error.

CONCLUSIONS

1. Answers will vary depending on which unknown students are given. Unknown in sample data is naphthalene.
2. Students should select the substance with the melting point closest to that of the unknown.

EXTENSION AND APPLICATION

Any soluble impurity will lower the melting point of any pure substance. The lowering of the melting point is one of the colligative properties of a substance. Colligative properties depend on the number of particles involved rather than on the kind of particle.

Copyright © Glencoe/McGraw-Hill, a division of The McGraw-Hill

Name

Date Class

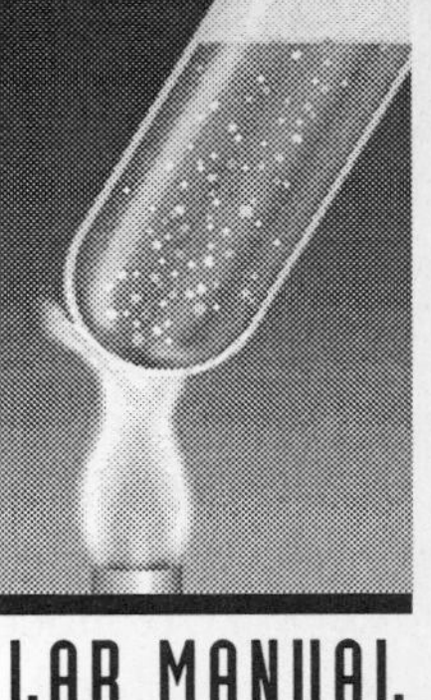

LAB MANUAL

4-2

Distinguishing Ionic and Covalent Compounds

When ionically bonded substances dissolve in water, the ions separate from each other enabling electric current to flow through the solution. An ionic compound can usually be identified by its high melting point. As a result of this high melting point, ionic compounds are crystalline solids at room temperature.

A covalent compound is made of molecules. Weak intermolecular forces hold molecules of covalent materials together. The weak forces are responsible for their relatively low melting points. Covalent compounds are usually liquids or gases at room temperature. With a few important exceptions, covalent compounds do not conduct electricity when dissolved in water, when they dissolve in water at all.

In this laboratory activity you will:

- **a.** build a simple conductivity tester to determine which substances produce ions in aqueous solution;
- **b.** examine various compounds to determine their state;
- **c.** research the literature to determine the melting points of these compounds;
- **d.** predict whether the compounds are more likely to be ionic or covalent.

OBJECTIVES

- **Construct** a conductivity tester.
- **Measure** conductivity of solutions.
- **Observe** the states of compounds.
- **Research** the melting temperatures of compounds.
- **Relate** properties to bond type.

MATERIALS

general purpose LED
9 V battery
battery clip
1k-ohm resistor
connecting wire
alligator clips (3)
96-well microplate
96-well template
soda straw
thin-stem pipet
wire stripper
apron
goggles
Handbook of Chemistry and Physics

PROCEDURE

Part 1: Building a Conductivity Tester

1. Refer to Figure A to assemble the LED, resistor, battery, and wires together.

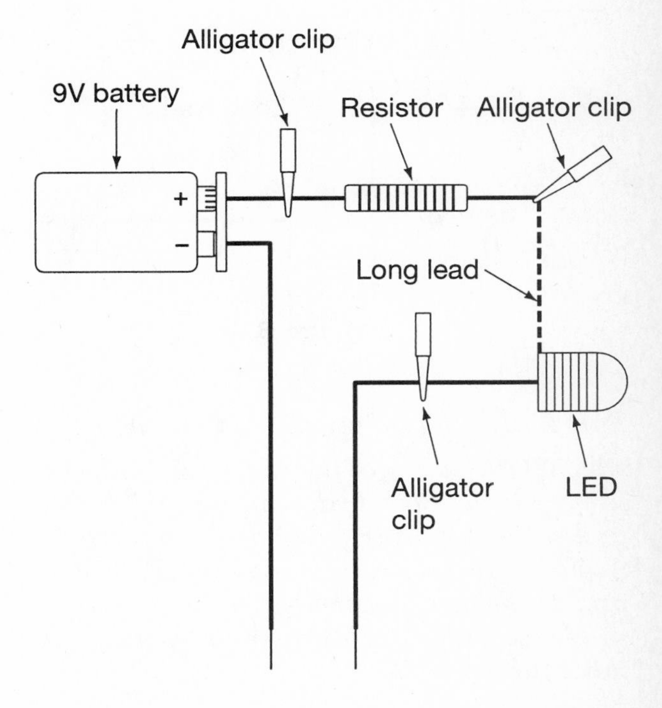

Figure A

Copyright © Glencoe/McGraw-Hill, a division of The McGraw-Hill Companies, Inc.

2. Connect the red lead of the battery to one end of the resistor with an alligator clip.
3. Using an alligator clip, connect the other end of the resistor to the positive lead of the LED (usually the longer lead).
4. Strip the insulation from both ends of two pieces of connecting wire, each 3–4 cm long.
5. Connect one piece of wire to the other lead of the LED with an alligator clip. Connect the second piece of wire to the lead of the battery clip.
6. Push the end of one of the connecting wires crosswise through a soda straw.
7. Push the end of the other wire through the straw about 0.5 cm from the first wire. (See Figure B.) The ends of the connecting wires will be the electrodes of the conductivity tester.

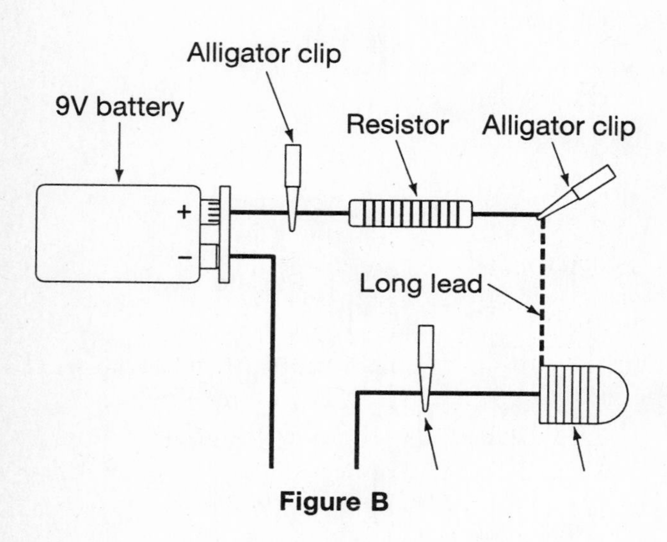

Figure B

Part 2: Testing Solutions for Conductivity

CAUTION: *Many of the chemicals you will use are toxic. Both hydrochloric acid (HCl) and sulfuric acid (H_2SO_4) are very corrosive. Avoid contact with skin and eyes. Follow proper chemical hygiene procedures. Wash your hands thoroughly after completing this laboratory activity.*

1. Place 10 drops of solution 1 in well A1 of your microplate. Record the name and formula of the solute on your template and Microplate Data Form. Place 10 drops of solution 2 in well A2 of your microplate. Record the name and formula of this solute on your template and Microplate Data Form. Continue in like manner until each solution available to you has been dispensed into your microplate. (Rinse the pipet both inside and outside after each use).
2. Insert the electrodes of the tester into a well that contains a solution. The electrodes must not touch each other.
3. Note the relative conductivity of the solution by looking at the brightness of the LED.
4. Record your observation on the Microplate Data Form. Use the following code: C = Conduction; NC = No Conduction; PC = Partial Conduction
5. Test all the solutions. Wash the leads with distilled water after each test. Record the results.

Part 3: Observing the Physical State

1. Inspect the flasks containing each of the undissolved substances.
2. In the corresponding box of your Microplate Data Form, record for each compound "s" if the compound is a crystalline solid, "l" if the material is liquid, and "g" if the substance is a gas.

Part 4: Finding the Melting Point

1. Use a reference source, such as the *Handbook of Chemistry and Physics*, to find the melting temperatures of all the substances used in this activity.
2. In the corresponding box of your Microplate Data Form record the melting temperature (in degrees Celsius) for each compound.

DATA AND OBSERVATIONS

Refer to the instructions in step 4 of Part 2, step 2 of Part 3, and step 2 of Part 4 to record your observations and information about each substance in the following table. Also, indicate your predictions about bond type.

Teacher Guide

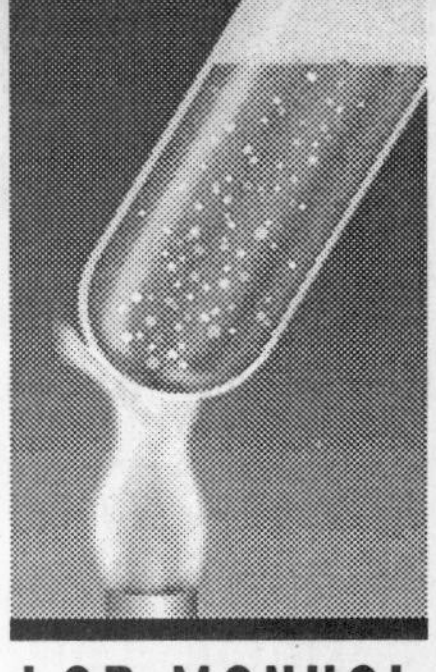

LAB MANUAL

4-2

Distinguishing Ionic and Covalent Compounds

The conductivity apparatus used in this laboratory activity is inexpensive, safe, and requires only a few drops of test solution. The use of electronic circuitry also introduces the students to the science of physics. If you or your students have already built the conductivity tester described on page 60T of the Teacher Wraparound Edition of the textbook, you may use that instead.

Misconception: Students often think of water as a good conductor of electricity. They make this assumption because tap water is usually a dilute solution of salts in water. Pure water, which students rarely encounter before studying chemistry, does not contain salts and does not conduct electricity.

PROCESS SKILLS

1. Students will construct a device to test the conductivity of various solutions.
2. Students will observe results of conductivity tests.
3. Students will examine compounds and observe their states.
4. Students will research melting points of various compounds.
5. Students will relate conductivity, state, and melting point of the test materials to the type of bonding in each substance.

PREPARATION HINTS

- You can use stock solutions prepared for other lab activities since the amounts required for this activity are very small. If solutions are tightly stoppered, they can be kept from year to year. Preparation amounts are given for one liter of solution. To prepare 500 mL of solution reduce both the mass of solute and the volume of solution by half.
- In addition to the solutions, display a sealed sample container of each substance. A small stoppered flask or vial sealed with transparent tape is appropriate. For gaseous hydrogen chloride display just a stoppered "empty" container.
- For each group, make a copy of the 96-well template on p. T8.

MATERIALS

The following solutions are all 0.1 *M* and are prepared, *using distilled water*, as follows.

$Al(NO_3)_3 \cdot 9H_2O$ – Aluminum nitrate
(37.5 g in 1 L of solution)
NH_4Cl – Ammonium chloride
(5.35 g in 1 L of solution)
CH_3COCH_3 – Acetone
(5.8 g in 1 L of solution)
C_2H_5OH – Ethyl alcohol
(4.6 g in 1 L of solution)
$C_3H_5(OH)_3$ – Glycerine
(9.2 g in 1 L of solution)
HCl – Hydrochloric Acid
(3.65 g in 1 L of solution) or (8.3 mL conc/L)
$MgBr_2$ – Magnesium bromide
(29.2 g in 1 L of solution)
CH_3OH – Methyl alcohol
(3.2 g in 1 L of solution)
NaCl – Sodium chloride
(5.85 g in 1 L of solution)
H_2SO_4 – Sulfuric Acid
(9.8 g in 1 L of solution) or (5.5 mL conc/L)
HOH – Distilled water
(Not necessary to dilute)

general purpose LED
9 V battery
battery clip
1 k-ohm resistor
connecting wire
alligator clips (3)
soda straw
96-well microplate
96-well template
thin-stem pipet
apron
goggles
Handbook of Chemistry and Physics

PROCEDURE HINTS

- Before proceeding, be sure students have read and understood the purpose, procedure, and safety precautions for this laboratory activity.
- Label the solutions by both name and formula. This will assist students in their formula writing and recognition. Numbering the solutions from 1 to 11 will assure that students' data sheets will be comparable.
- Inspect the completed conductivity tester before students begin. Ascertain that connections are properly made.
- Students may have trouble inserting the wire through the straw. A needle or pin can be used to puncture the straw before the wire is inserted.
- Set up three distribution/work locations: one for students to fill their microplates as they have completed their Microplate Data Form, a second location to examine the states of the materials, and a research center with the *Handbook of Chemistry and Physics*, *The Merck Index*, or similar references.
- In post-lab discussion, point out that distilled water served as a control during this activity.
- **Troubleshooting:** The most common problem with this activity arises from contamination. Wash bottles or dropping bottles may be effectively used as dispensing bottles to reduce the possibility of contamination as well as to reduce waste.
- **Disposal:** Dispose of solutions in accordance with the instruction in **Disposal** J. Save sample vials for next year.

DATA AND OBSERVATIONS

Students' summary table should be similar to the one shown. The actual bond type should be supplied by the teacher during the post lab discussion.

Substance	Conductivity	State	Melting Point (°C)	Predicted Bond type
$Al(NO_3)_3$ (circled)	C	s	73.5	ionic
NH_4Cl (circled)	C	s	340 (sublimes)	ionic
CH_3COCH_3	NC	l	−95.4	covalent
C_2H_5OH	NC	l	−117.3	covalent
$C_3H_5(OH)_3$	NC	l	17.8	covalent
HCl	C	g	−114.8	covalent (acid)
$MgBr_2$ (circled)	C	s	700	ionic
CH_3OH	NC	l	−93.9	covalent
NaCl (circled)	C	s	801	ionic
H_2SO_4	C	l	10.4	covalent (acid)
HOH	NC	l	0	covalent

ANALYSIS

1. Insert substances circled in chart separated by commas.
2. Insert substances that are NOT circled in chart separated by commas.

CONCLUSIONS

1. Low melting points, poor or nonconductivity in solution, and gaseous or liquid states are indicative of covalent bonds. Ionically bonded substances tend to be crystalline, conduct electricity in solution, and have high melting points.
2. Ionically bonded substances dissolve in water, and the ions separate from each other. These ions conduct an electric current through the solution.

EXTENSION AND APPLICATION

The equation shows the formation of ions. A covalent compound that forms ions, as does HCl in water, is capable of conducting an electric current.

Copyright © Glencoe/McGraw-Hill, a division of The McGraw-Hill Companies, Inc.

Name

Date Class

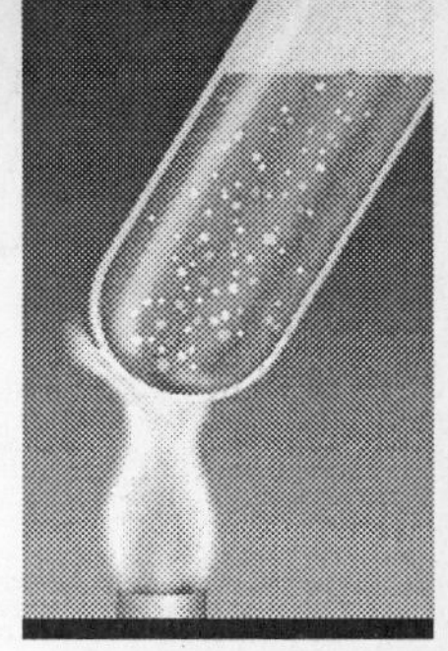

LAB MANUAL

5-1

Making Models of Compounds

At first glance, chemical formulas may look complicated. Once you know the rules for writing them, they look much more simple. Atoms or ions combine in small whole number ratios. By this we mean such combinations as 1 N with 1 O (NO) or 1 N with 2 O's (NO_2) or 2 N's with 4 O's (N_2O_4). Not every combination will do, however. For ionic compounds, there is just one important rule: the total positive charge must equal the total negative charge. For example, in sodium carbonate the charge of the sodium ion is 1+ (Na^+) and the charge of the carbonate ion is 2– (CO_3^{2-}). It takes two sodium ions with a total positive charge of 2+ to neutralize the 2– charge of just one carbonate ion, giving the formula Na_2CO_3. In this laboratory activity, you will practice combining monatomic or polyatomic ions to give the formulas for ionic compounds.

OBJECTIVES

- **Use models** of ions to assemble formulas.
- **List** correct formulas for ionic compounds.
- **Name** ionic compounds from their formulas.

MATERIALS

sheets of ion models (3)
scissors

PROCEDURE

1. Cut up the ion model sheets into individual ions.
2. Separate the ion models into two piles: positively charged ions and negatively charged ions.
3. Note that positive ion models have + charges on the right and negative ion models have – charges on the left. This arrangement allows you to match the charges so that they are neutralized. Note also that when the +'s and –'s are lined up, the symbols are in the correct order to write chemical formulas.
4. Prepare a table for formulas similar to the one under Data and Observations. Label it *Table 1* and include all the ion formulas listed under Data and Observations. In the order given, use the negative ion formulas for the vertical columns and the positive ion formulas for the rows.
5. Prepare a table for chemical names by repeating the labels described in step 4. Label it *Table 2*.
6. Using the ion models, assemble formulas for the compounds formed by each positive ion with each negative ion. Remember: positive charges must equal negative charges.
7. Count the number of ions of each charge in each formula and write the formula in the appropriate space in Table 1.
8. Write the name for each compound in Table 2.

DATA AND OBSERVATIONS

Negative ion formulas: Use these formulas for the vertical columns, left to right.

F^- Cl^- Br^- $C_2H_3O_2^-$ SO_4^{2-}
CO_3^{2-} NO_3^- PO_4^{3-} $Cr_2O_7^{2-}$

Positive ion formulas: Use these formulas for the horizontal rows, top to bottom.

NH_4^+ K^+ Li^+ Na^+ Ag^+
Ca^{2+} Mg^{2+} Ba^{2+} Fe^{2+} Al^{3+}

	F^-	Cl^-	Br^-
NH_4^+			
K^+			
Li^+			

ANALYSIS

1. Which of your positive ions is not a metal ion?

2. What is a binary compound? Which of your negative ions always form binary compounds with metal ions?

3. What kinds of positive and negative ions always combine in a one-to-one ratio? Give two examples of compounds with ions in a one-to-one ratio.

4. What is the largest number of ions in any of your formulas? What is the total positive charge and the total negative charge in each formula with this many ions? Give two examples of these formulas.

CONCLUSIONS

1. Even experienced chemists always double-check formulas after they have written them. Check all of your formulas for the correct numbers of positive and negative ions so that the charge equals zero. Also check to see that polyatomic ions have parentheses around them if there is more than one in a formula. Correct any errors.
2. Check the names in Table 2. In which of these names must a Roman numeral be used? Why?

EXTENSION AND APPLICATION

1. All the compounds in this activity are composed of charged ions. What class of compounds is composed of ions? What class of compounds is composed of atoms, not ions?
2. Two ionic compounds often react so that the positive and negative ions change places. For example, $AgNO_3$ and NaCl react to form AgCl and $NaNO_3$. Name the two compounds formed by the reaction between barium nitrate and potassium sulfate. What are the formulas for these compounds?
3. Find a bottle of vitamins or soft drink or a package of cookies. Check the ingredient names on the label and make a list of any that you think are ionic compounds. How could you tell?

Teacher Guide

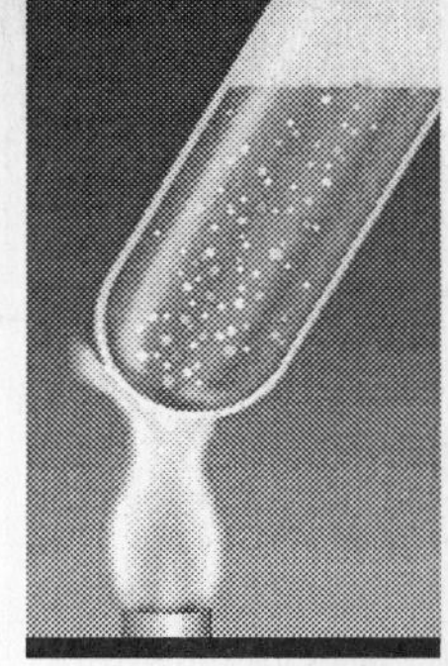

Making Models of Compounds

LAB MANUAL

5-1

Students will find this exercise a great help in writing chemical formulas. The ion models have been designed so that the positive charges are on the right side of the model and the negative charges are on the left side. Aligning the charges means that students automatically have the ions in the correct order to write an ionic chemical formula. Matching the positive and negative charges of the ions ensures that students will correctly neutralize the charges and make a correct model of the compound.

Misconception: Students should be reminded that these are not geometrically correct models of compounds. They should also be reminded that polyatomic ions stay together for the purposes of this activity but that in chemical reactions, the ions can break down to form other products.

PROCESS SKILLS

1. Students will assemble models of ionic compounds.
2. Students will write chemical formulas.
3. Students will name chemical compounds.
4. Students will recognize the patterns in which ions combine.

PREPARATION HINTS

Prepare a set of ion models (3 sheets per student or group), using the copy masters on pages T10–T12.

MATERIALS

sheets of ion models (3 per student group)
scissors

PROCEDURE HINTS

- Before proceeding, be sure students have read and understood the purpose, procedure, and safety precautions for this laboratory activity.
- An alternative way to run this activity is to have pairs of students play a game using the ions as cards. The students start with five ions each, selected at random. They take turns selecting ions from a box and matching the ions to form compounds. A student must name the compound correctly after matching the ions, otherwise the ions are returned to the pile. The student who has the greatest number of correctly named compounds when no ions remain in the pile is the winner.

DATA AND OBSERVATIONS

See Tables 1 and 2 on the next page.

ANALYSIS

1. The ammonium ion, NH_4^+, is not a metal ion.
2. A binary compound is made of atoms of two elements. The fluoride, chloride, and bromide ions form binary compounds with all metal ions.
3. Positive and negative ions that have equal positive and negative charges always combine in a one-to-one ratio, for example, KF and $CaSO_4$.
4. The largest number of ions in any formula is 5. In these formulas, the total + charge is 6+ and the total – charge is 6–, for example, $Ca_3(PO_4)_2$ or $Al_2(CO_3)_3$.

CONCLUSIONS

2. The Roman numeral II must be used in the compounds that contain Fe^{2+}. The reason is that iron also forms an ion with a 3+ charge.

Copyright © Glencoe/McGraw-Hill, a division of The McGraw-Hill Companies, Inc.

Table 1

	F^-	Cl^-	Br^-	$C_2H_3O_2^-$	SO_4^{2-}	CO_3^{2-}	NO_3^-	PO_4^{3-}	$Cr_2O_7^{2-}$
NH_4^+	NH_4F	NH_4Cl	NH_4Br	$NH_4C_2H_3O_2$	$(NH_4)_2SO_4$	$(NH_4)_2CO_3$	NH_4NO_3	$(NH_4)_3PO_4$	$(NH_4)_2Cr_2O_7$
K^+	KF	KCl	KBr	$KC_2H_3O_2$	K_2SO_4	K_2CO_3	KNO_3	K_3PO_4	$K_2Cr_2O_7$
Li^+	LiF	$LiCl$	$LiBr$	$LiC_2H_3O_2$	Li_2SO_4	Li_2CO_3	$LiNO_3$	Li_3PO_4	$Li_2Cr_2O_7$
Na^+	NaF	$NaCl$	$NaBr$	$NaC_2H_3O_2$	Na_2SO_4	Na_2CO_3	$NaNO_3$	Na_3PO_4	$Na_2Cr_2O_7$
Ag^+	AgF	$AgCl$	$AgBr$	$AgC_2H_3O_2$	Ag_2SO_4	Ag_2CO_3	$AgNO_3$	Ag_3PO_4	$Ag_2Cr_2O_7$
Ca^{2+}	CaF_2	$CaCl_2$	$CaBr_2$	$Ca(C_2H_3O_2)_2$	$CaSO_4$	$CaCO_3$	$Ca(NO_3)_2$	$Ca_3(PO_4)_2$	$CaCr_2O_7$
Mg^{2+}	MgF_2	$MgCl_2$	$MgBr_2$	$Mg(C_2H_3O_2)_2$	$MgSO_4$	$MgCO_3$	$Mg(NO_3)_2$	$Mg_3(PO_4)_2$	$MgCr_2O_7$
Ba^{2+}	BaF_2	$BaCl_2$	$BaBr_2$	$Ba(C_2H_3O_2)_2$	$BaSO_4$	$BaCO_3$	$Ba(NO_3)_2$	$Ba_3(PO_4)_2$	$BaCr_2O_7$
Fe^{2+}	FeF_2	$FeCl_2$	$FeBr_2$	$Fe(C_2H_3O_2)_2$	$FeSO_4$	$FeCO_3$	$Fe(NO_3)_2$	$Fe_3(PO_4)_2$	$FeCr_2O_7$
Al^{3+}	AlF_3	$AlCl_3$	$AlBr_3$	$Al(C_2H_3O_2)_3$	$Al_2(SO_4)_3$	$Al_2(CO_3)_3$	$Al(NO_3)_3$	$AlPO_4$	$Al_2(Cr_2O_7)_3$

Table 2

	F^-	Cl^-	Br^-	$C_2H_3O_2^-$	SO_4^{2-}	CO_3^{2-}	NO_3^-	PO_4^{3-}	$Cr_2O_7^-$
NH_4^+	ammonium fluoride	ammonium chloride	ammonium bromide	ammonium acetate	ammonium sulfate	ammonium carbonate	ammonium nitrate	ammonium phosphate	ammonium dichromate
K^+	potassium fluoride	potassium chloride	potassium bromide	potassium acetate	potassium sulfate	potassium carbonate	potassium nitrate	potassium phosphate	potassium dichromate
Li^+	lithium fluoride	lithium chloride	lithium bromide	lithium acetate	lithium sulfate	lithium carbonate	lithium nitrate	lithium phosphate	lithium dichromate
Na^+	sodium fluoride	sodium chloride	sodium bromide	sodium acetate	sodium sulfate	sodium carbonate	sodium nitrate	sodium phosphate	sodium dichromate
Ag^+	silver fluoride	silver chloride	silver bromide	silver acetate	silver sulfate	silver carbonate	silver nitrate	silver phosphate	silver dichromate
Ca^{2+}	calcium fluoride	calcium chloride	calcium bromide	calcium acetate	calcium sulfate	calcium carbonate	calcium nitrate	calcium phosphate	calcium dichromate
Mg^{2+}	magnesium fluoride	magnesium chloride	magnesium bromide	magnesium acetate	magnesium sulfate	magnesium carbonate	magnesium nitrate	magnesium phosphate	magnesium dichromate
Ba^{2+}	barium fluoride	barium chloride	barium bromide	barium acetate	barium sulfate	barium carbonate	barium nitrate	barium phosphate	barium dichromate
Fe^{2+}	iron(II) fluoride	iron(II) chloride	iron(II) bromide	iron(II) acetate	iron(II) sulfate	iron(II) carbonate	iron(II) nitrate	iron(II) phosphate	iron(II) dichromate
Al^{3+}	aluminum fluoride	aluminum chloride	aluminum bromide	aluminum acetate	aluminum sulfate	aluminum carbonate	aluminum nitrate	aluminum phosphate	aluminum dichromate

EXTENSION AND APPLICATION

1. Compounds composed of ions are ionic compounds. Compounds composed of atoms, not ions, are called molecular, or covalent, compounds.
2. The compounds formed by the reaction between barium nitrate and potassium sulfate are barium sulfate and potassium nitrate, $BaSO_4$ and KNO_3.
3. Students' lists will vary. The names of the ionic compounds can be recognized because they begin with the name of a metal ion and they end in *-ate* or *-ide*.

Name

Date Class

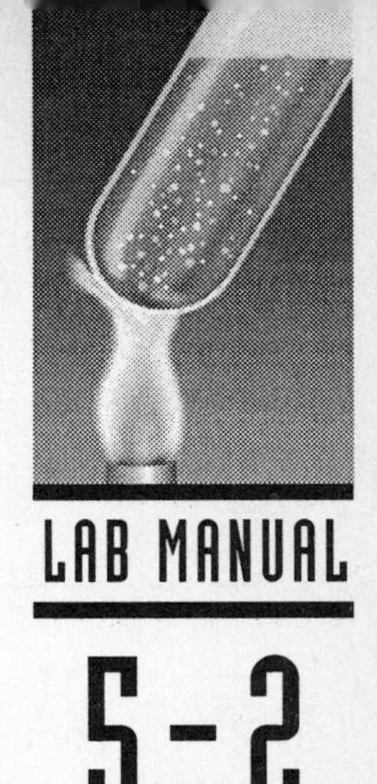

LAB MANUAL

5-2

Formulas and Oxidation Numbers

Oxidation numbers and the charges of ions give the information needed to write the formulas of many chemical compounds. Only a few guidelines are needed.

1. In a neutral compound, the charges on ions, or the oxidation numbers, balance out to zero.
2. One positive charge balances one negative charge.
3. Ions with positive charges or positive oxidation numbers are written first.
4. Subscripts show the relative numbers of atoms or ions in a compound.
5. To show more than one of a polyatomic ion, the symbol is enclosed in parentheses and the subscript follows; for example, $Al_2(SO_4)_3$.

In this activity, you will use paper models to show how chemical formulas are derived from oxidation numbers.

OBJECTIVES

- **Write** formulas of chemical compounds.
- **Name** chemical compounds.

MATERIALS 

scissors
pencil
paper
sheet of ion models

PROCEDURE

1. Cut out each of the ion squares on your sheet of ion models.
2. Assemble the ions for a compound containing nickel(II) and iodide ions. To do this, place the Ni^{2+} ion on a piece of paper. Place enough I^- ions alongside the Ni^{2+} ion to balance the charges.
3. Record the formula and name of the compound of nickel(II) and iodine in a data table similar to the one shown.
4. Use the rules listed in the introduction and in your textbook for writing formulas and naming compounds. Assemble the ions for five compounds from the following list and record their formulas and names in your data table. Nickel(III) and chlorite; zinc and oxide; copper(I) and sulfide; nickel(III) and chlorate; tin(II) and sulfite; copper(II) and iodide; tin(IV) and sulfide; nickel(III) and oxide; copper(II) and sulfite; zinc and hydrogen sulfite.

DATA AND OBSERVATIONS

Combining Ions	Chemical Formula	Name of Compound
nickel(II) and iodide	NiI_2	nickel(II) iodide

ANALYSIS

1. Some compounds are described as "binary compounds." What does this mean? List the formulas and names of any binary compounds you have constructed.

2. Which elements on your list form ions with two different oxidation numbers?

3. Parentheses must be used to show more than one of a polyatomic ion. List the formulas of any compounds on your list where this was necessary.

CONCLUSIONS

Should the formulas you have written be described as molecular formulas? Explain.

EXTENSION AND APPLICATION

1. Some elements have more than one oxidation number. To show the oxidation number of such elements in a compound, a Roman numeral is given in the name of the compound. Give names for the following compounds.

   a. UF_5 **c.** $PbCl_2$
   b. UF_6 **d.** $PbCl_4$

2. Manganese has an oxidation number of 4+ in a number of compounds. Write the formulas and names of compounds of manganese (IV) with oxygen and bromine.

3. Which of the following are molecular formulas?

   H_2O $C_4H_8O_4$
   NaBr $MnSO_4$
   CH_4

4. Hydrogen peroxide and water both contain the same two elements. Using reference materials, write the chemical formulas for these two substances and describe their properties and uses.

Copyright © Glencoe/McGraw-Hill, a division of The McGraw-Hill Companies, Inc.

Teacher Guide

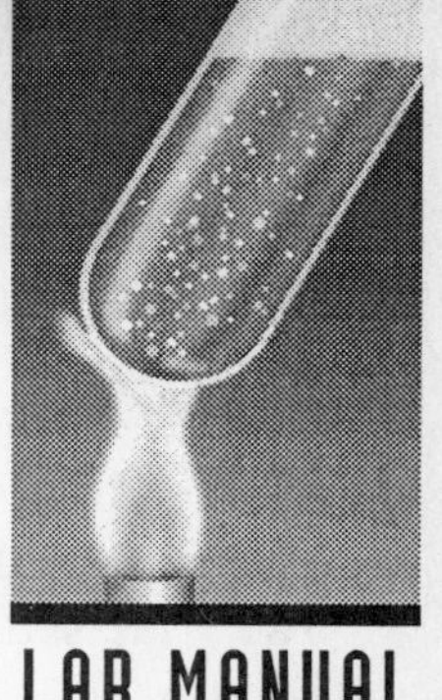

LAB MANUAL

5-2

Formulas and Oxidation Numbers

Only by using a universally agreed upon system can chemical formulas and names be written so that they will be understood by chemists everywhere. Ionic charges and oxidation numbers are fundamental to correctly writing formulas and names of inorganic compounds.

Misconception: Students may miss the point of charge balancing and think they can use the formula for one compound of an element as a model for the formulas for all other compounds of that element.

PROCESS SKILLS

1. Students will predict the formulas of ionic compounds.
2. Students will name ionic compounds.

PREPARATION HINTS

- Sheets of ion models can be mounted on cardboard or heavy paper and used over and over again.
- Have available a few reagent bottles to show students that both the names and formulas of chemicals are written on the labels.

MATERIALS

scissors
pencil
paper
sheet of ion models

DATA AND OBSERVATIONS

Combining Ions sample answers:	Chemical Formula	Name of Compound
nickel(II) and iodide	NiI_2	nickel(II) iodide
nickel(III) and chlorite	$Ni(ClO_2)_3$	nickel(III) chlorite
zinc and oxide	ZnO	zinc oxide
copper(I) and sulfide	Cu_2S	copper(I) sulfide
nickel(III) and chlorate	$Ni(ClO_3)_3$	nickel(III) chlorate
tin(II) and sulfite	$SnSO_3$	tin(II) sulfite
copper(II) and iodide	CuI_2	copper(II) iodide
tin(IV) and sulfide	SnS_2	tin(IV) sulfide
nickel(III) and oxide	Ni_2O_3	nickel(III) oxide
copper(II) and sulfite	$CuSO_3$	copper(II) sulfite
zinc and hydrogen sulfite	$Zn(HSO_3)_2$	zinc hydrogen sulfite

PROCEDURE HINTS

- Before proceeding, be sure students have read and understood the purpose, procedure, and safety precautions for this laboratory activity.
- Review the example in procedure step 2 so that students will be clear on what to do.
- You may wish to review naming compounds.

ANALYSIS

1. Binary compounds are composed of two elements only. Those on the list are NiI_2, ZnO, Cu_2S, CuI_2, SnS_2, and Ni_2O_3.
2. Nickel, copper, and tin
3. $Ni(ClO_2)_3$, $Ni(ClO_3)_3$, $Zn(HSO_3)_2$

CONCLUSIONS

The formulas written here all represent ionic compounds. They show formula units. They cannot be called "molecular formulas."

EXTENSION AND APPLICATION

1. a. uranium(V) fluoride
   b. uranium(VI) fluoride
   c. lead(II) chloride
   d. lead(IV) chloride
2. MnO_2 manganese(IV) oxide, $MnBr_4$ manganese(IV) bromide
3. H_2O, $C_4H_8O_4$, CH_4
4. The formulas for hydrogen peroxide and water are H_2O_2 and H_2O respectively.

Properties of H_2O_2 include very reactive, unstable (producing water and oxygen gas), polar covalent with hydrogen bonding, colorless, bitter taste (slightly acidic), miscible with water, soluble in ether, insoluble in petroleum ether, caustic to skin, density = 1.4422 g/mL, mp = –0.41°C, bp = 150.2°C.

Uses of H_2O_2 include oxidizer in rocket fuel (pure form); dilute solutions in disinfectants, for cleaning metals, refining oils, for bleaching in textile and plastic industry.

Properties of H_2O include very stable, very polar with hydrogen bonding and large dipole, colorless, tasteless (when pure), forms H^+ and OH^-, expands upon freezing so that the solid floats, density of solid at 0°C = 0.9168 g/mL, density of liquid at 4°C = 1.0000 g/mL, mp = 0.000°C, bp = 100.000°C.

Uses of H_2O: Water is essential to life. Chemically, water is known as the most universal solvent in that a great number of chemicals will dissolve in it. This solubility gives water many industrial and medical uses.

Name

Date Class

Types of Chemical Reactions

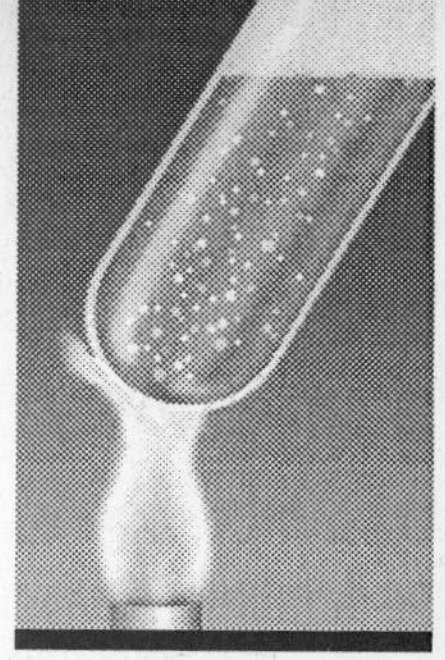

LAB MANUAL

6-1

Many chemical reactions are accompanied by observable physical changes. The appearance of a solid (precipitate) when solutions of substances are mixed is one such change. Others are generation of a gas and a change in the color of a solution. In this laboratory activity, you will observe what happens when a number of different substances are combined. Many of the reactions you will observe can be classified as single displacement, double displacement, decomposition, or synthesis reactions.

OBJECTIVES

- **Carry out** chemical reactions in a microplate.
- **Identify** the pairs of substances that react.
- **Classify** the reactions that occur.
- **Write** balanced equations for the reactions.

MATERIALS

microtip pipets	file
96-well microplate	drinking straw
96-well template	goggles
beaker	apron
deionized water	

PROCEDURE

1. The substances from Group A listed under Data and Observations will each be combined with the substances from Group B. Form a **hypothesis** about which reactants in Group B might take part in single displacement reactions with reactants in Group A.
2. Label a 96-well template and Microplate Data Form as illustrated under Data and Observations. Use rows A through D for Set 1. Use rows E through H for Set 2.
3. In preparation for the reactions in Set 1, place the microplate on your template so that the numbered columns are away from you and the lettered rows are at the left.
4. **CAUTION:** *Many of the chemicals you will use are toxic. Follow proper chemical hygiene procedures. Wash your hands thoroughly after completing this laboratory activity.* Place a few granules of MnO_2 in each of wells A1 and A2 of the microplate. These are the only two wells in row A to be used. The MnO_2 does not take part in any reaction here, but helps to make reactions occur. **CAUTION:** *MnO_2 is a strong oxident. Contact with organic material should be avoided.*
5. Using a microtip pipet, place 4 drops of $Cu(NO_3)_2$ solution in each of wells B1 through B11. Rinse the pipet in a beaker of deionized water. *Remember to rinse the microtip pipet every time you change chemicals.*
6. Place 4 drops of $Ni(NO_3)_2$ solution in each of wells C1 through C11.
7. Place 4 drops of $Pb(NO_3)_2$ solution in each of wells D1 through D11.

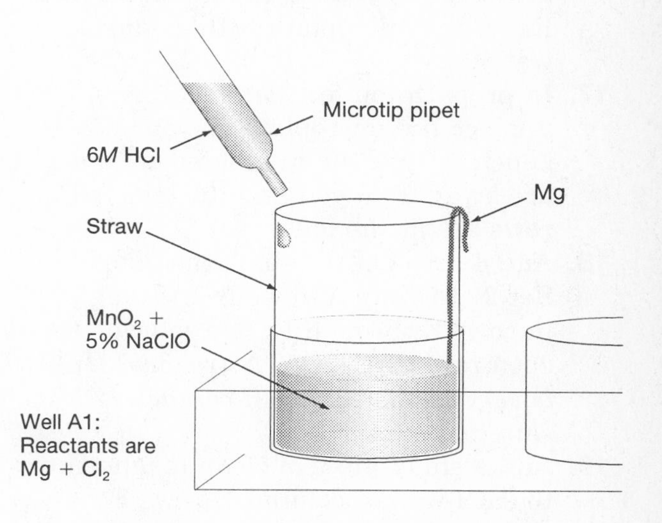

8. Place a 2-cm piece of drinking straw in well A1. Clean a 3-cm strip of magnesium metal with a file. Bend the magnesium so that most of it hangs inside the straw, as shown in the diagram.
9. Temporarily remove the straw-metal assembly. Add 5 drops of 5 percent NaClO solution to the MnO_2 in well A1. Carefully replace the straw-metal assembly in well A1 so that the solution does not come in direct contact with the magnesium.
10. Add 1 drop of 6*M* HCl to well A1 by letting it run down the inside of the straw. The result is generation of chlorine gas, Cl_2, to test the combination of magnesium and chlorine. **CAUTION:** *Hydrochloric acid (HCl) is very corrosive. Avoid contact with skin and eyes.*
11. Add 5 drops of H_2O_2 solution to well A2.
12. Add a small piece of clean magnesium to each well in column 1, rows B through D.
13. Add a small piece of zinc to each well in column 2, rows B through D.
14. Add 4 drops of solutions of each of the remaining substances in Group B, one to each column beginning with column 3 and continuing through column 11. For example, add 4 drops of Na_2CO_3 solution to each of wells B3, C3, and D3; add 4 drops of Na_2SO_4 solution to each of wells B4, C4, and D4.
15. Record your observations for each combination of reactants in the labeled Microplate Data Form.
16. Discard the mixtures in the microplate according to your teacher's instructions. Rinse the microplate with deionized water.
17. In preparation for the reactions in Set 2, place the microplate on your lab bench so that the numbered columns are away from you and the lettered rows are at the left.
18. Add 4 drops of the solutions listed in Set 2 of Group A to wells 1 through 11 in rows E through H, one solution to each row. **CAUTION:** *Nitric acid (HNO_3) is very corrosive. Avoid contact with skin and eyes.*
19. Add a small piece of clean magnesium to each well in column 1, rows E through H.
20. Add a small piece of zinc to each well in column 2, rows E through H.
21. Next, add 4 drops of each of the solutions listed for columns 3 through 11 in Group B to the wells in rows E through H, one solution to each column.
22. Record your observations in the Microplate Data Form.
23. Discard the mixtures in the microplate according to your teacher's instructions. Rinse the microplate carefully with deionized water.

DATA AND OBSERVATIONS

1. **Hypothesis:** ______________________

____

____

2. Label a Microplate Data Form as shown in the partial form. In Set 1, the reactants in well A1 are Mg and Cl_2; in well A2 the reactant is H_2O_2. In rows B–D in Set 1 and in rows E–H in Set 2, the row reactants listed in Group A are combined with the column reactants listed in Group B.

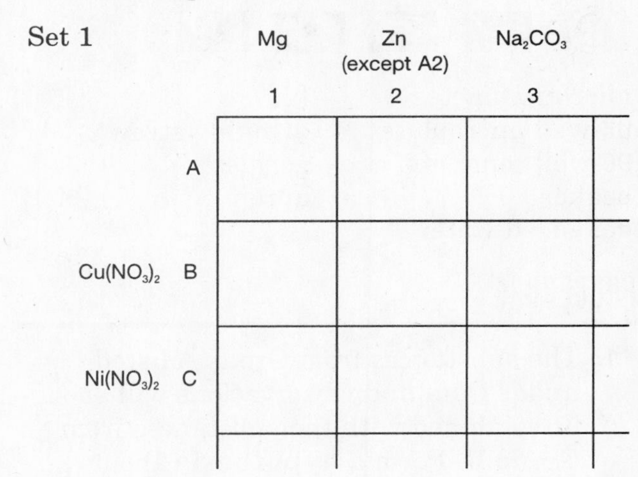

Group A: Label the lettered rows as follows.

A	—	Set 1
B	$Cu(NO_3)_2$	
C	$Ni(NO_3)_2$	
D	$Pb(NO_3)_2$	
E	HNO_3	Set 2
F	$AgNO_3$	
G	$Al(NO_3)_3$	
H	$Fe(NO_3)_3$	

Group B: Label the numbered columns as follows.

1 Mg
2 Zn (except well A2)
3 Na_2CO_3
4 Na_2SO_4
5 Na_2CrO_4
6 NaCl
7 NaI
8 NaSCN
9 $Na_2Cr_2O_7$
10 NaOH
11 H_2O

Enter your observations in the Microplate Data Form as shown in the partial form, using *NR* where there is no observable reaction and ppt for formation of a precipitate (include ppt color). *X* indicates a reaction occurred; you will add these marks as you analyze your observations.

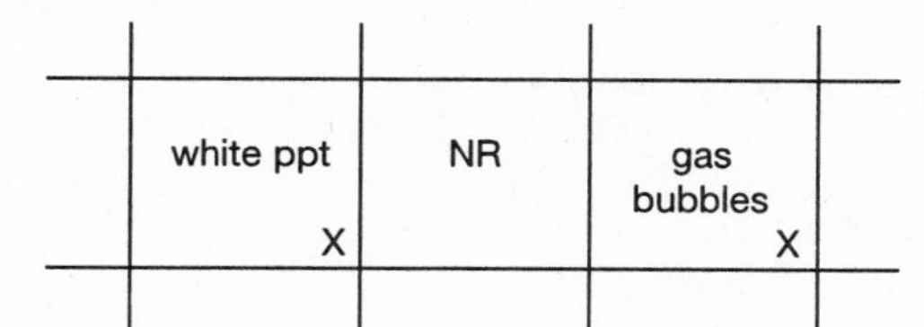

ANALYSIS

1. On your labeled Microplate Data Form, place an *X* in the lower right-hand corner of the square for each well where a reaction occurred.
2. In the space below and on page 52 or page 54, write a balanced equation for each combination of substances that resulted in a reaction, labeled as illustrated with the row letters and column numbers. Omit column numbers where no reaction occurred.

Set 1
Row A
Col 1 [balanced equation]
2 [balanced equation]

You need not write equations for any well where there was a color change but no precipitate formed. The color shows a reaction that formed a different kind of ion in solution.

CONCLUSIONS

1. Identify each balanced equation as representing a single displacement (SD), double displacement (DD), synthesis (S), or decomposition (D) reaction. Write the symbol for the type of reaction next to the equation on your list.
2. Was your hypothesis about the probable reactants in single displacement reactions correct? Explain.

____

____

____

EXTENSION AND APPLICATION

1. What type of reaction between two ionic compounds in water solution was most common?
2. Which of the substances that you tested most often reacted to form a complex ion, as shown by a color change but no precipitate?
3. From each of three different columns, choose one reaction that you classified as double displacement with precipitate formation. Using the Solubility Rules in Table 7 on page xxv and your balanced equations, identify which of the two products formed in each of the three reactions was the solid.

Teacher Guide

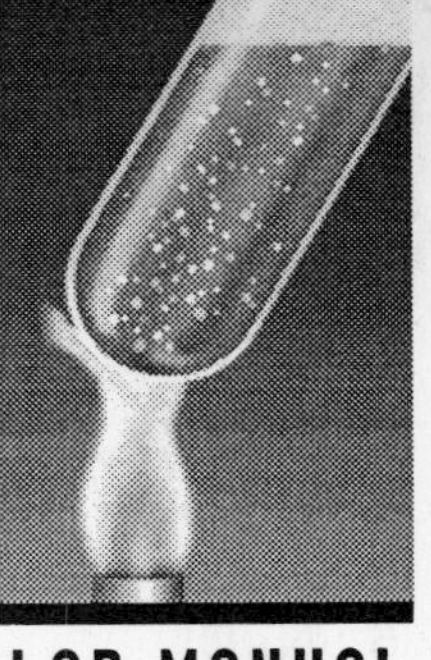

Types of Chemical Reactions

LAB MANUAL

6-1

In this experiment, students observe four simple types of inorganic reactions described in Chapter 6. Synthesis is represented by the reaction of magnesium with chlorine gas to produce magnesium chloride. Decomposition is shown by the reaction of hydrogen peroxide in the presence of MnO_2 catalyst to give water and oxygen. Single displacement reactions are demonstrated by several reactions of magnesium or zinc with soluble salts. Double displacement is shown by numerous precipitation reactions between soluble salts. Several reactions that do not fit these categories are included and make the point that not every reaction neatly fits one of these four patterns. In addition to allowing students to observe and classify a large number of reactions, this activity provides extensive practice in writing balanced chemical equations.

Misconception: Students may believe that all mixtures of substances lead to chemical reactions. On the other hand, they may also believe that if no visible change occurs, no reaction has occurred. Make clear to them that neither assumption is correct. *Proving* a reaction has occurred requires verifying that a new substance has formed.

PROCESS SKILLS

1. Students will combine substances in a microplate and identify substances that react.
2. Students will write balanced equations for the reactions.
3. Students will classify the reactions that occur.

PREPARATION HINTS

- All solutions except 3% hydrogen peroxide may be prepared months in advance of this experiment.
- To ensure good, reproducible results, 0.1*M* solutions should be used throughout, except as noted in the materials list.
- Be sure that an adequate supply of deionized or distilled water is available so that students can rinse pipets every time they change substances.
- Be sure to consult the labels of containers for the correct molecular masses of the compounds used in this experiment. Often, the chemicals are hydrated and a different mass will be needed from that given here.
- For each group, make a copy of the 96-well template on page T8.

MATERIALS

Group A reactants (for lettered rows)

Set 1

A MnO_2 (s)
5% sodium hypochlorite (household bleach)
6*M* HCl (Add concentrated HCl to water in a 1:1 ratio. **Caution:** *Add the acid to the water. Do not add the water to the acid.*)

B $Cu(NO_3)_2 \cdot 3H_2O$ (24.2 g in 1 L of solution)

C $Ni(NO_3)_2 \cdot 6H_2O$ (29.1 g in 1 L of solution)

D $Pb(NO_3)_2$ (33.1 g in 1 L of solution)

Set 2

E Nitric acid (6 mL of concentrated HNO_3 in 1 L of solution. (**CAUTION:** *Add acid to water.*)

F $AgNO_3$ (17 g in 1 L of solution)

G $Al(NO_3)_3 \cdot 9H_2O$ (37.5 g in 1 L of solution)

H $Fe(NO_3)_3 \cdot 9H_2O$ (40.4 g in 1 L of solution)

Group B reactants (for numbered columns)

1. magnesium
2. H_2O_2 (3% household hydrogen peroxide) zinc
3. Na_2CO_3 (10.6 g in 1 L of solution)
4. $Na_2SO_4 \cdot 10H_2O$ (32.2 g in 1 L of solution)
5. Na_2CrO_4 (16.2 g in 1 L of solution)
6. NaCl (5.8 g in 1 L of solution)
7. NaI (15.0 g in 1 L of solution)
8. NaSCN (8.1 g in 1 L of solution)
9. $Na_2Cr_2O_7 \cdot 2H_2O$ (29.8 g in 1 L of solution)
10. NaOH (4.0 g in 1 L of solution)
11. deionized water

microtip pipets	file
96-well microplate	drinking straw
96-well template	goggles
beaker	apron

PROCEDURE HINTS

- Before proceeding, be sure students have read and understood the purpose, procedure, and safety precautions for this laboratory activity.
- Point out that the wells in column 11, in which the substances are mixed with water, can be used for comparison, an aid in detecting whether a change has taken place in other wells.
- **Troubleshooting:** Students must be cautioned to observe carefully and for a reasonably long time. Reactions will not always be easy to detect, especially where gases are produced. It takes a while before enough gas has collected to be noticed.
- **Troubleshooting:** Students may have a difficult time determining whether a precipitate or a complex ion has formed. An easy criterion is that if a solution is cloudy, the reaction has produced a precipitate. If the reaction mixture is colored but clear, a complex ion has been formed.
- **Troubleshooting:** Some students may need help with writing the equations. Students will most likely not be able to write an equation or classify the reaction of nitric acid with sodium carbonate. You may wish to explain the formation of H_2CO_3 and its decomposition to H_2O and CO_2.
- **Safety Precaution:** Chromate and dichromate are proven carcinogens. Although the quantities used are small, appropriate precautions should be taken. You might prefer to omit these ions and have students carry out reactions with the remaining substances only.
- **Safety Precaution:** Heavy metal compounds, such as those containing lead, are generally extremely toxic. Remind students to follow proper chemical hygiene procedures when working with such compounds.
- **Disposal:** At the end of the experiment, collect all materials from the microplates and all rinse water in a wash basin or large beaker. Dispose of these chemicals according to the instructions in **Disposal** J.

DATA AND OBSERVATIONS

1. Students should be able to hypothesize that reactions involving magnesium or zinc as a reactant might be single displacement reactions. These are the only reactions with elements as reactants.
2. See Microplate Data Form on the next page.

ANALYSIS

1. See Microplate Data Form.
2. The equations that students should be able to complete are given, numbered according to the columns.

Set 1

Row A

Col 1. $Mg + Cl_2 \rightarrow MgCl_2$ (S)

Col 2. $2H_2O_2 \xrightarrow[\text{catalyst}]{MnO_2} 2H_2O + O_2$ (D)

Row B

Col 1. $Cu(NO_3)_2 + Mg \rightarrow Mg(NO_3)_2 + Cu$ (SD)

Col 2. $Cu(NO_3)_2 + Zn \rightarrow Zn(NO_3)_2 + Cu$ (SD)

Col 3. $Cu(NO_3)_2 + Na_2CO_3 \rightarrow CuCO_3 + 2NaNO_3$ (DD)

Col 5. $Cu(NO_3)_2 + Na_2CrO_4 \rightarrow CuCrO_4 + 2NaNO_3$ (DD)

Col 7. $Cu(NO_3)_2 + 2NaI \rightarrow CuI_2 + 2NaNO_3$ (DD)

Col 9. $Cu(NO_3)_2 + Na_2Cr_2O_7 \rightarrow CuCr_2O_7 + 2NaNO_3$ (DD)

Col 10. $Cu(NO_3)_2 + 2NaOH \rightarrow Cu(OH)_2 + 2NaNO_3$ (DD)

Row C

1. $Ni(NO_3)_2 + Mg \rightarrow Mg(NO_3)_2 + Ni$ (SD)
2. $Ni(NO_3)_2 + Zn \rightarrow Zn(NO_3)_2 + Ni$ (SD)

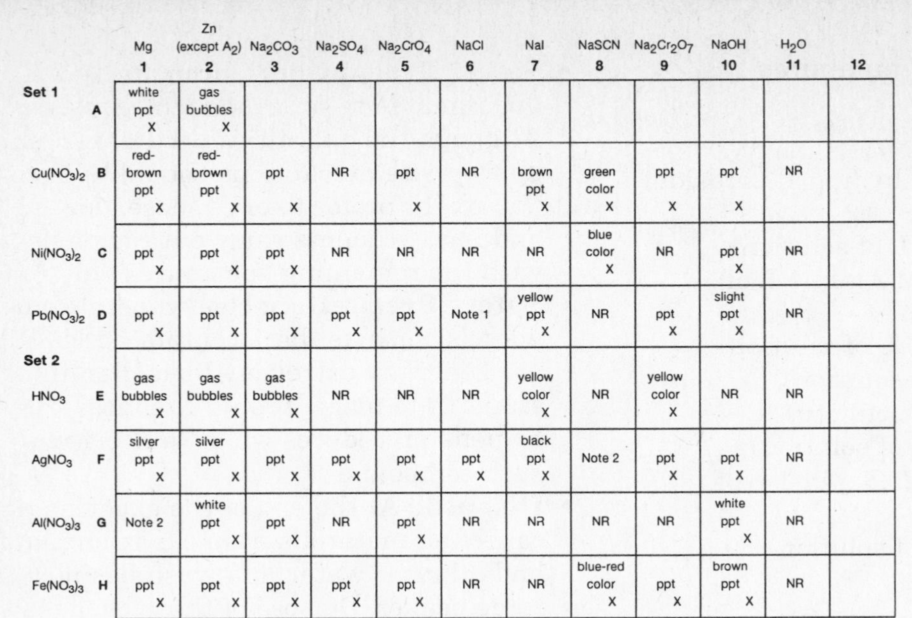

		Mg 1	Zn (except A_2) 2	Na_2CO_3 3	Na_2SO_4 4	Na_2CrO_4 5	NaCl 6	NaI 7	NaSCN 8	$Na_2Cr_2O_7$ 9	NaOH 10	H_2O 11	12
Set 1	A	white ppt X	gas bubbles X										
$Cu(NO_3)_2$	B	red-brown ppt X	red-brown ppt X	ppt X	NR	ppt X	NR	brown ppt X	green color X	ppt X	ppt X	NR	
$Ni(NO_3)_2$	C	ppt X	ppt X	ppt X	NR	NR	NR	NR	blue color X	NR	ppt X	NR	
$Pb(NO_3)_2$	D	ppt X	ppt X	ppt X	ppt X	ppt X	Note 1	yellow ppt X	NR	ppt X	slight ppt X	NR	
Set 2 HNO_3	E	gas bubbles X	gas bubbles X	gas bubbles X	NR	NR	NR	yellow color X	NR	yellow color X	NR	NR	
$AgNO_3$	F	silver ppt X	silver ppt X	ppt X	ppt X	ppt X	ppt X	black ppt X	Note 2	ppt X	ppt X	NR	
$Al(NO_3)_3$	G	Note 2	white ppt X	ppt X	NR	ppt X	NR	NR	NR	NR	white ppt X	NR	
$Fe(NO_3)_3$	H	ppt X	ppt X	ppt X	ppt X	ppt X	NR	NR	blue-red color X	ppt X	brown ppt X	NR	

Note 1: Sometimes a slight precipitate is seen and sometimes it is not.

Note 2: Sometimes a precipitate is seen and sometimes it is not.

3. $Ni(NO_3)_2 + Na_2CO_3 \rightarrow NiCO_3 + 2NaNO_3$ (DD)
10. $Ni(NO_3)_2 + 2NaOH \rightarrow Ni(OH)_2 + 2NaNO_3$ (DD)

Row D

1. $Pb(NO_3)_2 + Mg \rightarrow Mg(NO_3)_2 + Pb$ (SD)
2. $Pb(NO_3)_2 + Zn \rightarrow Zn(NO_3)_2 + Pb$ (SD)
3. $Pb(NO_3)_2 + Na_2CO_3 \rightarrow PbCO_3 + 2NaNO_3$ (DD)
4. $Pb(NO_3)_2 + Na_2SO_4 \rightarrow PbSO_4 + 2NaNO_3$ (DD)
5. $Pb(NO_3)_2 + Na_2CrO_4 \rightarrow PbCrO_4 + 2NaNO_3$ (DD)
6. $Pb(NO_3)_2 + 2NaCl \rightarrow PbCl_2 + 2NaNO_3$ (may not be observed) (DD)
7. $Pb(NO_3)_2 + 2NaI \rightarrow PbI_2 + 2NaNO_3$ (DD)
9. $Pb(NO_3)_2 + Na_2Cr_2O_7 \rightarrow PbCr_2O_7 + 2NaNO_3$ (DD)
10. $Pb(NO_3)_2 + 2NaOH \rightarrow Pb(OH)_2 + 2NaNO_3$ (DD)

SET 2

Row E

1. $2HNO_3 + Mg \rightarrow Mg(NO_3)_2 + H_2$ (SD)
2. $2HNO_3 + Zn \rightarrow Zn(NO_3)_2 + H_2$ (SD)
3. $2HNO_3 + Na_2CO_3 \rightarrow H_2CO_3 + 2NaNO_3$ (DD)
4. $H_2CO_3 \rightarrow H_2O + CO_2$ (D)

Row F

1. $2AgNO_3 + Mg \rightarrow Mg(NO_3)_2 + 2Ag$ (SD)
2. $2AgNO_3 + Zn \rightarrow Zn(NO_3)_2 + 2Ag$ (SD)
3. $2AgNO3 + Na_2CO_3 \rightarrow Ag_2CO_3 + 2NaNO_3$ (DD)
4. $2AgNO_3 + Na_2SO_4 \rightarrow Ag_2SO_4 + 2NaNO_3$ (DD)
5. $2AgNO_3 + Na_2CrO_4 \rightarrow Ag_2CrO_4 + 2NaNO_3$ (DD)
6. $AgNO_3 + NaCl \rightarrow AgCl + NaNO_3$ (DD)
7. $AgNO_3 + NaI \rightarrow AgI + NaNO_3$ (DD)
8. $AgNO_3 + NaSCN \rightarrow AgSCN + NaNO_3$ (may not be observed) (DD)
9. $2AgNO_3 + Na_2Cr_2O_7 \rightarrow Ag_2Cr_2O_7 + 2NaNO_3$ (DD)
10. $AgNO_3 + NaOH \rightarrow AgOH + NaNO_3$ (DD)

Row G

1. $2Al(NO_3)_3 + 3Mg \rightarrow 3Mg(NO_3)_2 + 2Al$ (may not be observed) (SD)
2. $2Al(NO_3)_3 + 3Zn \rightarrow 3Zn(NO_3)_2 + 2Al$ (SD)
3. $2Al(NO_3)_3 + 3Na_2CO_3 \rightarrow Al_2(CO_3)_3 + 6NaNO_3$ (DD)
5. $2Al(NO_3)_3 + 3Na_2CrO_4 \rightarrow Al_2(CrO_4)_3 + 6NaNO_3$ (DD)
10. $Al(NO_3)_3 + 3NaOH \rightarrow Al(OH)_3 + 3NaNO_3$ (DD)

Row H

1. $2Fe(NO_3)_3 + 3Mg \rightarrow 3Mg(NO_3)_2 + 2Fe$ (SD)
2. $2Fe(NO_3)_3 + 3Zn \rightarrow 3Zn(NO_3)_2 + 2Fe$ (SD)
3. $2Fe(NO_3)_3 + 3Na_2CO_3 \rightarrow Fe_2(CO_3)_3 + 6NaNO_3$ (DD)
4. $2Fe(NO_3)_3 + 3Na_2SO_4 \rightarrow Fe_2(SO_4)_3 + 6NaNO_3$ (DD)
5. $2Fe(NO_3)_3 + 3Na_2CrO_4 \rightarrow Fe_2(CrO_4)_3 + 6NaNO_3$ (DD)
9. $2Fe(NO_3)_3 + 3Na_2Cr_2O_7 \rightarrow Fe_2(Cr_2O_7)_3 + 6NaNO_3$ (DD)
10. $Fe(NO_3)_3 + 3NaOH \rightarrow Fe(OH)_3 + 3NaNO_3$ (DD)

CONCLUSIONS

1. Identifications of the type of reaction are given with the equations in the Analysis section.
2. Answers will vary, depending on students' hypotheses.

EXTENSION AND APPLICATION

1. Double displacement with formation of a precipitate was the most common reaction between two ionic compounds in water solution.
2. Sodium thiocyanate was the substance tested that most often formed complexes, as shown by clear, colored solutions.
3. Students' answers will vary. Once they recognize that sodium nitrate is water-soluble, they should be able to identify the other product as the precipitate.

Name

Date Class

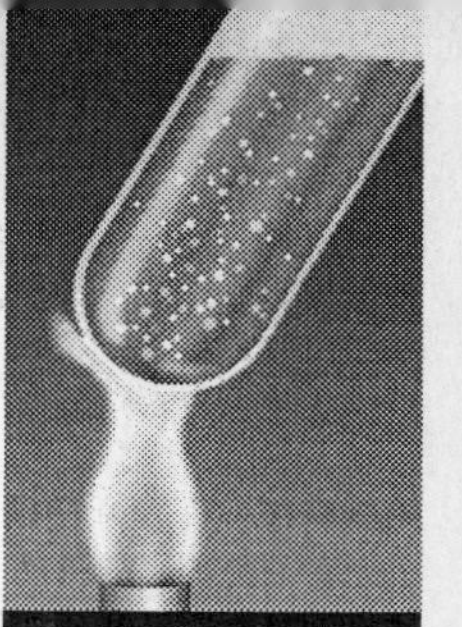

LAB MANUAL

6-2

Effects of Concentration on Chemical Equilibrium

In a chemical system at equilibrium, forward and reverse reactions are taking place at equal rates, and concentrations of reactants and products remain constant. The concentrations of reactants and products need not be equal, however. Equilibrium can be reached at any relative concentrations of reactants and products, depending upon the conditions and the specific reaction.

LeChatelier's principle is a very useful guide to predicting or explaining what happens when changes are made to a system at equilibrium. In this activity, you will observe the reaction between hydrated cobalt ion and chloride ion. With this reaction, the outcome of changes in equilibrium conditions are visible.

$$Co(H_2O)_6^{2+}(aq) + 4Cl^-(aq) \rightleftarrows CoCl_4^{2-}(aq) + 6H_2O(l)$$

pink *blue*

The reactant is pink and the product is blue. When the equilibrium conditions are such that roughly equal amounts of reactant and product are present, a violet color is seen. You will first observe how changing concentrations of reactants and products affects equilibrium. Then you will investigate the effects of different temperatures on the equilibrium.

OBJECTIVES

- **Set up** an equilibrium system.
- **Observe** the effects of changing concentrations and temperature on the equilibrium.

MATERIALS

apron
goggles
96-well microplate
96-well template
microtip pipets (4)
hot plate
thermal mitt
shallow pan or pneumatic trough
toothpicks
distilled water
clear tape
thermometer
400-mL beaker

PROCEDURE

Part 1: Effects of Concentration

1. Based on LeChatelier's principle, form a **hypothesis** about the direction in which increasing the concentration of Cl^- will shift the reaction of hydrated cobalt ion and chloride ion. Write your hypothesis under Data and Observations.
2. Label a 96-well template and a Microplate Data Form as shown under Data and Observations. Place the microplate on the template with the numbered columns away from you and the lettered rows to the left.
3. **CAUTION:** *Many of the chemicals you will use are toxic. Follow proper chemical hygiene procedures. Wash your hands thoroughly after completing this laboratory activity.* Place 4 drops of $Co(NO_3)_2$ solution in each well A1 through A8.
4. Add 1 drop of concentrated HCl to well A1. **CAUTION:** *Hydrochloric acid (HCl) is very corrosive. Avoid contact with skin and eyes.*
5. Add 2 drops of concentrated HCl to well A2.
6. Continue to add HCl to each succeeding well, increasing the amount for each well until you reach 8 drops in well A8. Stir each well with the same toothpick, beginning with well A1.
7. Record the color of the solution in each well in row A of the Microplate Data Form.

8. Add 4 drops of concentrated HCl to each well.
9. Stir each well with the same toothpick, as before. Record the color for the solution in each well in the second row of the data form.
10. Add 5 drops of distilled water to each well.
11. Again, stir each well with the same toothpick. Record the color for the solution in each well in the third row of the data form.
12. Add 2 drops of 1*M* $AgNO_3$ to each well.
13. Stir each well with a single toothpick. Record the color for the solution in each well in the fourth row of the data form.
14. Discard the solutions in wells A1 through A8 according to the directions of your teacher. Rinse the wells with distilled water and discard the rinse water in the same manner as the solutions.
15. Rinse the microtip pipets with distilled water.

Part 2: Effects of Temperature

1. Set up another set of equilibrium reactions by repeating steps 2–6 of the procedure for Part 1.
2. Label row G of the Microplate Data Form "Raise Temperature." Label Row H "Lower Temperature."
3. Fill a 400-mL beaker about one-half full with tap water. Heat the water to 10°C – 15°C above room temperature.
4. Prepare a water bath by pouring the heated water into a shallow pan or pneumatic trough.
5. Cover the tops of wells A1 through A8 with a piece of clear tape. Seal each well by running your finger over the top.
6. Float the microplate on the surface of the hot water.
7. Allow 2 minutes for the plate to be heated by the water. Record the colors of the solutions in the wells.
8. Replace the water in the trough with water that is 10°C below room temperature.
9. Float the microplate on the surface of the cold water.
10. Allow 2 minutes for the plate to be cooled by the water. Record the colors in the wells.
11. Discard the solutions in wells A1 through A8 according to the directions of your teacher. Rinse the wells with distilled water and discard the rinse water in the same manner as the solutions.
12. Rinse the microtip pipets with distilled water.

DATA AND OBSERVATIONS

Hypothesis: ______________________________

____

____

____

Label a Microplate Data Form as shown and use it to record your observations. Use *P* to indicate a pink color, *V* for violet, and *B* for blue.

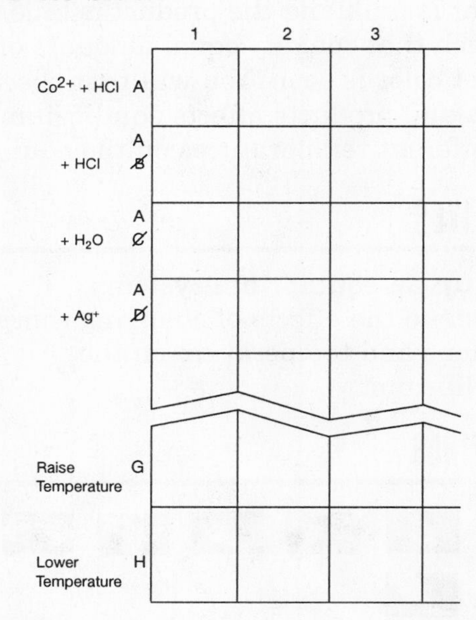

ANALYSIS

Part 1: Effects of Concentration

1. How does the concentration of Cl^- vary in the initial solutions prepared in row A? In which well is the indicator violet? In how many wells is the $CoCl_4^{2-}$ concentration higher than the $Co(H_2O)_6^{2+}$ concentration?

____

____

____

2. What condition changed when excess concentrated HCl was added? How did this change affect the position of the violet indicator well and the number of wells with higher $CoCl_4^{2-}$ concentration?

3. What condition changed when water was added? How did this change affect the position of the indicator well and the number of wells with higher $CoCl_4^{2-}$ concentration?

4. What condition changed when 1*M* $AgNO_3$ was added? How did this change affect the position of the indicator well and the number of wells with higher $CoCl_4^{2-}$ concentration?

Part 2: Effects of Temperature

1. What effect did a higher temperature have on the relative numbers of pink and blue wells?

2. What effect did a lower temperature have on the relative numbers of pink and blue wells?

CONCLUSIONS

1. Using LeChatelier's principle and your analysis, explain the results of adding concentrated HCl, H_2O, and $AgNO_3$ to the reaction of $Co(H_2O)_6^{2+}$ with Cl^- at equilibrium.

2. According to LeChatelier's principle, an increase in temperature shifts an equilibrium in the direction of the endothermic reaction and a decrease in temperature shifts it in the opposite direction of the exothermic reaction. In the reaction of $Co(H_2O)_6^{2+}$ with Cl^-, which direction of the equilibrium is exothermic and which is endothermic?

3. Was your hypothesis about the effect of Cl^- concentration on the equilibrium correct? Explain.

EXTENSION AND APPLICATION

1. Sodium chloride would have been a safer source of Cl^- for the reaction in this experiment, but it could not be used. Explain why you think concentrated HCl was necessary.

2. The active ingredient in swimming pool disinfectant is hypochlorous acid, HClO, which is most effective in a narrow pH range.

$$HClO \rightleftarrows H^+ + ClO^-$$

How is the concentration of HClO affected when the concentration of H^+ is increased and decreased? Sometimes baking soda, $NaHCO_3$, is added to swimming pools. Why?

Teacher Guide

LAB MANUAL

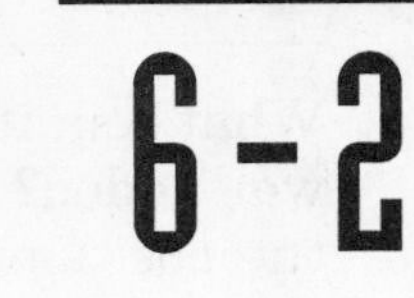

Effects of Concentration on Chemical Equilibrium

In this activity, the different colors of the pink reactant $Co(H_2O)_6^{2+}$ and the blue product $CoCl_4^{2-}$ allow the students to see the shift in the equilibrium under varying conditions.

$$\underset{pink}{Co(H_2O)_6^{2+}(aq)} + 4Cl^-(aq) \rightleftarrows \underset{blue}{CoCl_4^{2-}(aq)} + 6H_2O(l)$$

Students should be able to explain all of the observed changes by applying LeChatelier's principle. The concentrations of the reactants and the products, as well as the temperature, are varied in the experiment.

Misconception: Students often do not recognize that equilibrium is a dynamic state because they see so many reactions that appear to go only one way. Emphasize to the students that all chemical reactions reach an equilibrium, although most of the reactions they have witnessed reach equilibrium with the product so heavily favored that the establishment of equilibrium is not apparent.

PROCESS SKILLS

1. Students will establish an equilibrium system.
2. Students will compare the effects of variations in the concentrations of reactants and products and in temperature.
3. Students will explain their observations by using LeChatelier's principle.

PREPARATION HINTS

- Students should handle concentrated HCl as little as possible. It is a good idea to prepare pipets containing the concentrated HCl and, if you wish, cobalt nitrate solutions in advance. Minimize the amount of HCl by providing only 1 mL in each pipet. It is important the HCl be stored in the bulb. Do not allow concentrated HCl to be stored in the stem of the pipets. In case the students run out of the acid supplied, prepare extra HCl pipets.
- Provide clear tape to seal the tops of the wells. Do not use translucent tape. The HCl will make the tape opaque, in any case.
- Silver nitrate should be stored in a brown or black bottle.
- For each group, make a copy of the 96-well template on page T8.

MATERIALS

12*M* HCl (stock concentrated HCl)
0.2*M* $Co(NO_3)_2$ (Add 5.81 g of $Co(NO_3)_2 \cdot 6H_2O$ to enough water to make 100 mL of solution.)
1*M* $AgNO_3$ (Add 17.0 g $AgNO_3$ to 100 mL of distilled water.)

apron	thermometer
goggles	shallow pan or pneumatic trough
96-well microplate	toothpicks
96-well template	distilled water
microtip pipets (4)	clear tape
400-mL beaker	thermal mitt
hot plate (or laboratory burner, ring stand, ring, and wire gauze)	ice (if needed)

PROCEDURE HINTS

- Before proceeding, be sure students have read and understood the purpose, procedure, and safety precautions for this laboratory activity.

- If your lab has hot water, it can be used to make the hot water bath. If your tap water is not colder than room temperature, you will need some ice to make the cold water bath.
- **Safety Precaution:** It is extremely important to insist that students observe safety rules when performing this experiment. Concentrated HCl is corrosive. It causes skin burns and damages clothing.
- **Disposal:** Dispose of used solutions according to the instructions in **Disposal** J.

DATA AND OBSERVATIONS

Hypotheses will vary, but students should be able to predict, based on LeChatelier's principle, that increasing the concentration of Cl^- will shift the reaction in the forward direction and increase the concentration of the product, $CoCl_4^{2-}$. See the Microplate Data Form for Parts 1 and 2 below.

ANALYSIS

Part 1: Effects of Concentration

1. The Cl^- concentration increases across row A. The violet indicator is in well A3. The $CoCl_4^{2-}$ concentration is higher than the $Co(H_2O)_6^{2+}$ concentration in 5 wells (blue wells).
2. The addition of HCl increased the Cl^- concentration and shifted the violet indicator well toward well A1, resulting in 6 blue wells instead of 5.
3. Adding water decreased the concentrations of the reactant ions. The violet indicator well moved toward well A8. Here there were 4 wells with higher $CoCl_4^{2-}$ concentration than $Co(H_2O)_6^{2+}$ concentration.
4. Addition of $AgNO_3$ decreases the Cl^- concentration by forming AgCl. The violet indicator well moves much farther toward well A8, giving only one well where $CoCl_4^{2-}$ concentration is higher than $Co(H_2O)_6^{2+}$ concentration.

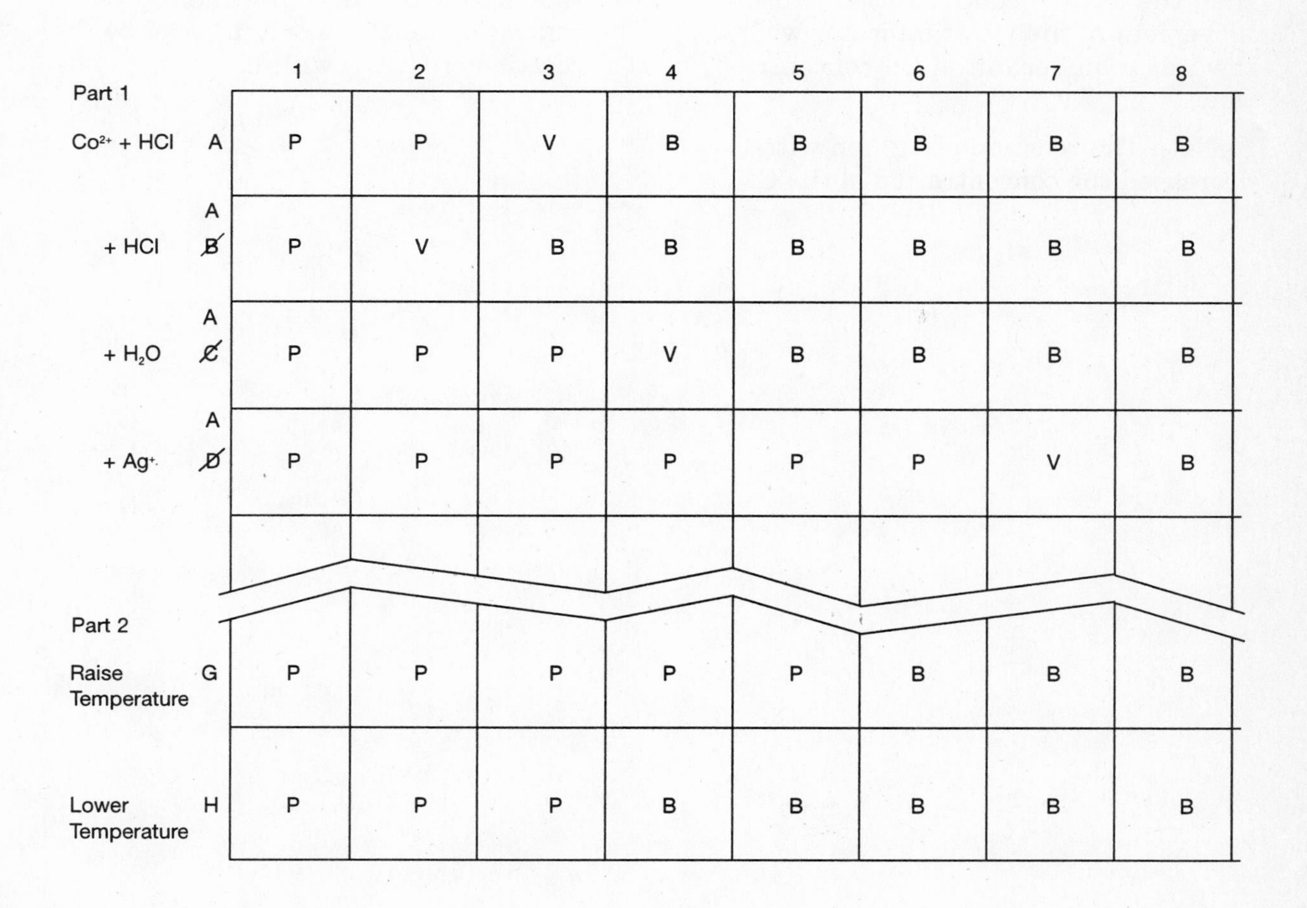

		1	2	3	4	5	6	7	8
Part 1									
Co^{2+} + HCl	A	P	P	V	B	B	B	B	B
+ HCl	A (~~B~~)	P	V	B	B	B	B	B	B
+ H_2O	A (~~C~~)	P	P	P	V	B	B	B	B
+ Ag^+	A (~~D~~)	P	P	P	P	P	P	V	B
Part 2									
Raise Temperature	G	P	P	P	P	P	B	B	B
Lower Temperature	H	P	P	P	B	B	B	B	B

Part 2: Effects of Temperature

1. A higher temperature increased the number of pink wells and decreased the number of blue wells.
2. A lower temperature gave the same number of blue wells as in the initial solutions, but the violet well was replaced by a pink one.

CONCLUSIONS

1. HCl: Adding the HCl increased the Cl^- concentration, driving the reaction forward and increasing the concentration of blue $CoCl_4^{2-}$. The change in position of the violet indicator well showed that equal concentrations of reactant and product were reached at a well with a lower initial concentration of Cl^-.

   H_2O: Adding H_2O decreases the concentration of both reactants, favoring the formation of pink $Co(H_2O)_6^{2+}$. The result was more wells where the $Co(H_2O)_6^{2+}$ concentration was higher than the $CoCl_4^{2-}$ concentration and the movement of the violet indicator well towards a higher initial Cl^- concentration.

   $AgNO_3$: The reaction of Ag^+ ion with Cl^- decreased the concentration of the Cl^- significantly. The result was the movement of the violet indicator well much farther toward a higher initial concentration of Cl^-.
2. The higher temperature favored the pink reactant, $Co(H_2O)_6^{2+}$. Therefore, the reverse reaction in this equilibrium is the endothermic reaction and the forward reaction is the exothermic reaction.
3. Answers will vary, depending on students' hypotheses.

EXTENSION AND APPLICATION

1. Concentrated HCl is 12*M*. Sodium chloride is not soluble enough to give a 12*M* solution. The high concentration of Cl^- is needed to give a visible blue color for the product because the equilibrium favors the reverse reaction.
2. Increasing H^+ favors formation of HClO (the concentration increases) and decreasing H^+ favors ionization of HClO (the concentration decreases). Baking soda is added to decrease the H^+ concentration in the swimming pool by the reaction of HCO_3^- with H^+.

Name

Date Class

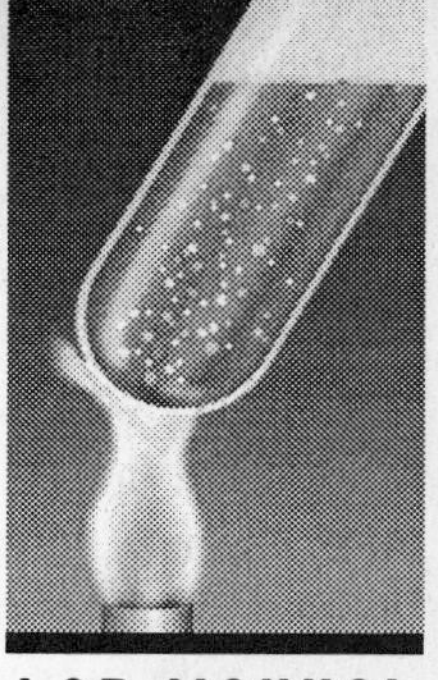

LAB MANUAL

7-1

Atomic Spectra

Each element in the Periodic Table has a specific number of electrons in specific energy levels. When atoms of any element absorb energy, electrons move from lower to higher energy levels. As these electrons fall back into their original energy levels, the excess energy is released as electromagnetic radiation.

The characteristic patterns of radiation for different elements are called atomic spectra. For some metallic elements, this radiation is in the visible region of the spectrum. The spectra can be seen when compounds containing these elements are heated in a flame. The purpose of this activity is to compare the atomic spectra of different metallic elements.

OBJECTIVES

- **Observe** the characteristic atomic emission spectra of some metallic elements.
- **Diagram** the observed spectra.
- **Analyze** a mixture from its spectrum.

MATERIALS

laboratory burner
hand spectroscope
standard spectra chart

PROCEDURE

1. Form a **hypothesis** about the appearance of the spectra of the metallic elements in the three different compounds used in this activity.
2. Work in pairs. Place a laboratory burner in a fume hood. Light and adjust the flame of the burner until the flame is all blue.
3. Observe the burner flame through the slit in the hand spectroscope.
4. Shake the covered plastic bottle containing copper carbonate. **CAUTION:** *Copper carbonate is toxic if ingested.*
5. Loosen the top of the bottle, hold the bottle close to the collar (air inlet) of the burner, and remove the lid.
6. Record the color of the flame. Then, observe the flame with the hand spectroscope and sketch the specific color bands that you see. For example, to record a line at 500 nm, draw a scale like the one below and place a vertical line at 500 nm:

350 (nm) 450 550 650 750

If your spectroscope does not have a built-in scale, compare your observed bands with those on a standard spectral chart like the one below to estimate wavelengths.

Continuous Spectrum (See your textbook.) Wavelength, λ(nm)

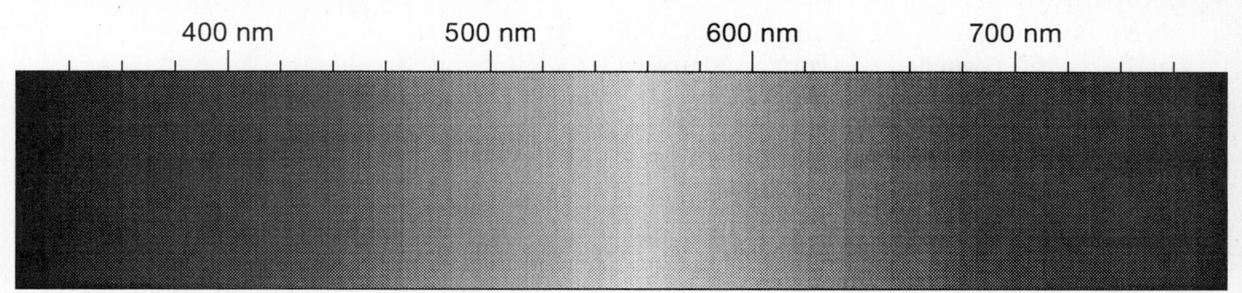

7. Repeat steps 4–6 with the plastic bottles containing lithium carbonate and sodium carbonate. **CAUTION:** *Sodium carbonate can irritate the skin. Lithium carbonate dissolves in water to form a strong base.*
8. Repeat steps 4–6 with your unknown mixture of salts.

DATA AND OBSERVATIONS

Hypothesis: ______________________

____

____

____

Element:	Copper	Sodium	Lithium
Flame color:			

Spectral lines of copper:

Spectral lines of sodium:

Spectral lines of lithium:

Spectral lines of unknown:

ANALYSIS

1. Compare the bright lines you observed for copper, sodium, and lithium with the lines you recorded for your mixture. Identify which elements are in your mixture.

____

____

____

2. Were any of your spectra identical? Similar? Was your hypothesis correct?

____

____

____

CONCLUSIONS

1. Why did each element give a different observed flame color?

____

____

____

2. How can you identify the composition of a mixture of any two salts?

____

____

____

EXTENSION AND APPLICATION

1. Distance between the observer and the glowing element does not play a major factor in the examination of spectra. How could atomic spectra be useful to an astronomer observing stars?
2. As the concentration of an element in a mixture increases, the intensity of the color of that element's lines in the spectrum increases. Investigate the principle of an Atomic Absorption Spectrometer. How does this laboratory instrument make it possible to determine the concentrations in mixtures?

Teacher Guide

LAB MANUAL

7-1

Atomic Spectra

Elements that absorb energy release the energy absorbed by emitting photons of electromagnetic radiation. The emission spectrum of each element in the Periodic Table is unique. The emission spectra of many metals fall in the visible range and can be observed when metal salts are ionized in a flame.

PROCESS SKILLS

1. Students will observe spectra of different elements.
2. Students will analyze a mixture by observing its spectrum.

PREPARATION HINT

Have available a copy of *The Handbook of Chemistry and Physics*, or similar guidebook, for student reference.

MATERIALS

plastic bottles containing anhydrous carbonates of copper, lithium, and sodium
plastic bottle containing an unknown mixture of two of the above salts
laboratory burner
hand spectroscope
standard spectra chart

PROCEDURE HINTS

Before proceeding, be sure students have read and understood the purpose, procedure, and safety precautions for this laboratory activity.

- The salts selected should be dry and as powdered as possible.
- Carbonates and oxides of metals are preferred because they usually do not form hydrated salts.
- Small plastic containers are perfect for use in this activity. Snap cap film containers are ideal.
- Film containers may be labeled and kept from year to year by placing in large, self-sealing plastic bags.
- If necessary, the film containers can be emptied and refilled with salts that have been dried overnight in a desiccator or laboratory oven set on low.
- An empty shoe or goggle box with a slit on one end and a diffraction-grating-covered opening on the other makes an inexpensive spectroscope.
- **Safety Precaution:** Adequate ventilation is necessary for this activity.
- **Disposal:** Dispose of the chemicals according to the instructions in **Disposal** F.

DATA AND OBSERVATIONS

Students' hypotheses may vary. A correct hypothesis should say that the appearances of the spectra will differ.

Element:	Copper	Sodium	Lithium
Flame color:	Green	Yellow	Red

Copper

350 (nm) 450 550 650 750

Sodium

350 (nm) 450 550 650 750

Lithium

350 (nm) 450 550 650 750

Unknown mixture (Answer will depend on the mixture)
For example, Cu and Na

ANALYSIS

1. The lines of the elements in the mixture are combined in the spectrum. Answer should identify the lines that show the presence of each different element identified.
2. None of the observed spectra are identical. Some of the lines on the spectra of two or more elements may be similar or even appear to be identical. No two elements, however, will have all lines the same in their spectra because no two elements have the same energy levels.

CONCLUSIONS

1. The observed flame color is the sum of the lines in the spectrum of each element.
2. By observing the spectrum of a mixture and comparing the distinctive line with known atomic spectra, the elements in the mixture can be identified. For example, only sodium gives a double line at 589 nm. If a sample shows a double line at this location in the spectrum, then sodium is in the sample.

EXTENSION AND APPLICATION

1. By observing and measuring the spectra of distant stars, astronomers can tell the elements that are present on a star.
2. In an atomic absorption spectrometer a special lamp emits light at the same wavelength of the emission line of an element. The light passes through an atomized sample containing that element and some of the light is absorbed. A detector records the amount absorbed, which indicates the concentration of the element in the sample. By using different lamps for different elements, mixtures can be analyzed. Atomic absorption spectrometers can detect 1 atom of a particular element per billion atoms of other elements.

Name

Date Class

Transition Metals

The elements with atomic numbers 21 through 30 exhibit many properties different from those of other elements in the same period. The elements directly below in the next two periods, however, have similar properties. Together, these elements in Groups 3 to 12 are known as the transition metals. The purpose of this activity is to compare the chemical reactions of some transition metal ions with those of non-transition metals from the same period.

OBJECTIVES

- **Observe** physical and chemical properties of transition metal ions in aqueous solution.
- **Observe** results of mixing three different chemicals (NH_3, KSCN, HCl) with metal ions.
- **Compare** chemical reactions of transition metal ions with those of other metal ions.

MATERIALS

96-well microplate
96-well template
microtip pipets
toothpicks

PROCEDURE

1. Review the list of metals that you will be testing. Form a **hypothesis** about which three metals are likely to have properties different from the others.
2. Label a 96-well template as shown on the diagram below. Place a 96-well microplate on your template with the numbered columns away from you and the lettered rows to the left.
3. **CAUTION:** *Many of the chemicals you will use are toxic. Follow proper chemical hygiene procedures. Wash your hands thoroughly after completing this laboratory activity*. Place 5 drops of KNO_3 in each of wells A1, B1, C1, and D1.
4. Place 5 drops of $Ca(NO_3)_2$ in each of wells A2, B2, C2, and D2.
5. Repeat this process of placing 5 drops of solutions of metal compounds in subsequent columns of the microplate. Use chemicals in the order in which they are listed on the microplate template. The last drops will be in column 10.
6. Add 5 drops of 6*M* NH_3 to each well in row A. Mix well with a toothpick. **CAUTION:** *Ammonia solution is caustic.*
7. Add 5 drops of KSCN solution to each well in row B. Mix well with a toothpick.
8. Finally, add 5 drops of HCl to each well in row C. **CAUTION:** *Hydrochloric acid*

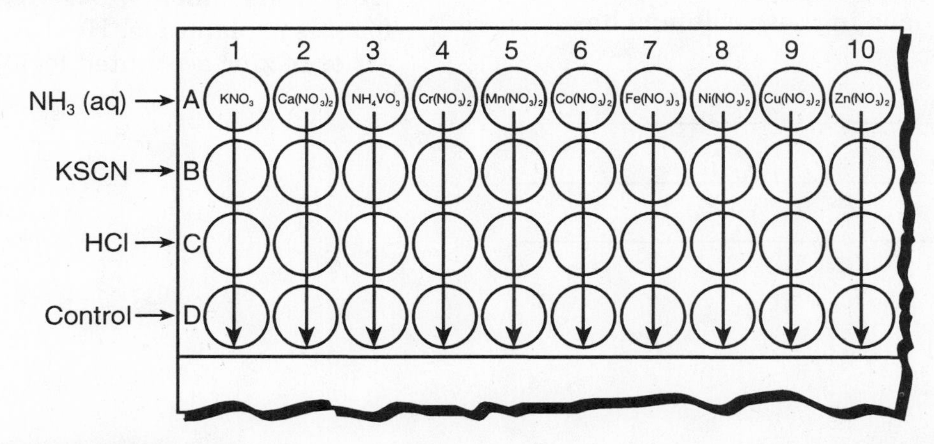

(HCl) is very corrosive. Avoid contact with skin and eyes. Do not mix HCl and KSCN. Clean up all spills immediately. Mix well. Use row D as a control for comparing with the other rows.

DATA AND OBSERVATIONS

1. **Hypothesis:** ____________________

   ____

   ____

2. Label the rows and columns of your Microplate Data Form the same way as your template. Observe row D of your microplate. Record your observations in the Data Form.
3. Compare the solutions in the wells in rows A, B, and C in each column to the solution in row D in that column. Where there was a change, record what you observe.

ANALYSIS

1. Compare the order of the metal ions in the wells from left to right in the rows of your microplate to the order of metals in Period 4 of the periodic table. Note the relationship of your metal ions to the positions of the metals in the periodic table.

   ____

   ____

   ____

2. What did you observe in columns 1, 2, and 10 in your microplate? What do the metal ions in these columns have in common?

   ____

   ____

   ____

CONCLUSIONS

1. Review what you observed in columns 3 through 9 of your microplate. What do the results in these columns have in common? What do the metal ions in these columns have in common? Is there any evidence for the occurrence of more than one type of chemical reaction?

   ____

   ____

   ____

   ____

   ____

   ____

2. Was your hypothesis correct? Why or why not?

   ____

   ____

   ____

   ____

   ____

EXTENSION AND APPLICATION

1. What physical or chemical properties would help identify a salt as containing a transition metal ion?
2. Modern atomic theory provides an explanation for the similarities of the transition elements. Describe the basis for this explanation. How are the properties of zinc accounted for?

Teacher Guide

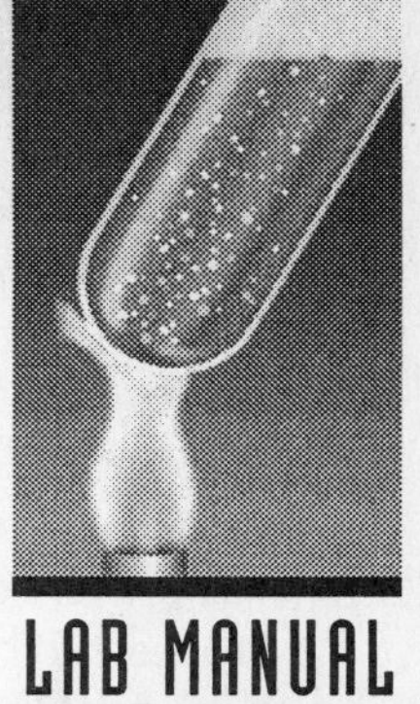

LAB MANUAL

7-2

Transition Metals

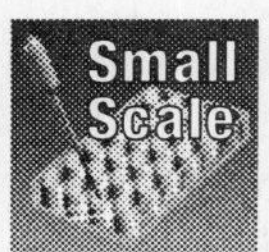

Most of the salts of the transition elements are brightly colored, and most transition metal ions in water solution have a distinctive color. The color of the solution is due to the interaction of the metal ions with water molecules and the formation of hydrated ions. Similar complex ions with characteristic colors are also formed by transition metal ions with a variety of reagents, including NH_3, SCN^-, and Cl^-.

Misconception: Students tend to focus their attention on the reactivity differences and similarities in the periodic table groups of elements, particularly the main-group elements. This activity emphasizes differences across the table and the properties of the transition metals.

PROCESS SKILLS

Students will compare the properties and reactions of transition metal ions and non-transition metal ions.

PREPARATION HINTS

For each group, make a copy of the 96-well template on page T8.

MATERIALS

O.1*M* KNO_3 (10.1 g dissolved in enough distilled water to make 1 L of solution)
O.1*M* $Ca(NO_3)_2 \cdot 4H_2O$ (23.6 g dissolved in enough distilled water to make 1 L of solution)
O.1*M* $Cr(NO_3)_3 \cdot 9H_2O$ (40 g dissolved in enough distilled water to make 1 L of solution)
O.1*M* $Mn(NO_3)_2$ (Comes commercially as a 50 percent solution. Prepare by dissolving 36 mL of $Mn(NO_3)_2$ in 964 mL distilled water)
O.1*M* $Co(NO_3)_2 \cdot 6H_2O$ (29.1 g dissolved in enough distilled water to make 1 L of solution)
O.1*M* $Fe(NO_3)_3 \cdot 9H_2O$ (40.4 g dissolved in enough distilled water to make 1 L of solution)
O.1*M* $Ni(NO_3)_2 \cdot 6H_2O$ (29.1 g dissolved in enough distilled water to make 1 L of solution)
O.1*M* $Cu(NO_3)_2 \cdot 3H_2O$ (24.2 g dissolved in enough distilled water to make 1 L of solution)
O.1*M* $Zn(NO_3)_2 \cdot 6H_2O$ (29.7 g dissolved in enough distilled water to make 1 L of solution)
O.1*M* NH_4VO_3 (Dissolve 4 g NaOH in 500 mL of distilled water. Stirring constantly, add 10 g NH_4VO_3. When the NH_4VO_3 has dissolved completely, add 10 mL of concentrated H_2SO_4. Then add distilled water to make a total volume of 1 L.)
6*M* NH_3 (400 mL of NH_4OH to 600 mL of distilled water)
1*M* KSCN (97.2 g dissolved in enough distilled water to make 1 L of solution)
6*M* HCl (500 mL of HCl to 500 mL of distilled water)
96-well microplate
96-well template
microtip pipets
toothpicks

PROCEDURE HINTS

Before proceeding, be sure students have read and understood the purpose, procedure, and safety precautions for this laboratory activity.

- Substitution for the compounds shown in the Materials list should be avoided since many of the transition elements cited have multiple oxidation states and the results will vary.
- If available, the lanthanide ions (*f* sublevel fillers) provide an interesting contrast to the reaction of elements whose ions have incomplete *d* sublevels.

- **Troubleshooting:** If the concentrations of HCl and NH_3 are too small, the desired effects are not noted.
- In particular, the formation of precipitates or complex ions with ammonia can vary with concentrations or oxidation states of the metals.
- **Safety Precaution:** Care should be taken when handling hydrochloric acid and ammonia.
- **Disposal:** Waste material should be handled as heavy metal waste. Collect all wastes in a fume hood. Dispose according to instructions in **Disposal** J.

DATA AND OBSERVATIONS

Students' hypotheses will vary. A correct hypothesis is that K^+, Ca^{2+}, and Zn^{2+} will have different properties.

	1	2	3	4	5	6	7	8	9	10
A	NR	NR	Orange cpx	Green ppt.	White ppt.	Green ppt.	Brown ppt.	Pale blue cpx	Blue cpx	NR
B	NR	NR	Pale green cpx	Gray cpx	NR	NR	Red cpx	Pale green cpx	Green cpx	NR
C	NR	NR	NR	Gray cpx	NR	Blue cpx	Orange cpx	Light green cpx	Green cpx	NR
D	Colorless		Yellow soln.	Blue gray soln.	Colorless soln.	Pink soln.	Dark yellow soln.	Emerald green soln.	Blue soln.	Colorless soln.

ANALYSIS

1. The ions in solution are in the same order as the elements in the periodic table. The elements scandium and titanium are omitted, however, because the salts of these elements are not usually found in the high school chemistry stockroom. The ions are those of the first two nontransition metals and all the transition metals in Period 4.
2. There are no visible reactions of these ions. The ions in these columns are K^+, Ca^{2+}, and Zn^{2+}. The first two are not transition metals. Zinc, although it is a transition metal, is more like the nontransition metals in its properties because it has a full *d* sublevel.

CONCLUSIONS

1. All of the observable changes indicating possible chemical reactions occurred in these columns. All of the ions in these columns are those of transition metals. Two types of changes—color changes and formation of precipitates—indicate that two different types of reactions have occurred. Color changes without precipitation are due to the formation of complex ions, a type of reaction common for transition metal ions and unusual for main-group metals. Precipitation indicates formation of hydroxides with ammonia solution.
2. Answers will vary, depending on students' hypotheses

EXTENSION AND APPLICATION

1. Properties that would indicate the presence of a transition metal ion are a brightly colored solid, a colored water solution, and changes in color with NH_3, KSCN, HCl or other reagents that form complex ions.
2. The transition elements are those elements in which a *d* sublevel is being filled with electrons across each period. Such similarities in electron configurations lead to similarities in properties. Zinc, as the final transition element in Period 4, has a filled *d* sublevel and two electrons in the outer *s* sublevel, making it more like a main-group element in electron configuration.

Name

Date Class

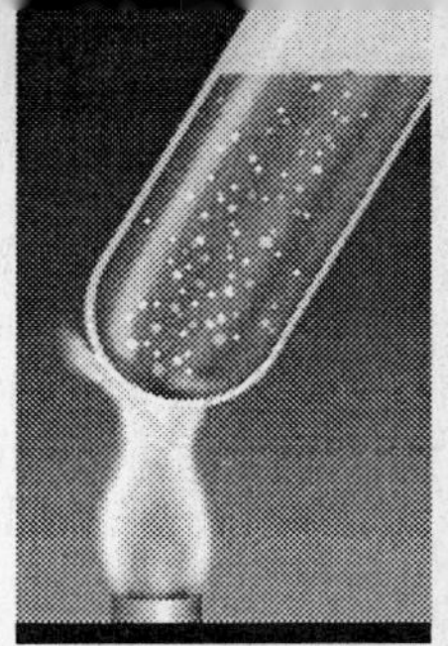

LAB MANUAL

8-1

Periodicity and Chemical Reactivity

The periodic table, arranged according to the atomic structures of the elements, can be used to predict the physical and chemical properties of elements and their compounds. The vertical columns of the table are referred to as groups or families. The halogen family, for example, is Column 17, or Group 17. The horizontal rows are periods. Period 3, for example, contains sodium, magnesium, aluminum, and so on.

This laboratory activity investigates two properties as functions of an element's location within a family on the periodic table. You will work with compounds of Group 2, the alkaline earth family, to determine the variation and the pattern in the solubility and reactivity within the family.

OBJECTIVES

- **Observe** the results of reactions involving Group 2 metal ions.
- **Determine** patterns of reactivity and solubility of the elements of a family.

MATERIALS

apron
goggles
96-well microplates (3)
black construction paper
thin stem pipets
toothpicks
distilled water
copies of Microplate Data Form (2)

PROCEDURE

Part 1

1. Form a **hypothesis** to describe the pattern of solubility and reactivity of the alkaline earth metals. Write your hypothesis under Data and Observations.
2. Place a microplate on a piece of black construction paper. The letters should be along the left side and the numbers away from you.
3. Place exactly 10 drops of Mg^{2+} solution in well A1, 10 drops of Ca^{2+} solution in well B1, 10 drops of Sr^{2+} solution in well C1, and 10 drops of Ba^{2+} solution in well D1. Make sure you use a different pipet for each solution. Also be sure to hold the thin stem pipets vertically so that all the drops will be of the same size. **CAUTION:** *Barium compounds are toxic. Follow proper chemical hygiene procedures. Wash your hands thoroughly after completing this laboratory activity.*
4. Prepare Microplate Data Form 1. Title the data form "Reactions of Group 2 Metal Ions with Sulfate Ions." Label the rows to show the metal ions that are in the first well of each row.
5. Add 5 drops of distilled water to each of the wells *except well 1* in rows A–D.
6. Transfer 5 drops from well A1 to well A2, and mix thoroughly with a toothpick.
7. Transfer 5 drops of this diluted solution in well A2 to well A3.
8. Continue transferring 5 drops from one well to the next through well A12. Be sure to mix the solution thoroughly each time before transferring some of it to the next well.
9. Remove 5 drops from well A12 and discard them.
10. Repeat steps 6 through 9 for each of rows B, C, and D. The plate now holds a series dilution of the metal ions from Group 2 of the periodic table.
11. Now add 5 drops of the sodium sulfate (Na_2SO_4) solution to each well that contains a solution of a metal ion. Observe any reactions.
12. Record your observations in Microplate Data Form 1.

Copyright © Glencoe/McGraw-Hill, a division of The McGraw-Hill Companies, Inc.

Part 2

1. Prepare Microplate Data Form 2, labeling the rows with the metal ions as you did in Part 1. Title this data form "Reactions of Group 2 Metal Ions with Oxalate Ions."
2. Using a clean microplate and the same thin stem pipets you used in Part 1 to transfer the metal ion solutions, repeat steps 3 through 10 of Part 1.
3. Repeat step 11 of Part 1, adding 5 drops of sodium oxalate ($Na_2C_2O_4$) solution to each well in place of the sodium sulfate.
4. Record your observations in Microplate Data Form 2.

Part 3

1. Prepare Microplate Data Form 3 and label it as you labeled forms 1 and 2. Title this form "Reactions of Group 2 Metal Ions with Carbonate Ions."
2. Using a clean microplate and the same thin stem pipets you used in Parts 1 and 2 to transfer the metal ion solutions, repeat steps 3 through 10 of Part 1.
3. Repeat step 11 of Part 1, adding 5 drops of sodium carbonate (Na_2CO_3) solution to the wells in place of sodium sulfate.
4. Record your observations in Microplate Data Form 3.

DATA AND OBSERVATIONS

Hypothesis: ______________________

____

____

Use your Microplate Data Forms to record your observations.

ANALYSIS

1. What pattern can you describe for the precipitation of the sulfate salts ?

   ____

   ____

2. What pattern can you describe for the precipitation of the oxalate salts?

   ____

   ____

3. What pattern can you describe for the precipitation of the carbonate salts?

   ____

   ____

CONCLUSIONS

1. From the data and observations recorded in your Microplate Data Forms, determine which of the sulfate, oxalate, and carbonate salts are the most soluble.

   ____

2. Which of the salts are the least soluble?

   ____

3. How are the reactivity tendencies observed with the sulfate and Group 2 ions supported by the reactions with oxalate and carbonate ions?

   ____

   ____

4. Write a general statement to describe the reactivity of Group 2 ions.

   ____

   ____

   ____

5. Did your results of this experiment support your hypothesis about the properties of Group 2 metals? Explain.

   ____

   ____

EXTENSION AND APPLICATION

Suggest a way that this experiment could be extended in order to place each element within a family in the periodic table. What types of reaction would be most useful in characterizing the elements? Which family of elements would be the hardest to characterize?

Copyright © Glencoe/McGraw-Hill, a division of The McGraw-Hill Companies, Inc.

Teacher Guide

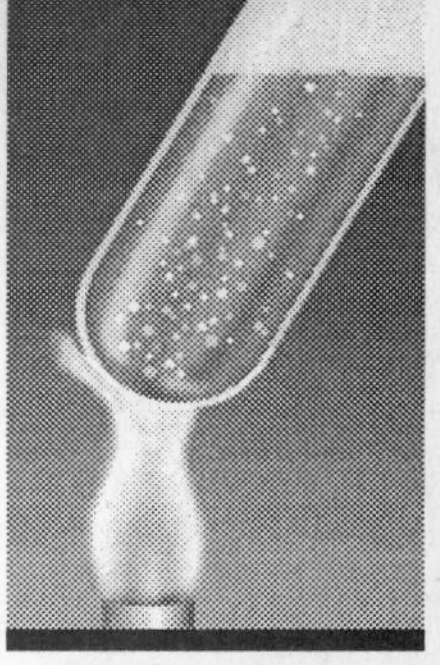

LAB MANUAL

8-1

Periodicity and Chemical Reactivity

This activity gives students a demonstration of the periodic law. By observing the reactivity of the elements within a group, students can form generalizations concerning properties of the group.

Misconception: The pattern of precipitation of the compounds formed when Group 2 ions react does show a regularity. However, students may think that this pattern is repeated with every anion. Although the anions selected for this experiment show a periodic pattern, not all anions will show the same pattern.

PROCESS SKILLS

1. Students will prepare three series dilutions of Group 2 metal ions.
2. Students will observe any precipitation that occurs when various anion solutions are added to Group 2 metal ions.
3. Students will determine a pattern of reactivity for the members of Group 2.

PREPARATION HINTS

- Prepare all solutions very carefully. If you cannot obtain a specific compound mentioned in the materials list, you may substitute another, provided the concentration of the active ion is the same.
- Beryllium compounds are not recommended for this activity because of their highly toxic nature.
- Use distilled water to dilute all solutions.
- Clean microplates will help students observe the formation of precipitates.
- For each student make two additional copies of the Microplate Data Form on page T9.

MATERIALS

0.1*M* solutions of the following:

$Mg(NO_3)_2 \cdot 6H_2O$	(25.6 g in 1 L of solution)
$Ca(NO_3)_2 \cdot 4H_2O$	(23.6 g in 1 L of solution)
$Sr(NO_3)_2 \cdot 4H_2O$	(28.4 g in 1 L of solution)
$Ba(NO_3)_2$	(26.1 g in 1 L of solution)
$Na_2SO_4 \cdot 10H_2O$	(32.2 g in 1 L of solution)
$Na_2C_2O_4$	(13.4 g in 1 L of solution)
Na_2CO_3	(10.6 g in 1 L of solution)

apron
goggles
96-well microplates (3)
black construction paper
thin-stem pipets (7)
toothpicks
distilled water
copies of Microplate Data Form (2)

PROCEDURE HINTS

- Before proceeding, be sure students have read and understand the purpose, procedure, and safety precautions for this laboratory activity.
- Make sure students use a different thin-stem pipet for each solution. Also stress that they must hold the pipets vertically so that drop sizes will be uniform.
- Students must count drops accurately.
- **Disposal:** Collect the solutions and dispose of them according to the instructions in **Disposal** J.

Copyright © Glencoe/McGraw-Hill, a division of The McGraw-Hill Companies, Inc.

DATA AND OBSERVATIONS

Students' hypotheses will vary. A correct hypothesis should say that each family of elements shows regular patterns of reactivity and solubility.

Microplate Data Form 1
Reactions of Group 2
Metal Ions with Sulfate Ions

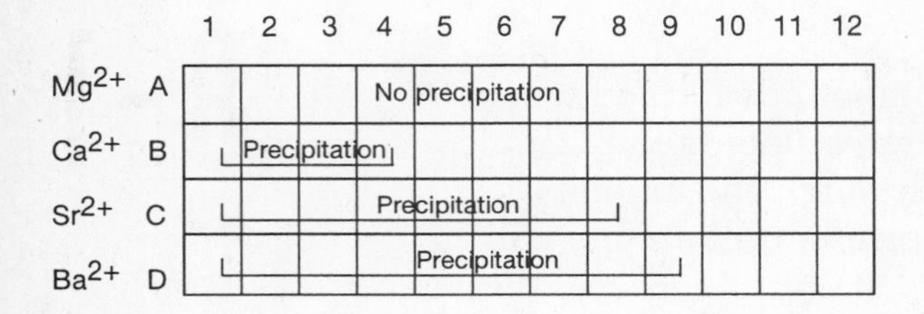

Microplate Data Form 2
Reactions of Group 2
Metal Ions with Oxalate Ions

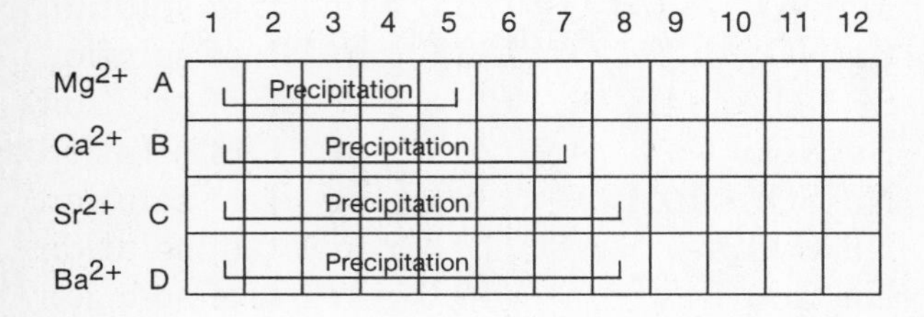

Microplate Data Form 3
Reactions of Group 2
Metal Ions with Carbonate Ions

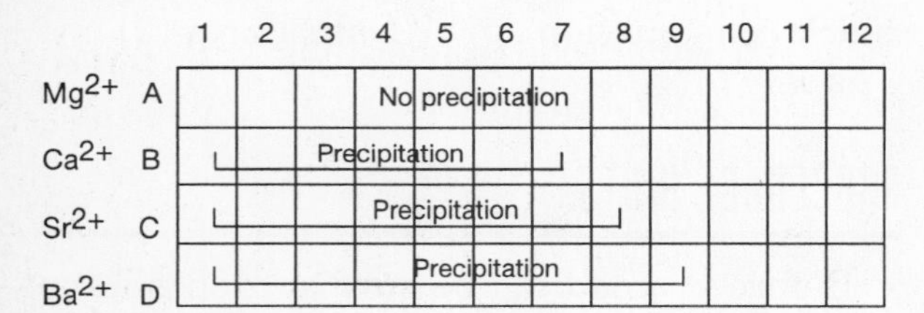

ANALYSIS

1. The amount of precipitation of the sulfate salts increases as the atomic mass of the metal ion increases.
2. The amount of precipitation of the oxalate salts generally increases as the atomic mass of the metal ion increases.
3. The amount of precipitation of the carbonate salts increases as the atomic mass of the metal ion increases.

CONCLUSIONS

1. The magnesium salts are the most soluble.
2. The barium salts are the least soluble.
3. The reactivity of the Group 2 ions with the sulfate ion has a pattern that is repeated with the other ions.
4. Students' answers will vary, but should indicate that properties within a family can be predicted.
5. Answers will vary depending on students' hypotheses.

EXTENSION AND APPLICATION

Each family of elements would produce regular reactivity patterns. The pattern might be seen in reaction rate, formation of a precipitate, or other physical and chemical properties. The most useful reactions would be those that are most easily monitored and observed. The noble gases would be most difficult to characterize by reactivity because these elements do not react readily with any other element or compound.

Copyright © Glencoe/McGraw-Hill, a division of The McGraw-Hill Companies, Inc.

Name

Date Class

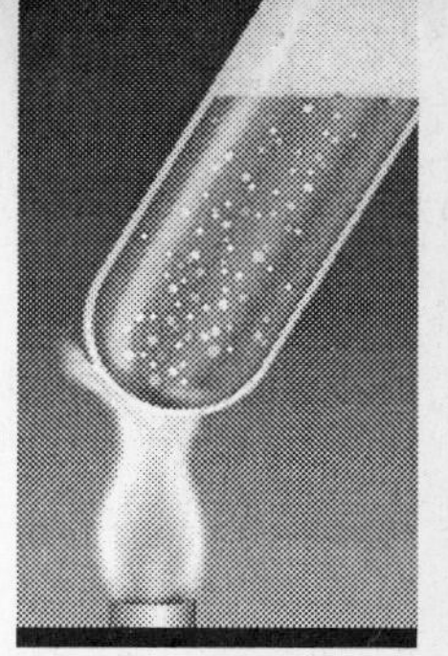

Comparing Activities of Selected Metals

LAB MANUAL

8-2

Metals are elements that tend to give up electrons and form positive ions in solution. They usually have a positive oxidation state and form ionic bonds with nonmetals. Some metals, for example, react with oxygen to form oxides. Metal activity depends on the tendency of the metal atom to lose electrons. The more readily a metal forms ionic bonds with nonmetals, the higher its activity will be. Some metals are so active that they react with water. In this experiment, you will observe and compare the activities of several metals.

OBJECTIVES

- **Observe** the reactions of metals.
- **Arrange** the metals in order of activity.
- **Compare** results with an activity series table.

MATERIALS

apron
goggles
96-well microplate
96-well template
microtip pipet
forceps or small tongs
distilled water

PROCEDURE

Part 1

1. You will observe the reactivity of copper, lead, nickel, iron, magnesium, aluminum, zinc, and silver. Which of these metals do you think is the most active? Which is the least active? Form a **hypothesis** about the reactivity of the metals and write it under Data and Observations.
2. Label a 96-well template as shown on the diagrams on this page and the next. Place a 96-well microplate on the template so that the lettered rows are at the left and the numbered columns are away from you.
3. **CAUTION:** *Many of the chemicals you will use are toxic. Follow proper chemical hygiene procedures. Wash your hands thoroughly after completing this laboratory activity.* With the microtip pipet, place 10 drops of 0.1*M* $Cu(NO_3)_2$ in wells A1 through A8. Rinse the pipet with distilled water.
4. Place 10 drops of 0.1*M* $Pb(NO_3)_2$ in wells B1 through B8. Rinse the pipet with distilled water after this and every other use.
5. Add $Ni(NO_3)_2$ solution to wells C1 through C8 and the $Fe(NO_3)_2$ solution to wells D1 through D8.
6. Add a small piece of copper metal or copper wire to each well in column 1, rows A through D.
7. Add a small piece of lead metal or wire to each well in column 2, rows A through D.
8. Continue to add pieces of metal or wire to each column in the order shown in this diagram.

Solutions of:		Cu 1	Pb 2	Ni 3	Fe 4	Mg 5	Al 6	Zn 7	Ag 8
$Cu(NO_3)_2$	A	○	○	○	○	○	○	○	○
$Pb(NO_3)_2$	B	○	○	○	○	○	○	○	○
$Ni(NO_3)_2$	C	○	○	○	○	○	○	○	○
$Fe(NO_3)_3$	D	○	○	○	○	○	○	○	○

9. While you wait 10 minutes for the reactions to take place, label a Microplate Data Form, using symbols and formulas to show the metal and the solution in each well of the microplate.

Copyright © Glencoe/McGraw-Hill, a division of The McGraw-Hill Companies, Inc.

10. After 10 minutes, use forceps or small tongs to remove the metal from each well. Some of the metals are active enough to have reacted and replaced the metal ions of the solution. Observe any change in the metals or the solutions and record the changes in your Microplate Data Form. Use a *+* to show that a reaction occurred and a *0* to indicate that no reaction occurred. Recall that formation of a precipitate or a gas or a change in the color of the solution may indicate that a reaction took place.
11. Discard the solutions and rinse the microplate according to the directions of your teacher.

Part 2

Repeat steps 2–11 of Part 1 using solutions of Mg, Al, Zn and Ag in rows E through H. Add the same metals in the same order as in Part 1. The diagram shows the metal and the solution that should be in each well.

Metals:

Solutions of:		Cu 1	Pb 2	Ni 3	Fe 4	Mg 5	Al 6	Zn 7	Ag 8
$Mg(NO_3)_2$	E	○	○	○	○	○	○	○	○
$Al(NO_3)_3$	F	○	○	○	○	○	○	○	○
$Zn(NO_3)_2$	G	○	○	○	○	○	○	○	○
$AgNO_3$	H	○	○	○	○	○	○	○	○

DATA AND OBSERVATIONS

Hypothesis: ______________________

____

____

Record all observations in your Microplate Data Form.

ANALYSIS

1. Did all the metals react in the same way? Explain.

____

____

2. Were there any metals that did not react at all? Explain.

____

____

3. Determine the number of reactions of each metal.

Metal	Cu	Pb	Ni	Fe	Mg	Al	Zn	Ag
Number of Reactions								

CONCLUSIONS

1. Arrange the metals in order of reactivity, from the highest to the lowest. Which metal is the most active? Which is the least active?

____

____

2. Did your results support your hypothesis about the reactivity of the metals? Explain.

____

____

3. How do your results compare with a table of metal reactivities in a reference book or provided by your teacher?

____

____

EXTENSION AND APPLICATION

A magnesium rod is usually found inside a water heater. What purpose do you think the rod serves?

Copyright © Glencoe/McGraw-Hill, a division of The McGraw-Hill Companies, Inc.

Teacher Guide

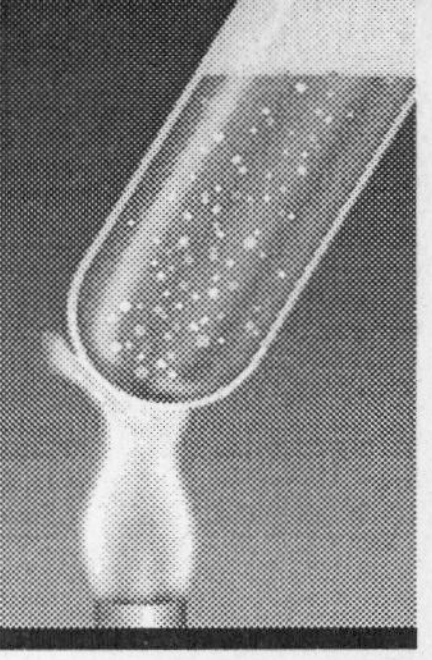

Comparing Activities of Selected Metals

The activity of a metal is not totally dependent on the location of the metal in the periodic table. While it is true that metal activity within a family increases as the atomic mass increases, some active metals are found at the top of the table (Al), while the most inactive metals (Au, Ag) are much farther down. The ionization energy of each metal is a better indication of activity than is its atomic mass. The easier it is to remove an electron from the outer energy level of a metal atom, the more active is the metal.

Misconception: Some students think that all metals behave in the same way as sodium and potassium. This misconception leads them to believe that there is no difference in activity among metals.

PROCESS SKILLS

1. Students will observe the reactions of selected metals with ions in solution.
2. Students will arrange the metals in order of their activity.
3. Students will compare their results with an activity series table.

PREPARATION HINTS

- This laboratory activity will take a double period. If you cannot complete the entire process in one day, schedule Part 1 for one day and Part 2 for the next day.
- For each group, make a copy of the 96-well template on page T8.
- Have on hand reference materials providing Activity Series Tables.
- In preparing the salt solutions, be sure to check the hydration of your stock salt and adjust the masses needed, if necessary.

MATERIALS

0.1*M* solutions of the following:

$Cu(NO_3)_2 \cdot 3H_2O$ (24.2 g in 1 L of solution)
$Pb(NO_3)_2$ (33.1 g in 1 L of solution)
$Ni(NO_3)_2 \cdot 6H_2O$ (29.1 g in 1 L of solution)
$Fe(NO_3)_3 \cdot 9H_2O$ (40.4 g in 1 L of solution)
$Mg(NO_3)_2 \cdot 6H_2O$ (25.6 g in 1 L of solution)
$Al(NO_3)_3 \cdot 9H_2O$ (37.5 g in 1 L of solution)
$Zn(NO_3)_2 \cdot 6H_2O$ (29.7 g in 1 L of solution)
$AgNO_3$ (17.0 g in 1 L of solution)

The following metals, supplied as wire or 1 mm × 10 mm strips: copper, lead, nickel, iron, magnesium, aluminum, zinc, silver

apron
goggles
96-well microplate
96-well template
microtip pipet
forceps or small tongs
distilled water

PROCEDURE HINTS

- Before proceeding, be sure students have read and understood the purpose, procedure, and safety precautions for this laboratory activity.
- Make sure that students have all materials ready before they begin the activity.
- Some of the reactions will take some time to show evidence of change.
- **Safety Precaution:** Heavy metal compounds, such as those containing lead, are generally extremely toxic. Remind students to follow proper chemical hygiene procedures when working with such compounds.
- **Disposal:** Collect all solutions in a dishpan or another receptacle and dispose of according to the instructions in **Disposal** J.
- Silver and silver compounds are very costly. Closely monitor their use.

DATA AND OBSERVATIONS

Students' hypotheses will vary. Correct hypotheses should agree with the Activities Series Table.

Metals:

Solutions of:	Cu 1	Pb 2	Ni 3	Fe 4	Mg 5	Al 6	Zn 7	Ag 8	9	10	11	12
$Cu(NO_3)_2$ A	0	+	+	+	+	+	+	0				
$Pb(NO_3)_2$ B	0	0	+	+	+	+	+	0				
$Ni(NO_3)_2$ C	0	0	0	0	+	+	+	0				
$Fe(NO_3)_3$ D	0	0	+	0	+	+	+	0				
$Mg(NO_3)_2$ E	0	0	0	0	0	0	0	0				
$Al(NO_3)_3$ F	0	0	0	0	+	0	0	0				
$Zn(NO_3)_2$ G	0	0	0	0	+	+	0	0				
$AgNO_3$ H	+	+	+	+	+	+	+	0				

ANALYSIS

1. Each metal showed its own characteristic pattern of reaction.
2. Silver was the only metal that did not react. Silver is very inactive and does not displace any of the other metals from solution by single displacement.
3.

Metal	Cu	Pb	Ni	Fe	Mg	Al	Zn	Ag
Number of Reactions	1	2	4	3	7	6	5	0

CONCLUSIONS

1. Metals in order of activity from the highest to the lowest:
Mg, Al, Zn, Ni, Fe, Pb, Cu, Ag
2. Answers will vary, depending on students' hypotheses.
3. Answers will vary, depending on students' results.

EXTENSION AND APPLICATION

Magnesium is a very active metal and provides electrons to ions in the water more readily than do the metals that the tank is made of. For a time, the magnesium helps to prevent the corrosion of the metal water tank.

Name

Date Class

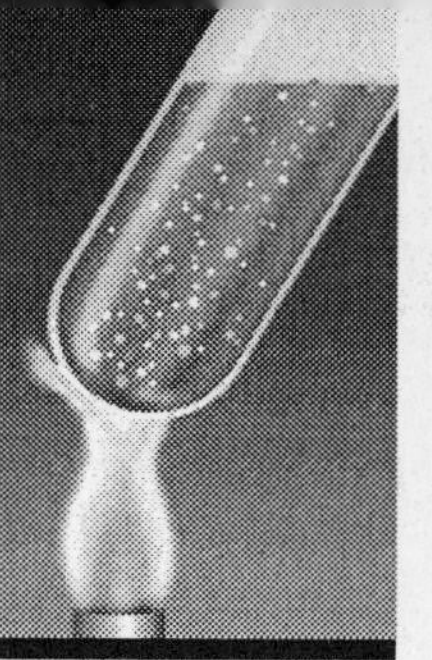

LAB MANUAL

9-1

Diagnostic Properties of Bonds

The millions of chemical substances found in nature are the result of approximately 100 chemical elements bonding together. These substances can be divided into three major groups: metals; salts, or ionic compounds; and covalent substances, the last making up the overwhelming majority of chemicals. Metallic substances, such as copper and aluminum, hold their atoms together by communal sharing of outer electrons. Salts, such as common table salt, NaCl, have two or more charged particles, which are ions resulting from the transfer of electrons. Covalent substances, such as oxygen and water, consist of molecules held together by electron sharing. In the laboratory, these three major groups display different properties in their appearance, relative melting or boiling points, solubilities in different solvents, and ability to conduct electricity.

In this activity, you will explore differences in properties, whereby millions of substances can be classified into one of three major groups. These groups are metals, such as copper (Cu) and aluminum (Al), salts, such as potassium chloride (KCl) and potassium iodide (KI), and covalent compounds, typified by naphthalene and benzoic acid.

OBJECTIVES

- **Make laboratory observations** of six solid substances, each of which exhibits the properties of one of the three types of chemical bonds.
- On the basis of your observations, **list** the distinguishing properties of the three types of chemical bonds.

MATERIALS

can lid with indentations
ring stand with ring
watch glass
deionized water
laboratory burner
chemical scoop
beaker (50 or 100 mL)
glass stirring rod
paper towel
conductivity tool

PROCEDURE

CAUTION: *Some of the chemicals you will use are toxic. Follow proper chemical hygiene procedures. Wash your hands thoroughly after completing this laboratory activity.*

Part 1: Appearance and Odor

1. Obtain a very small quantity of KCl with a chemical scoop. Place it on your watch glass on the lab bench top.
2. Describe the appearance of the solid. Notice any odor.
3. Record your observations in Table 1 under Data and Observations. Leave the solid on the watch glass for later work.

Part 2: Effects of Heat

1. Set up your ring stand as shown in Figure A. On the ring, place a can lid prepared with indentations.

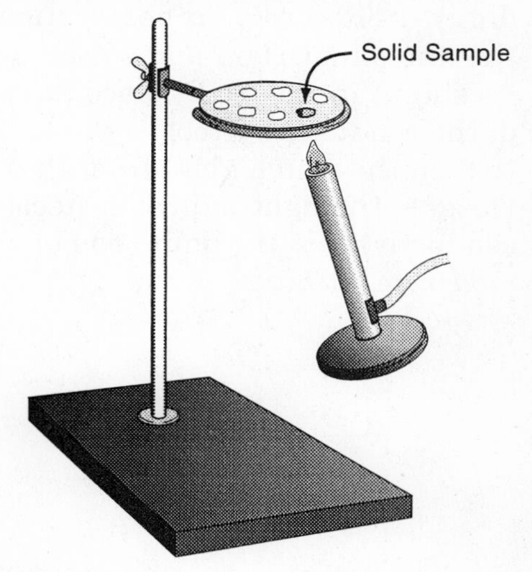

Figure A

2. Place in one of the indentations the same amount of KCl you used in Part 1.
3. Light the laboratory burner and, handling the base of the burner, gently heat the *underside* of the can lid at the location of the solid. Observe the solid. If no change occurs, heat more intensely. If there is still no change, heat the solid from above, lightly at first then intensely. To do so, hold the burner and pass the flame over the material.
4. Observe whether the solid melts, boils, evaporates, burns, or shows no change. Record your observations in Table 1.

Part 3: Solubility in Water

1. Add 3 mL of deionized water to the KCl on the watch glass (from Part 1.) Stir with a glass rod and let stand about 3 minutes.
2. Compare the remaining undissolved solid with the original quantity. Decide whether the solid is *soluble* (most dissolved in the water), *insoluble* (most remains at the center bottom of the glass) or *partly soluble* (some remains undissolved).
3. Record your observations in Table 1. Save the watch glass and solution or water and solid for the next procedure.

Part 4: Conductivity of Water Solutions

1. Pour 10 to 20 mL of deionized water into a beaker. Place the prongs of your conductivity tool into the water.
2. Deionized water should show little or no conductivity. Depending on the sensitivity of your conductivity tool, the light indicator should be off, blinking, or dimly lit. Record your observation on the line under Data and Observations.
3. As shown in Figure B, place the prongs of the conductivity tool in the liquid portion on the watch glass from Part 3. Observe the light indicator. Record the conductivity of the liquid as *none, low, moderate,* or *high*.
4. Wipe the prongs with a paper towel, immerse in the beaker of deionized water, and wipe again. Set aside the conductivity tool for the next solid to be tested. Discard the water solution or solvent-solid mix according to your teacher's instructions.

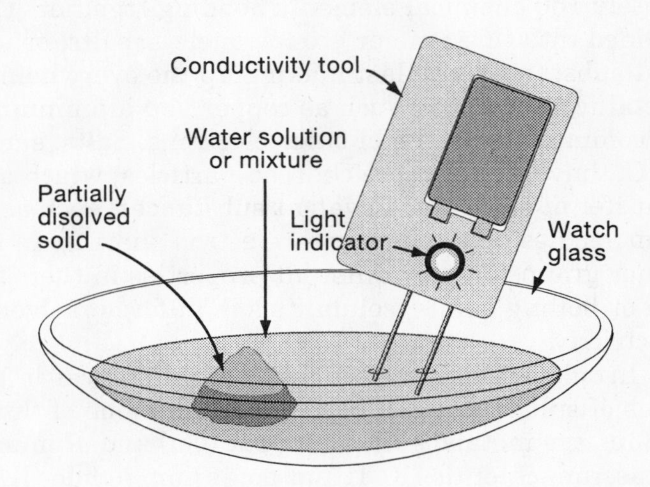

Figure B

Part 5: Solubility in Paint Thinner

1. **CAUTION:** *This test must be performed in the fume hood because of the volatility and flammability of the solvent.*
2. Repeat the solubility test as in Part 3, using paint thinner as the solvent instead of water.
3. Record in Table 1 whether the solid is soluble, insoluble, or partly soluble in paint thinner.
4. Discard the solution or solvent-solid mix into the appropriate container in the fume hood.

Repeat Parts 1 through 5 for each of the remaining five solids and complete Table 1.

DATA AND OBSERVATIONS

Conductivity of pure water

____

Table 1

Observations of Solids					
Solid	Appearance/ odor	Effects of heating	Solubility in water	Conductivity of water solution	Solubility in paint thinner
KCl					
KI					
Al					
Cu					
Benzoic acid					
Naphthalene					

ANALYSIS

1. Complete Table 2, using the observations you recorded in Table 1. Use the words *high, low,* or *none* to compare the properties of the three bonding types.

Table 2

Comparison of Relative Properties				
Bond type	Melting point	Solubility in water	Conductivity of water solution	Solubility in paint thinner
Metallic				
Ionic				
Covalent				

2. Which solid has the highest melting point?

____

3. Which solid has the lowest melting point?

____

CONCLUSIONS

1. Which chemical bond type is found in substances with high melting point, water solubility, and aqueous conductivity?

____

2. Which bond type is found in substances that have low melting point, are flammable, and tend to dissolve in nonpolar solvents?

____

3. Which bond type is found in substances that have luster and are insoluble in water and nonpolar solvents?

____

4. Most of the changes you observed in this experiment are considered physical, except for burning, which is a chemical change. The formula for naphthalene is $C_{10}H_8$. Write a combustion equation for the complete reaction of naphthalene with oxygen in air to form carbon dioxide and water. Indicate whether each reactant and product is a solid (s), liquid (l), or gas (g).

____

5. Consider chemical substances in your daily life, or look for examples in your chemistry book. In the table below, list two more examples of each of the three bonding types.

Ionic	Covalent	Metallic

6. Metals and nonmetals form substances that can be divided into three major chemical bond types. For each of the three bonding types, which types of elements—metallic or nonmetallic—bond?

Metallic: ________________________

Ionic: ________________________

Covalent: ________________________

EXTENSION AND APPLICATION

There are exceptions to the generalizations above, as there are to many scientific generalizations. Repeat the experiment with sugar, a covalent substance. Contrast your results with the conclusions you made above.

Teacher Guide

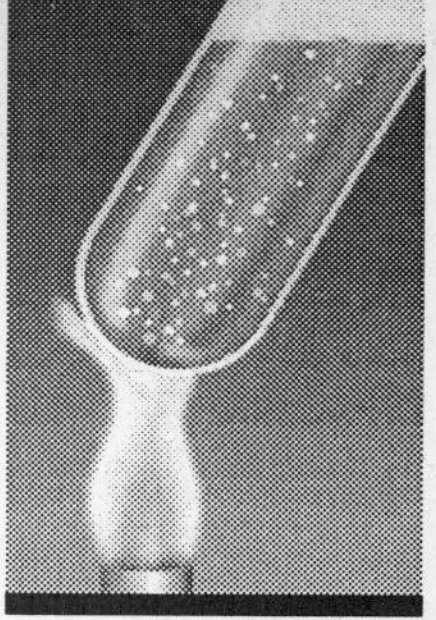

LAB MANUAL

9-1

Diagnostic Properties of Bonds

Chemistry courses emphasize bonding between atoms. Concepts of quantum mechanics, orbital theory, electron pairing and transfer, and electronegatives are presented in detail, yet the observational nature of chemical compounds must also be used as a way to distinguish the major kinds of chemical bonds. This laboratory experiment points out the high melting points of salts and metals. Students also observe that covalent compounds and the majority of organic compounds melt, and boil, at relatively low temperatures. Solubility in water generally identifies salts. Covalent and other organic compounds tend to dissolve in nonpolar solvents. Salt and acid solutions conduct electricity.

PROCESS SKILLS

1. Students will perform tests of solubility in different solvents, relative melting points, and conductivity of aqueous solutions on six samples, each of which is an example of one of the three kinds of chemical bonds.
2. Based on their findings, students will relate definite properties to substances containing each of the three major chemical bonds.

PREPARATION HINTS

- If possible, have students bring from their home clean lids from paint cans. With a hammer and an inverted nail, 10d to 16d common, make six nail indentations around the edge of the lid. Use a soft pine block of wood for a base when making the indentations.
- Paint thinner is an inexpensive nonpolar solvent. Because paint thinner is extremely flammable, dispense only in small quantities. Have students work in a fume hood while handling the paint thinner.
- Deionized water is important for a reference point of the conductivity tool so that students may compare relative conductivities.

MATERIALS

0.5 L deionized water
0.5 L paint thinner or TTE
15 g each of the solids KI, KCl, benzoic acid, and naphthalene
15 pieces of 0.5 cm squares of Al and Cu metal (cut from 22–30 gauge sheets)
paint can lid with indentations
ring stand and ring
chemical scoop
watch glass
glass stirring rod
conductivity tool
laboratory burner
beaker (50 or 100 mL)
paper towel

PROCEDURE HINTS

- Before proceeding, be sure students have read and understood the purpose, procedure, and safety precautions for this laboratory activity.
- Because the experiment is primarily qualitative and only relative comparisons are made, precise quantities of solids are not necessary. About 0.1 g or less of solid in a chemical scoop is sufficient for sample size. You might display a sample during prelab discussion.
- To obtain an approximate volume of 3 mL of solvent, students can measure 3 mL of solvent into a test tube, using a graduated cylinder. Direct students to use the height of the solvent in the test tube as a reference when measuring out other 3 mL volumes of solvents.
- Deciding on relative solubility will be chal-

lenging to students. Emphasize that to determine relative solubility students should compare the undissolved quantity of solid to its original quantity.

- Degree of conductivity may be indicated by the relative brightness or rate of blinking of the indicator light on the conductivity tool.
- Provide a collection container in the fume hood for the paint thinner.
- **Safety Precaution:** A tiny flake of naphthalene should be used on the can lid to avoid a smoky fire. You may wish to demonstrate the heating of this substance as a prelab demonstration.
- **Safety Precaution:** Remind students that the paint thinner solubility test must be performed in the fume hood. They should make sure that no burner flames are near the paint thinner.
- **Disposal:** Dispose of potassium chloride and potassium iodide according to the instructions in **Disposal** C; copper, aluminum, and naphthalene according to **Disposal** F; benzoic acid according to **Disposal** D; and paint thinner according to **Disposal** A.

DATA AND OBSERVATIONS

Table 1

Conductivity of pure water: *none* or *very low*

Observations of Solids					
Solid	Appearance/odor	Effects of heating	Solubility in water	Conductivity of water solution	Solubility in paint thinner
KCl	white crystalline, no odor	did not melt	soluble	high	insoluble
KI	white crystalline, no odor	started to melt when high flame directed at solid	soluble	high	insoluble
Al	lustrous, no odor	did not melt, appeared to react with hot flame from above	insoluble	NA	insoluble
Cu	some luster., no odor	did not melt or react with hot flame	insoluble	NA	insoluble
Benzoic acid	powdery, slight odor	burned slowly with flame from above	partly soluble	low	partly soluble
Naphthalene	shiny flakes, strong odor	melted readily and caught fire when heated from above	insoluble	NA	partly soluble

Table 2

Comparison of Relative Properties				
Bond type	Melting point	Solubility in water	Conductivity of water solution	Solubility in paint thinner
Metallic	high	none	N/A	none
Ionic	high	high	high	none
Covalent	low	low to none	low to none	low

2. KCl or Cu has the highest melting point.
3. Naphthalene has the lowest melting point.

CONCLUSIONS

1. Ionic
2. Covalent
3. Metallic
4. $C_{10}H_8(s) + 12O_2(g) \rightarrow 10CO_2(g) + 4H_2O(g)$
5. Common examples of the bonding types:
   ionic: common table salt, limestone, bicarbonate of soda (baking soda), magnesium sulfate (Epsom salt)

   covalent: sugar, gasoline, alcohol, wax, fats, carbohydrates, proteins

   metallic: iron, steel, chromium, nickel, gold, silver
6. Metallic substances form from metals, ionic from metals and nonmetals, and covalent from nonmetals.

EXTENSION AND APPLICATION

Sugar shows high water solubility and low or no aqueous conductivity.

Name

Date Class

Electron Clouds

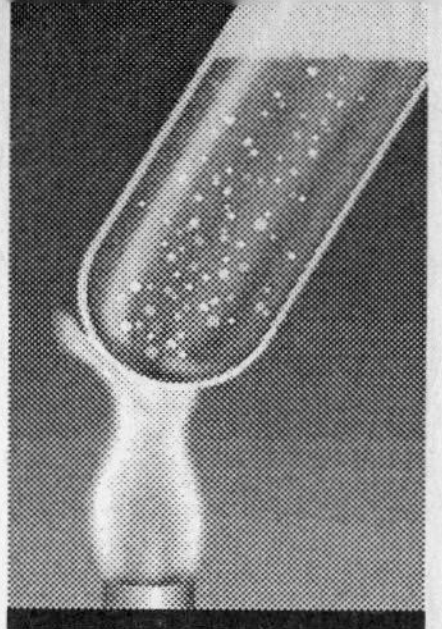

LAB MANUAL

9-2

Electrons occupy space about the nucleus. The exact path or location of electrons in an atom cannot be known. Instead, we speak about the probability of locating an electron in an electron cloud—a volume determined by the energy of the electron. Balloons allow us to model the electron clouds of bonding electron pairs whose arrangement determines the shapes of molecules.

Round balloons will be used to represent the spherical volume occupied by one or two *s* electrons. Pear-shaped balloons will be used to represent the volume occupied by pairs of bonding electrons. The purpose of this activity is to provide you with a model of how atoms join together to form molecules.

OBJECTIVES

- Use balloons to **model** the arrangement of electrons around the nucleus of an atom.
- **Observe** a model of bonding of atoms in three dimensions.

MATERIALS

round balloon (5)
pear-shaped balloons (4)
clear tape
string

PROCEDURE

1. Form a **hypothesis** about the shape of a molecule formed from 2, 3, and 4 atoms, respectively.
2. Blow up five round balloons to the same size. Tie their ends closed.
3. Make a loop of tape on each of the balloons to represent an electron. It takes a pair of loops to make a chemical bond.
4. Stick two of the balloons together at the two tapes to form a model of a covalent bond as shown in Figure A. This is a model of two hydrogen atoms and an example of linear bonding.
5. Blow up four pear-shaped balloons. Tie three ballons together at their knotted ends, as shown in Figure B.
6. Place a loop of tape at the end of each balloon opposite where they are knotted together. Attach a round balloon by its loop of tape on the end of each pear-shaped balloon, as shown in Figure B. This is a model of BCl_3 and an example of planar bonding.
7. Add the fourth pear-shaped balloon to the other three pear-shaped balloons by tying it where the others are knotted together.
8. Place a loop of tape at the end of the fourth pear-shaped balloon.
9. Place a round balloon at the end of the fourth pear-shaped balloon, as shown in Figure C. This is a model of CH_4 and an example of tetrahedral bonding.

Figure A

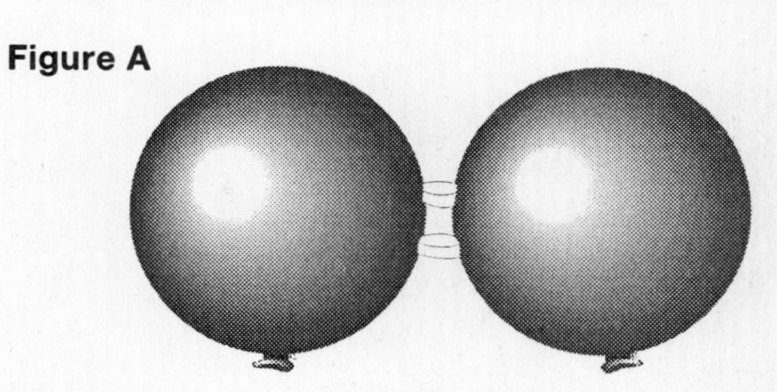

Linear bonding

Figure B

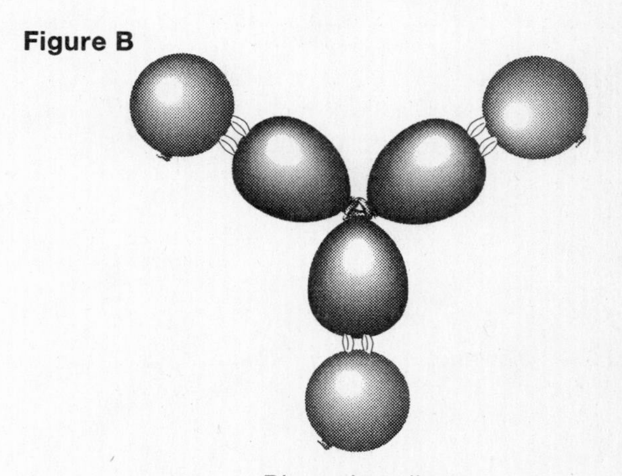

Planar bonding

Figure C

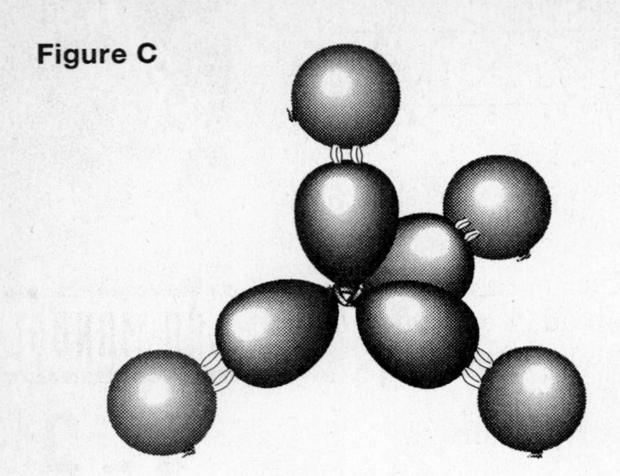

Tetrahedral bonding

DATA AND OBSERVATIONS

1. **Hypothesis:** ______________________

____

____

2. Compare the location of the bonded atoms in your three models. Using atomic symbols and straight lines for the bonds, diagram each type of molecule.

3. What are the angles between the bonds in your models of BCl_3 and CH_4?

____

____

____

____

ANALYSIS

Explain in your own words how the balloons represent the forces that shape molecules.

____

____

____

CONCLUSIONS

Which of the molecules listed below will be linear? planar? tetrahedral?

a. CH_3Cl **b.** Cl_2 **c.** CCl_4
d. HCl **e.** BF_3

____

____

EXTENSION AND APPLICATION

1. Use your round and pear-shaped balloons to make a model of $BeCl_2$. What are the bond angle and the molecular shape of this molecule?
2. Blow up two round balloons four times larger than before. Replace two round balloons in the tetrahedral molecule. What happens? When does something similar happen in a real molecule?
3. In determining molecular shape by the repulsion of electron pairs, double or triple bonds have the same effect as single bonds. Are ethylene, $H_2C = CH_2$, and acetylene, $H-C \equiv C-H$, planar or three-dimensional? What will be the bond angles of the carbon-carbon-hydrogen bonds in each molecule?

Teacher Guide

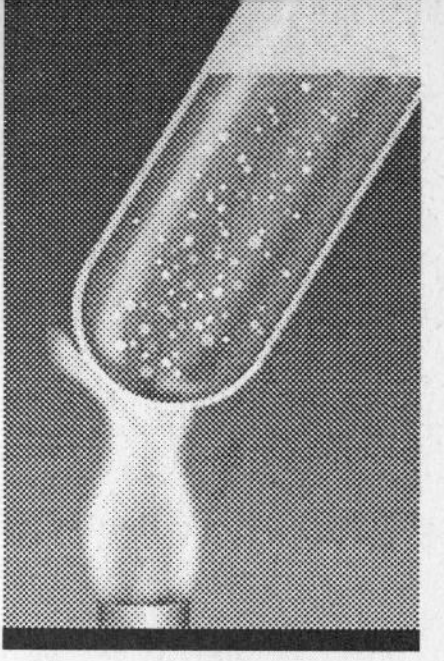

LAB MANUAL

9-2

Electron Clouds

The repulsion of electron pairs is a simple model that shows the arrangement of electron pairs in molecules. This activity focuses on the shapes of molecules in which there are no unshared electron pairs. It also demonstrates how the linear, trigonal, and tetrahedral arrangements of bonds arise naturally by electron-pair repulsion.

Misconception: Most students picture a molecule as flat or planar based on the pictures they see in textbooks. Students often picture electrons as standing still or having limited motion.

PROCESS SKILLS

1. Students will construct models that show electron-pair repulsion.
2. Students will demonstrate the bond angles in molecules with no unshared electron pairs.

MATERIALS

round and pear-shaped balloons
clear tape
string

PREPARATION HINTS

Have available styrofoam model materials for those students who wish to extend beyond this activity to VSEPR models.

PROCEDURE HINTS

Before proceeding, be sure students have read and understood the purpose and procedure for this laboratory activity.

- Use only round or pear-shaped balloons.
- Do not allow students to use double-sided tape to connect the balloons. Double-sided tape sticks to the balloons without any give. This leads to a large number of broken balloons.

DATA AND OBSERVATIONS

1. Students should be able to hypothesize a bond angle of 180° for two atoms, 120° for three atoms, and 109.5° for four atoms.
2.
   a. H — H
   b. Cl — B (bonded to Cl, Cl, Cl)
   c. H — C (bonded to H, H, H, H)
3. In the planar molecule, BCl_3, the B-Cl bonds are at 120° from each other. In the tetrahedral molecule, CH_4, the C-H bonds are at 109.5° from each other.

ANALYSIS

Electron pairs in molecules stay as far away from each other as possible because their negative charges repel each other. The tied, pear-shaped balloons naturally assume the same geometry in which they are as far apart as possible.

CONCLUSIONS

a. tetrahedral
b. linear
c. tetrahedral
d. linear
e. planar

EXTENSION AND APPLICATION

1. The $BeCl_2$ molecule is linear, with a 180° angle between the bonds.
2. The larger balloons create a strain on the tetrahedral arrangement of the pear-shaped balloons. The bond angle between the balloons supporting the large round balloons becomes larger than 109.5°, and the other bond angles become smaller. This happens when there are unshared electrons on the central atom in a molecule.
3. Each carbon atom in acetylene has two bonds that stay away from each other in a flat, linear arrangement. The bond angles will be 180°. Each carbon atom in ethylene has three bonds that stay away from each other in a flat, triangular arrangement. The molecule will be planar with 120° bond angles.

Copyright © Glencoe/McGraw-Hill, a division of The McGraw-Hill Companies, Inc.

Name

Date Class

LAB MANUAL

10-1

Crystal Shapes

In the last few years, many large, colorless "gems" have appeared in jewelry. The next time you see someone wearing a large "diamond" ring, look again. He or she may be wearing a cubic zirconium—a simulated stone.

Although the crystal structure in the cubic zirconium is quite different from the crystal structure of a diamond, they have one thing in common. Their particles are arranged in a characteristic, orderly way. The simple pattern or arrangement of particles in any crystal has a special name, the unit cell. In a crystal, the unit cells form a repeating pattern.

Although there are only fourteen different kinds of cells, the shape and relative size of the unit cell of each solid is unique. Chemists often use X-ray crystallography to study crystal structure.

As part of this laboratory, you will have an opportunity to grow and observe several different crystals. Time, temperature, pressure, and other conditions in the laboratory will affect the growth of your crystals.

OBJECTIVES

- **Construct** three-dimensional models of several different unit cells.
- **Demonstrate** growth of crystals by evaporation from saturated solutions under different conditions.
- **Compare** crystals of several pairs of compounds grown under different conditions.

MATERIALS

foam balls
toothpicks
glass microscope slides
microtip pipet
microscope or hand lens
apron
goggles

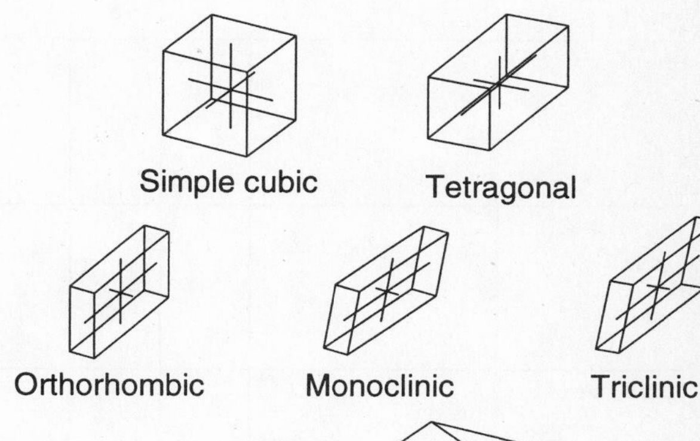

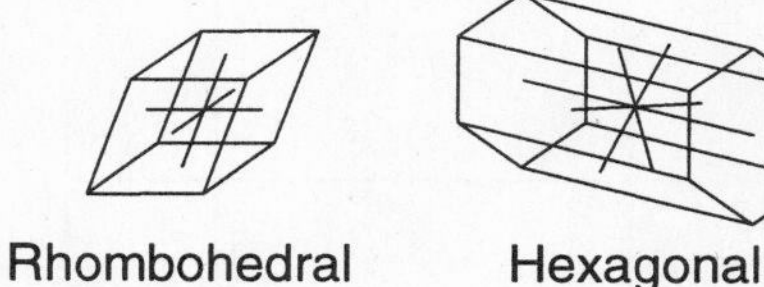

PROCEDURE

Part 1

1. Using foam balls and toothpicks, construct the unit cells of the types shown in the diagrams.
2. Draw a three-dimensional diagram of each unit cell in the table under Data and Observations.
3. Locate one corner of each crystal model.
4. Notice that the corner is made up of three angles, which can be designated as angles a, b, and c. Determine for each model whether each of the angles at a corner is a right angle, less than a right angle, or greater than a right angle. Record the size of each angle.

Part 2

1. Your teacher will provide you with sample tubes containing saturated solutions of three different compounds. **CAUTION:** *Do not taste any of the samples.*
2. Place 1 or 2 small drops of each sample solution on a separate glass slide. Observe the growth of crystals with a microscope or hand lens.
3. How does the length of time required for evaporation affect crystal formation? Form a **hypothesis** about the answer to this question. Write your hypothesis under Data and Observations.

4. Repeat step 2, but this time evaporate the solvent more rapidly by fanning the solution with paper.
5. Compare the size of the crystals that were allowed to grow slowly with the crystals that grew quickly.
6. Rinse the slides with water into the container provided by your teacher.

DATA AND OBSERVATIONS

Hypothesis: ______________________

____

____

Crystal Structures

Crystal Type	Diagram	Corner Angles		
		a	*b*	*c*
Simple cubic				
Tetragonal				
Orthorhombic				
Monoclinic				
Triclinic				
Rhombohedral				
Hexagonal				

ANALYSIS

1. Look at the model of the simple cubic crystal. How many atoms will each of the atoms in the crystal touch if the unit cell is surrounded completely by other unit cells?

____

2. Sodium chloride is a face-centered cubic crystal. How many chloride ions surround a sodium ion in crystalline NaCl?

____

CONCLUSIONS

1. When you were making your observations of the growth of crystals, where did crystal growth begin? Explain why.

____

____

2. Describe the differences between the crystals that were grown quickly and those that were allowed to grow slowly. Did your results support your hypothesis about the effect of evaporation time on crystal formation? Explain.

____

____

3. Based on your observations of crystal growth, suggest conditions or events that could cause deviation from the unit cell in real crystals.

____

____

EXTENSION AND APPLICATION

1. Using large and small foam balls to represent chlorine and sodium ions respectively, make a three-dimensional model of a sodium chloride unit cell.
2. Modern jet engines have blades made out of a single crystal of titanium alloy. What is the advantage of using a single crystal for an engine blade?

Teacher Guide

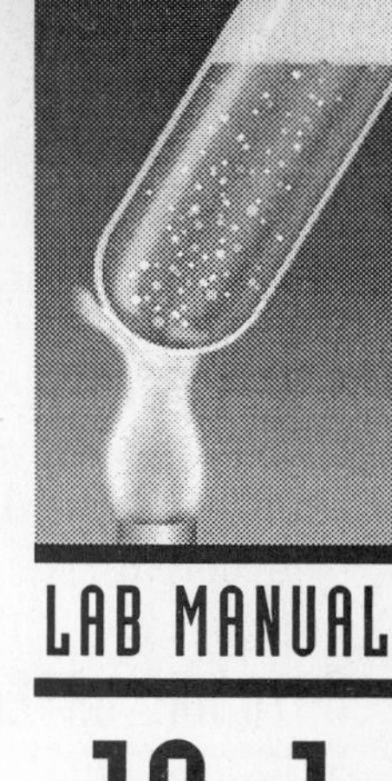

LAB MANUAL

10-1

Crystal Shapes

Crystals are usually grown from a melt or saturated solution. Industrial crystals include silicon wafers used in integrated circuits and semiconductors. Such industrial crystals are usually the product of zone refining done in clean rooms where technicians wear environmental suits for maximum cleanliness. These precautions must be taken to ensure the purest form of crystal. Of course, the perfect crystals of silicon or germanium used in transistors must be doped with specific impurities, but first the crystals must be pure.

The procedures students use when observing crystal growth under the microscope do not meet the rigorous standards required in industry. However, the experiment does afford students the opportunity to observe crystal growth in a very short time.

Misconceptions: Students tend to think that all crystals are either cubic or diamond-shaped. They also fail to associate purity of a compound with a crystalline form. Remind students that all true solids must be crystalline.

PROCESS SKILLS

1. Students will construct three-dimensional models of various unit cells in crystal structures.
2. Students will prepare slides of saturated solutions for growing crystals.
3. Students will observe and analyze differences between pairs of crystals grown at different rates.

PREPARATION HINTS

- Plastic foam balls and toothpicks for model building can be obtained from a chemical supply house. The foam balls may also be purchased at craft shops.
- Place small amounts of each saturated solution in labeled covered vials or test tubes.
- When preparing the saturated solutions, use only the solids and liquids suggested. Do not use acetone, ether, or gasoline as solvents because these compounds have a low flash point and their vapors may be harmful if students are exposed to them for long periods of time.
- Use pure ethanol. Any denaturing agent could affect crystal growth. **CAUTION:** *Ethanol is highly flammable.*

MATERIALS

saturated solutions of naphthalene, salicylic acid, and sodium chloride. Prepare as follows.

Solution A: Saturated solution of naphthalene in ethanol

Solution B: Saturated solution of salicylic acid in ethanol

Solution C: Saturated solution of sodium chloride in distilled water

vials or test tubes
foam balls
toothpicks
glass microscope slides
microtip pipet
microscope or hand lens
apron
goggles

PROCEDURE HINTS

- Before proceeding, be sure students have read and understood the purpose, procedure, and safety precautions for this laboratory activity.
- Students need not wear aprons and goggles in Part 1.
- Students usually work in pairs.
- Since many foam balls are used in this experiment, it may be necessary for students to assemble and disassemble their models rather than keeping them for a complete set.

- **Safety Precaution:** Caution students not to taste the solutions. The naphthalene is harmful; the ethanol should not be consumed.
- **Disposal:** Have students rinse the solid crystals from their slides into a dishpan. Dispose of the contents of the dishpan according to the instructions in **Disposal** J.

DATA AND OBSERVATIONS

Students' hypotheses can vary, but should indicate that crystals will be larger when grown slowly.

Crystal Structures

Crystal Type	Diagram	Corner Angles		
		a	*b*	*c*
Simple cubic	Diagrams should resemble those in Procedure, Part 1.	90°	90°	90°
Tetragonal		90°	90°	90°
Orthorhombic		90°	90°	90°
Monoclinic		90°	90°	>90°
Triclinic		<90°	>90°	<90°
Rhombohedral		>90°	<90°	>90°
Hexagonal		90°	60°	60°

ANALYSIS

1. When the unit cell is surrounded completely by other unit cells, each atom in the crystal touches six atoms.
2. Each sodium ion will be surrounded by six chloride ions in the crystalline form of sodium chloride.

CONCLUSIONS

1. Crystal growth began at the edges of the drop. The liquid at the edge of the drop evaporated fastest.
2. The crystals that were grown slowly were much larger than those that were grown quickly. Students' hypotheses about the effect of evaporation time on crystal formation could vary; their answers will depend on their hypotheses.
3. Deviations in real crystals are caused by uneven heating and cooling, foreign substances within the solution, and water or insoluble material trapped as the crystal forms.

EXTENSION AND APPLICATION

1. Cl^- Na^+

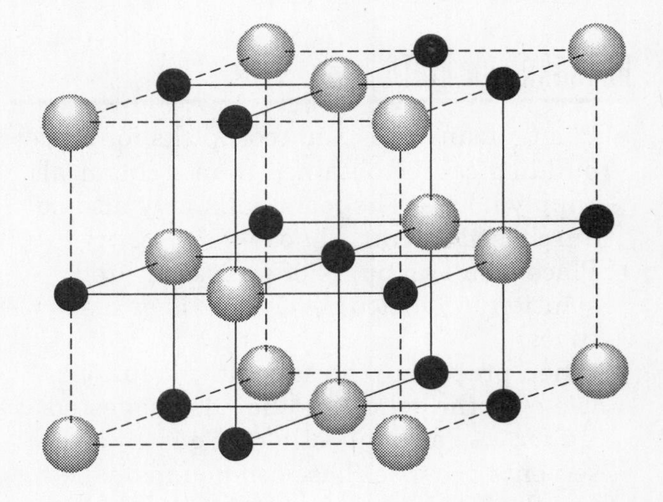

2. A single crystal of any metal is superior to many crystals joined together because a single crystal is stronger. A single crystal of titanium alloy used in jet engine blades must have no flaws or faults that would fail under stress.

Name ______________________________

Date ______________ Class ______________

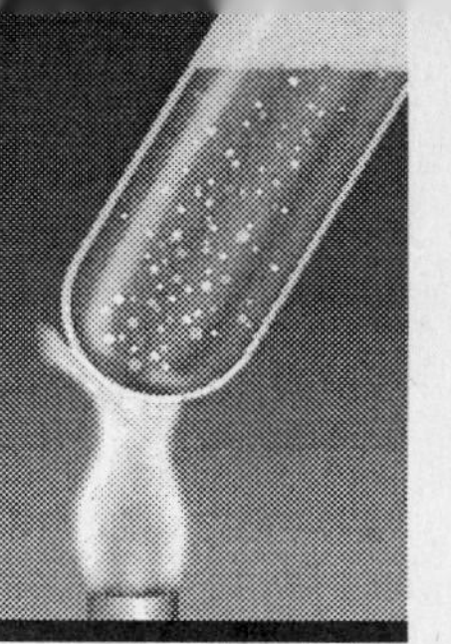

LAB MANUAL

10-2

Relating Gas Temperature and Pressure

An automobile tire contains air under pressure. When the tire moves over the surface of a highway, the temperature of the tire increases due to friction with the road. What happens to the pressure in the tire?

The molecules of a substance in the gaseous state have the greatest amount of mobility of the three major states of matter. Molecules of gas rotate, vibrate, and translate (move from place to place) with the least interference of all the states of matter. When the molecules of gas are heated, their kinetic energy increases. The warmer the gas, the faster the molecules of the gas move. Since the temperature of a substance is a measure of its average kinetic energy, temperature is a measurement of molecular motion in that substance.

If a gas is confined in a container, the molecules of the gas strike the walls of the container at a certain rate. If the gas in the container is heated, the velocity of the molecules increases. If the container will not expand, this increase in molecular action causes an increase in the internal pressure of the container. Continuing to heat the gas in the container may cause the container to explode.

In this activity, you will determine the change in pressure in a confined gas. The gas is confined in a test tube, which is joined by tubing to the space above water in an Erlenmeyer flask. As the gas in the test tube is heated, the molecules move faster and the pressure increases. The increased pressure is transmitted through the tubing to the surface of the water in the flask. The increased pressure forces water up into a tube in the flask.

OBJECTIVES

- Using a closed system, **measure** the change in the height of a column of air as the temperature of a confined gas in the system changes.
- **Relate** these height differences to a change in pressure as the temperature changes.
- **Graph** data to determine the relationship between temperature and pressure of a confined gas.

MATERIALS

apron
goggles
400-mL beaker
large test tube, heat-resistant
500-mL Erlenmeyer flask
hot plate
1.25-m length of glass tubing
one-hole rubber stopper to fit test tube
two-hole rubber stopper to fit flask
glass droppers, straight (2)
50-mm length of 5-mm rubber or plastic tubing
500-mm length of 5-mm rubber or plastic tubing
ring stands (2)
thermometer
utility clamps (2)
thermometer clamp
pinch clamp
metric ruler
beaker tongs or thermal mitts
test-tube holder
hot pad
glycerol
barometer

PROCEDURE

1. Form a hypothesis about how a decrease in temperature would affect the pressure of a confined gas. Record your hypothesis under Data and Observations.
2. Prepare a water bath. Fill the 400-mL beaker two-thirds full with tap water.

3. Use a clamp to attach a thermometer to the ring stand. Lower the thermometer into the beaker, as shown in Figure A. The thermometer should not touch the bottom or side of the beaker.

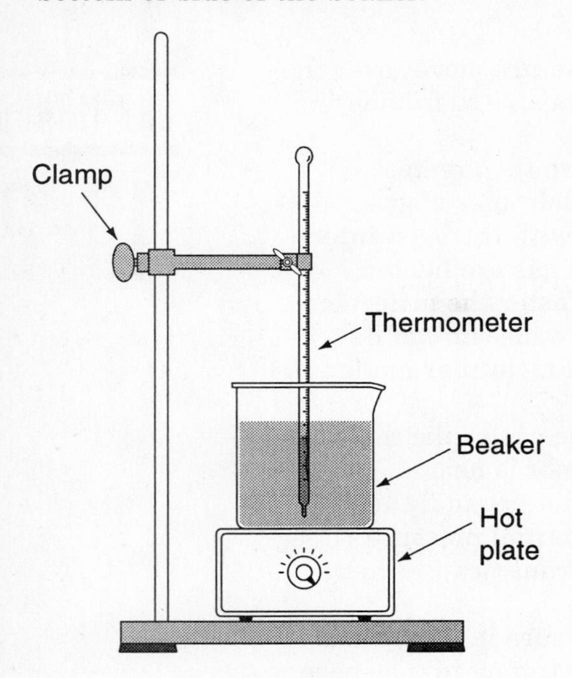

Figure A

4. Begin heating the water bath with a hot plate while you go on to steps 5–14.
5. Read the directions on page xv for inserting glass tubing into a stopper. You will carry out this procedure with both glass tubing and glass droppers. **CAUTION:** *Follow the directions carefully. The ends of the glass must be fire polished. Be sure to lubricate the end of the tubing or the dropper with glycerol or soapy water before inserting it into the stopper. Do not force the glass. It may shatter in your hand. If you have any difficulty, ask your teacher for help.*
6. Following the directions on page xv and observing the caution stated in step 5, insert the glass tubing into the two-hole stopper. Push the tubing through the hole until it can reach almost to the bottom of the Erlenmeyer flask when the stopper is in place in the flask.
7. Attach the shorter piece of rubber tubing to the end of the longer part of the glass tubing. Use a pinch clamp to close this piece of rubber tubing.
8. Fill the flask to the 300-mL mark with tap water.
9. Insert the pointed end of a glass tube from a straight dropper into the other hole in the stopper with the long glass tubing.
10. Insert the two-hole stopper assembled with the glass tubes into the Erlenmeyer flask.
11. Clamp both the long glass tubing and the flask to the second ring stand.
12. Insert the pointed end of the glass tube from the other dropper into the one-hole stopper and place the stopper in the test tube.
13. Attach one end of the long rubber tubing to the dropper tube in the test tube set-up. Attach the other end of the rubber tubing to the dropper tube in the flask set-up. These connections are shown in Figure B.

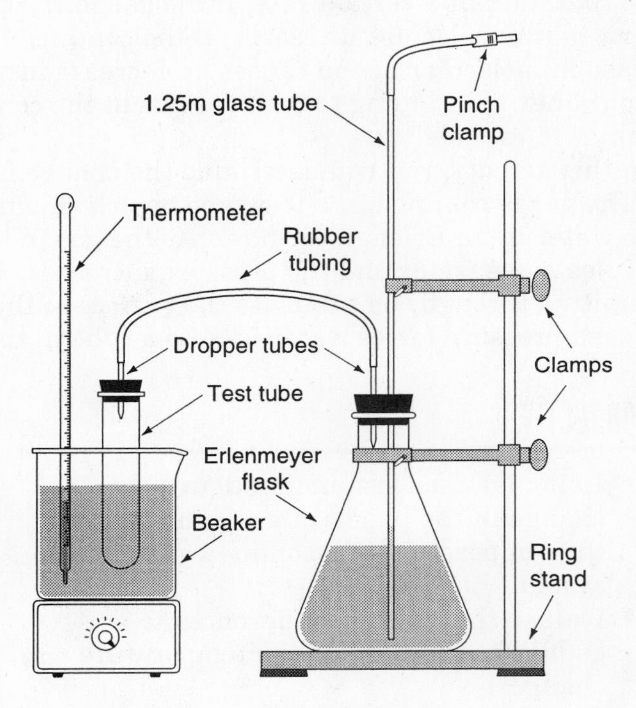

Figure B

14. Measure the height of the column of air trapped in the glass tubing (the distance from the water level in the long glass tubing to the upper end of the glass tubing). Read the room temperature from the classroom thermometer or obtain the reading from your teacher. Record the room temperature and the height of the air column in the first two columns in the table under Data and Observations.

Copyright © Glencoe/McGraw-Hill, a division of The McGraw-Hill Companies, Inc.

15. When the water in the water bath reaches a rolling boil, use a test-tube holder to place the test tube into the boiling water. Hold the test tube so that it is immersed in the water up to the level of the bottom of the stopper. Wait 5 minutes for the temperature of the gas in the test tube to reach the temperature of the boiling water.
16. Measure the height of the column of air trapped in the glass tubing. Record this distance and the temperature in the table.
17. Raise the thermometer on the ring stand so that you can move the beaker.
18. Using beaker tongs or thermal mitts, remove the beaker containing the test tube from the hot plate and set it on a heat-resistant pad. Keep the test tube immersed in the hot water. Immediately return the thermometer to the water bath, adjusting the position of the ring stand and clamp as necessary.
19. Allow the water bath to cool slowly. You may add small quantities of cool water to speed the cooling process. Remove an equal quantity of warm water from the beaker each time you add cool water.
20. With each 10°C (approximate) decrease in the temperature of the water bath, take a new measurement of the height of the column of trapped air. Record the height and the temperature in the table.
21. Stop taking measurements when the water bath reaches room temperature, about 20°C.
22. Read the atmospheric pressure on the barometer. Convert the pressure to kilopascals and record the value in the Data and Observations section.

DATA AND OBSERVATIONS

Hypothesis: ______________________________

____

____

Atmospheric pressure ________ kPa

Changes in Height and Pressure with Temperature Changes

Temperature (°C)	Height (cm)	Pressure (kPa)

ANALYSIS

1. Before heating takes place, the air confined in the test tube and the air trapped in the glass tube are at atmospheric pressure. As the temperature increases during heating, the pressure of the air confined in the test tube increases and is transmitted to the water in the flask. Water is forced into the long glass tubing, and the pressure on the column of air trapped above the water in the tubing increases. After heating, as the temperature decreases from 100°C, the pressure of the air confined to the test tube decreases and exerts less pressure on the water in the flask. The water level in the long glass tubing falls, and the pressure on the column of air trapped above the water in the tube decreases. The change in pressure at each temperature reading can be calculated by multiplying the original pressure by a factor made up of the original height of the air column divided by the new height of the air column. For example, if the original height of the air column at 20°C and atmospheric pressure is 45 cm and the

height is reduced to 22.5 cm at 100°C, then the pressure is twice as great. Use your data to calculate the pressure in the test tube in kilopascals. Use the atmospheric pressure as the original pressure. Record the pressures in the table.

2. Make a graph of your results. Assume that the temperature of the gas confined in the test tube is the same as the temperature of the water bath. Plot the temperature on the x-axis and the pressure in kilopascals on the y-axis. Draw the best-fitting curve for these points.

CONCLUSIONS

1. What was the relationship between temperature and the height of the column of air above the water that was forced up the glass tube?

2. What is the relationship between the temperature of a confined gas and the pressure exerted by the gas?

3. What would happen to the pressure exerted by the gas if the temperature were lower than room temperature?

4. Did your results support your hypothesis? Explain.

EXTENSION AND APPLICATION

Consider a situation that is the opposite of this experiment. A can contains a gas under pressure. What happens to the temperature of the gas and its container if the gas is suddenly released from the can?

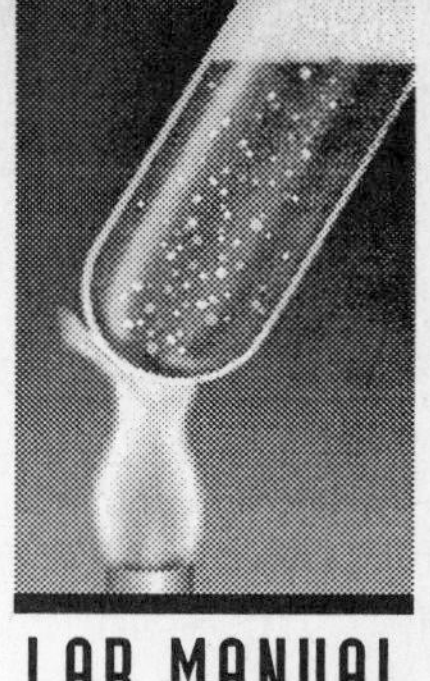

LAB MANUAL

10-2

Relating Gas Temperature and Pressure

When the temperature of an automobile tire increases, the internal pressure of the air in the tire also increases. This is due to the fact that a tire is fairly rigid and does not expand. If the tire were made of thinner material, the tire would expand and the volume of the tire rather than the pressure in the tire would increase. This is an example of the direct relationship between the temperature of a confined gas and the pressure exerted by the gas.

Misconception: Students almost never consider the reverse phenomenon when an explanation is given for the behavior of matter. The cooling that occurs when the pressure of a confined gas is released is a good example.

PROCESS SKILLS

1. Students will measure temperatures and corresponding heights of a column of air.
2. Students will infer that the changes in height are a measure of pressure changes.
3. Students will analyze data by means of a graph to determine the relationship between temperature and pressure of a confined gas.

PREPARATION HINTS

- It is a good idea to prepare the flask, test tube, and rubber stoppers with glass tubing for each group of students. This preparation ensures that the experimental setups will be done properly and provides a safer working environment.
- If students assemble the glass tubes and stoppers themselves, make sure they review the directions for this procedure on page 90 and supervise them closely to ascertain that they use the proper, safe technique for inserting a glass tube into a stopper.
- The part of the long glass tubing that is inside the flask should be about 1 cm shorter than the height of the flask.
- If available, use a tall-form beaker to keep the test tube in step 18 vertical.
- Tygon or latex tubing is preferable to rubber tubing. Make sure all stoppers are firmly seated in order to resist the small increase in pressure inside the apparatus.

MATERIALS

apron
goggles
400-mL beaker
18 mm × 150 mm test tube, heat resistant
one-hole rubber stopper to fit test tube
two-hole rubber stopper to fit flask
glass droppers, straight (2)
50-mm length of 5-mm rubber or plastic tubing
500-mm length of 5-mm rubber or plastic tubing
500-mL Erlenmeyer flask
1.25-m length of glass tubing
hot plate (or burner with ring and wire gauze)
ring stands (2)
utility clamps (2)
thermometer clamp
pinch clamp
metric ruler
beaker tongs or thermal mitts
test-tube holder
hot pad
glycerol
barometer

PROCEDURE HINTS

- Before proceeding, be sure students have read and understood the purpose, procedure, and safety precautions for this laboratory activity.
- If necessary, a laboratory burner can be used as a heat source. Students will have to set up a ring and wire gauze on the ring stand to support the water bath. After heating, the burner can simply be turned off. It is not necessary to move the beaker, as in step 17 of the Procedure.
- Neither the test tube nor the thermometer should touch the bottom of the beaker.

Copyright © Glencoe/McGraw-Hill, a division of The McGraw-Hill Companies, Inc.

- **Troubleshooting:** If no changes in the height of the air column can be measured with changes in temperature, there is a leak in the system. The most likely places for leaks are the rubber stopper-glass tube interface and the rubber stopper-dropper interface. Tighten these connections to correct the problem. Or, a leak can occur in the rubber tubing, especially if the tubing is old or cracked. Replace the tubing. Such leaks are less likely in Tygon, or latex tubing.
- **Safety Precaution:** Caution students to be careful as they work with boiling water. They must avoid spillage.
- **Disposal:** Since no hazardous substances are used in this experiment, there are no disposal problems.

DATA AND OBSERVATIONS

Hypotheses will vary but should indicate that in a confined gas, decreases in temperature and in pressure are directly related.

Atmospheric pressure 99.8 kPa

Changes in Height and Pressure with Temperature Changes

Temperature (°C)	Height (cm)	Pressure (kPa)
18	48.0	99.8
100	41.0	116.8
95	41.4	115.7
85	43.6	109.9
78	44.4	107.9
70	45.7	104.8
56	46.6	102.8
46	47.5	100.9
36	47.6	100.6
18	48	99.8

ANALYSIS

1. See the table.
2. See graph in right-hand column.

CONCLUSIONS

1. As the temperature decreased, the height of the column of air increased.
2. There is a direct relationship between the temperature of a confined gas and the pressure exerted by the gas.
3. The pressure would continue to drop. The pressure would fall below the pressure of the room. The container would appear to contain a partial vacuum.
4. Answers will vary, depending on students' hypotheses.

EXTENSION AND APPLICATION

If gas in a container is confined under pressure and the pressure is released suddenly, the gas and the container are cooled. This is the reason that an aerosol can gets cold when the contents are allowed to escape. This phenomenon is known as the Joule-Thomson effect.

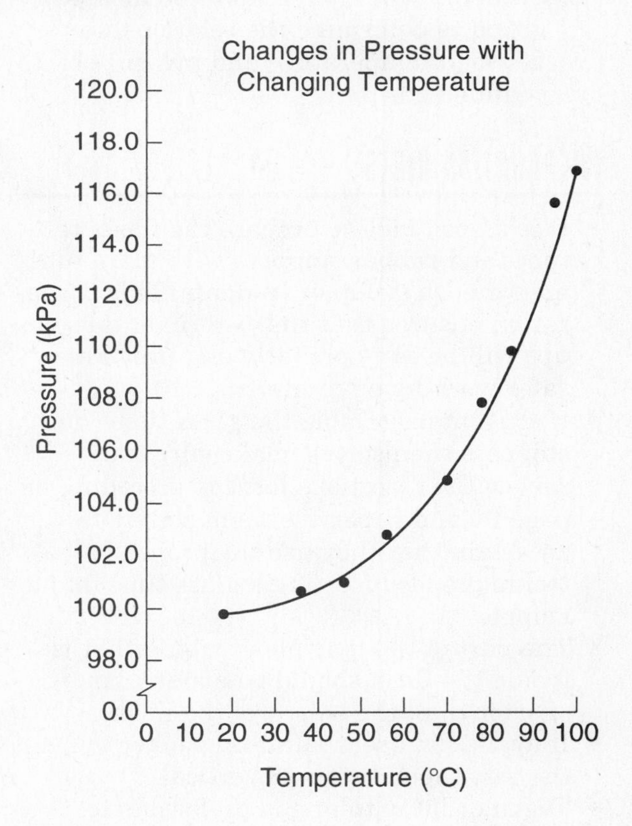

Copyright © Glencoe/McGraw-Hill, a division of The McGraw-Hill Companies, Inc.

Name

Date Class

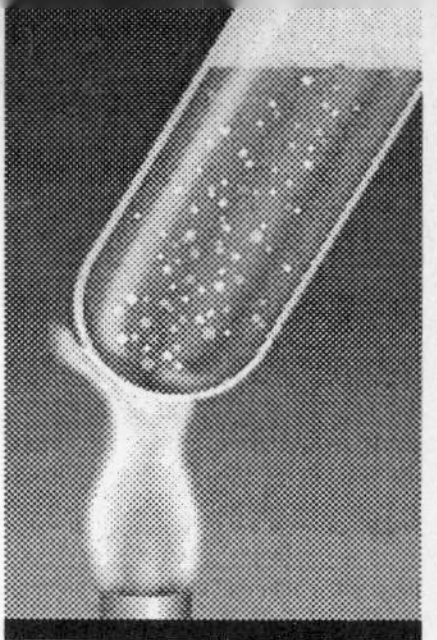

LAB MANUAL

11-1

Determining Absolute Zero

One of the variables that affects the volume of gas is the temperature of the gas and its surroundings. This volume-temperature relationship is quantified in Charles's law. The law states that as the temperature of a gas decreases, the volume of the gas decreases proportionately. An ideal gas at 273 K, for example, would decrease in volume by 1/273 of its original volume for each kelvin that the temperature decreases. If the temperature decreases to 0 K, the volume should decrease to zero. Real gases, however, liquefy and solidify long before this theoretical limit, called absolute zero, is reached. By using air and measuring its volume at various temperatures, it is possible to estimate the temperature that would correspond to absolute zero.

OBJECTIVES

- **Demonstrate** the relationship between the temperature of a gas and its volume.
- **Graph** the relationship.
- **Estimate** the temperature of absolute zero by extrapolation.

MATERIALS

apron
goggles
400-mL beakers (2)
thin-stem pipet
thermometer
hot plate
paper towel

PROCEDURE

1. Fill two 400-mL beakers half full with tap water.
2. Begin heating the water in one beaker to a temperature that is 10°C above room temperature.
3. Fill a thin-stem pipet completely with room-temperature water. To make sure the pipet is filled, first draw in as much water as possible. Then, holding the pipet upward, squeeze the bulb slightly to eject any air left in the bulb and stem. Keeping this pressure on the bulb, insert the tip of the stem into the water, as shown in Figure A. Release the pressure on the bulb, and the pipet will fill completely.
4. Dispense the water from the pipet, counting the total number of drops it takes to empty the pipet. (The number should be about 100 drops.)

Figure A

5. Record in the table under Data and Observations the room temperature and the number of drops you dispensed.
6. Stop heating the water when a temperature 10°C above the room temperature is reached.
7. Holding the thin-stem pipet by the stem, immerse the bulb in the warm water, as shown in Figure B on the next page.
8. Hold the pipet in the warm water for a few minutes so that the air in the pipet reaches the temperature of the water.
9. Pinch the stem of the pipet to seal off the bulb. Place the bulb in the other beaker of water, which is at room temperature.
10. Still pinching the stem, immerse the entire pipet, including the stem, in the water.

Release the stem underwater. A small amount of water should be drawn up into the pipet. This water is equal in volume to the amount that the gas expanded at atmospheric pressure when the pipet bulb was heated.

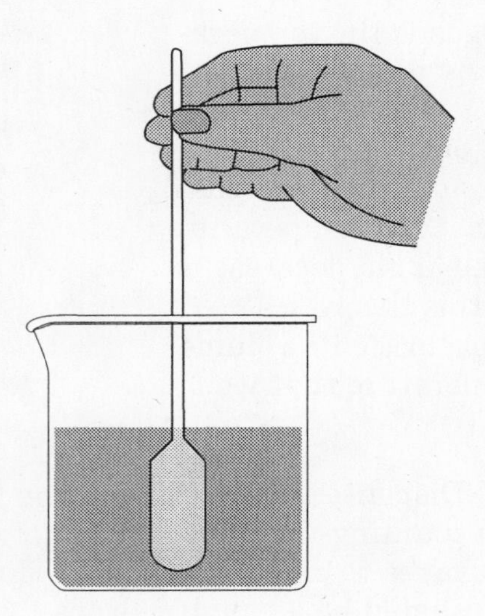

Figure B

11. Remove the pipet from the water bath. Dry the outside of the pipet with the paper towel. Expel the water, counting the number of drops of water that were drawn into the pipet.
12. Add the number of drops to the initial volume noted in the table. Record the total volume and the temperature of the warm water bath.
13. Dry the inside of the pipet by drawing in air and then releasing the bulb several times.
14. Repeat steps 6–13 at a temperature that is 10°C above the previous trial.
15. Continue this procedure until the water temperature is approximately 75°C. Record all your results in the table.

DATA AND OBSERVATIONS

Temperature/Volume Changes of Air

Temperature (°C)	Total volume (drops)	Temperature (°C)	Total volume (drops)

ANALYSIS

1. Make a graph of your data. Plot the temperature on the x-axis and the volume on the y-axis. The scale of the x-axis should go from –300°C to 100°C.
2. Connect the data points with a straight line. Then extend the line until it intersects the x-axis. This point is your estimate of absolute zero.

CONCLUSIONS

1. What is your estimate of absolute zero?

2. Calculate the percentage error for your experimental value. Use the formula

% error =

$$\frac{\text{theoretical temp.} - \text{experimental temp.}}{\text{theoretical temp.}} \times 100\%$$

3. What is the relationship between temperature and volume as shown by the graph?

EXTENSION AND APPLICATION

The theoretical limit known as absolute zero is also the point where all molecular motion stops. Investigate the subject of cryogenics, the science of low temperatures. What is the value of materials that are "supercold"?

Copyright © Glencoe/McGraw-Hill, a division of The McGraw-Hill Companies, Inc.

Teacher Guide

LAB MANUAL

11-1

Determining Absolute Zero

This experiment demonstrates Charles's law, the relationship between the volume and the temperature of a gas. The law states that the volume of a gas varies directly with the temperature. According to theory, an ideal gas would continue to decrease in volume as the temperature decreased until the temperature reached absolute zero. At that temperature, the volume of the gas would theoretically be zero. Under normal laboratory conditions, real gases behave as if they were ideal. So,, by measuring the volume of a sample of air as temperature is changed, students can arrive at a fair estimation of absolute zero.

Misconception: Students may think that gases differ in their response to temperature change. Charles's law, in fact, applies to all gases.

PROCESS SKILLS

1. Students will heat a sample of air and determine the change in its volume.
2. Students will prepare a graph of their results.
3. Students will estimate the point of absolute zero by extrapolation of the graph.

PREPARATION HINTS

You may want to divide the class into small groups and assign each group a specific temperature at which to keep their water bath.

MATERIALS

apron
goggles
400-mL beakers (2)
thin-stem pipet
thermometer
hot plate (or burner with ring stand, ring, and wire gauze)
paper towel

PROCEDURE HINTS

- Before proceeding, be sure students have read and understood the purpose, procedure, and safety precautions for this laboratory activity.
- If possible, use a hot plate in place of a burner to heat the water bath.
- Make sure students pinch the stem of the pipet very tightly when they transfer the bulb from the warm water bath to the water bath at room temperature.
- Pipets must be dry for each trial. If droplets of water remain in the pipet, have students partially refill the pipet with water, collect the droplets in the larger volume of water, and expel water with the stem pointed downward.
- Make sure water temperatures do not rise above 75°C. At about 80°C, the plastic pipets will begin to melt.

DATA AND OBSERVATIONS

Temperature/Volume Changes of Air

Temperature (°C)	Total volume (drops)	Temperature (°C)	Total volume (drops)
25	105	55	115
35	107	65	120
45	110	75	121

Copyright © Glencoe/McGraw-Hill, a division of The McGraw-Hill Companies, Inc.

ANALYSIS

You may want to have students graph temperature and volume to show changes of air so that they can observe the linearity of the points and establish an accurate slope. Then they can make the extrapolated graph more accurately.

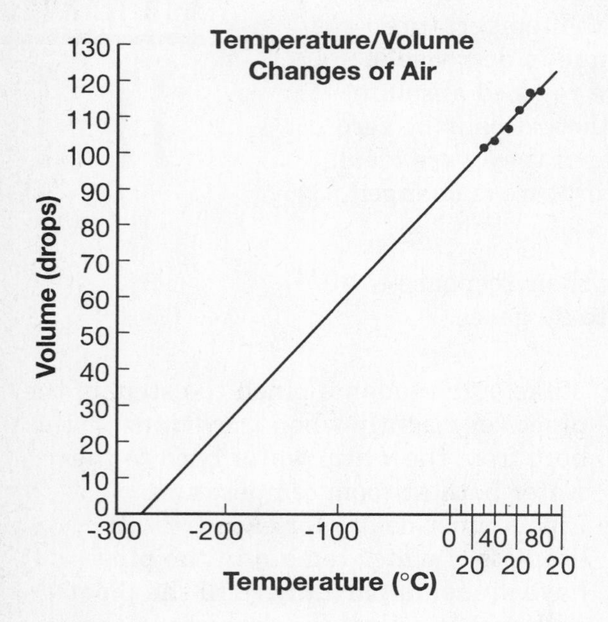

CONCLUSIONS

1. Answers will vary depending on students' data. Using the sample data, absolute zero is –276°C.
2. Answers will vary. Percentage error of sample data is 1 percent.
3. The volume of a gas changes directly with the temperature.

EXTENSION AND APPLICATION

The supercold condition has many applications. Large quantities of gas can be stored in small volumes, for example. Hospitals often keep tanks of liquid nitrogen and liquid oxygen outside; piping brings the gases inside for use. Oxygen is delivered directly to patients. Liquid nitrogen is used in transporting food under refrigeration. Liquefied gases are used as refrigerants in air conditioners and refrigerators.

Name

Date Class

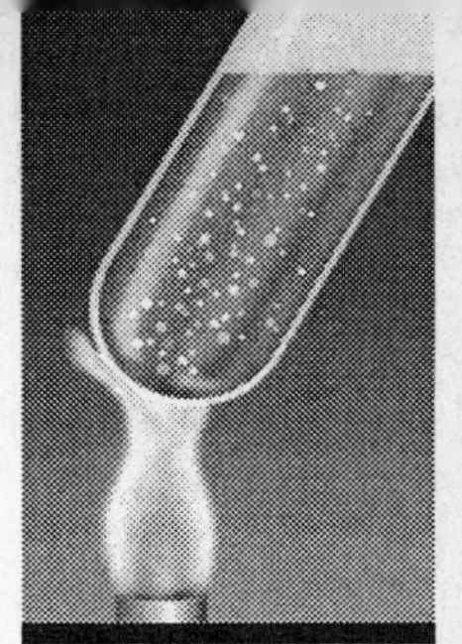

LAB MANUAL

11-2

Charles's Law

The gas laws, Boyle's law and Charles's law collectively, treat gases as if they were ideal. An ideal gas is one that is composed of massless and volumeless particles. Ideal gases do not condense into liquids. Hydrogen and helium are two gases that most closely follow the gas laws. If we think of molecules of gas as the ideal model, the kinetic energy of a molecule is directly proportional to the absolute temperature. The hotter the gas, the more kinetic energy the molecules have, the more they move, and the more volume they occupy at constant pressure.

In this activity, you will observe the relationship between the temperature and volume of a gas.

OBJECTIVES

- **Show** the relationship between temperature and volume of a gas.
- **Relate** the kinetic energy of a gas to the temperature of the gas.
- **Calculate** the volume of a gas at various temperatures.

MATERIALS

thermometer
1-L beaker
hot plate
ice
small balloon, round
string or thread
iron ring, smaller than the diameter of the beaker and balloon
tongs
metric ruler

PROCEDURE

1. Blow up a round balloon so that the diameter of the balloon is about one-half the diameter of a large beaker. It is not necessary to blow up the balloon all the way.
2. Wrap a piece of thread or string around the widest part of the balloon. Use a ruler to determine the length of the thread or string. Record the circumference of the balloon in Table 1. Also record the temperature in your classroom.
3. Prepare an ice-water bath in the beaker. Record the temperature of the water in Table 1.
4. Place the balloon in the ice water. Use the iron ring to hold the balloon below the water for about two minutes.
5. Remove the balloon from the ice water and quickly wrap a piece of thread or string around the widest part of the balloon. Use a ruler to determine the length of the thread or string. Record the circumference of the balloon in Table 1. Dispose of the ice water.
6. Prepare a hot-water bath in the beaker. The water should reach 90°–99°C. Record in Table 1 the temperature of the water. **CAUTION:** *Handle the hot-water bath carefully.*
7. Tape the balloon to the iron ring and place the balloon in the hot water. Use the iron ring to hold the balloon below the water for about two minutes.
8. Using tongs, remove the balloon from the hot water. Wrap a piece of thread or string around the widest part of the balloon. Use a ruler to determine the length of the thread or string. Record the circumference of the balloon in Table 1.

DATA AND OBSERVATIONS

Table 1

Temperature	Circumference of balloon
1. ________ °C	________ mm
2. ________ °C	________ mm
3. ________ °C	________ mm

Table 2

Circumference (C) of balloon (mm)	Diameter (D) of balloon (mm) $C / \pi = D$	Radius (r) of balloon (mm) $D / 2 = r$	Volume (V) of gas in the balloon (mm^3) $4/3 \pi r^3 = V$
1.			
2.			
3.			

ANALYSIS

1. Use the equations in Table 2 to calculate the volume of the gas in the balloon at each temperature.
2. Graph your results on graph paper. Plot temperature (°C) on the x-axis and volume of the gas (mm^3) on the y-axis. Draw the best-fitting straight line to connect your data points. Extend the line to 400°C.

CONCLUSIONS

1. Remember that the volume of a gas is proportional to the absolute temperature (K). On a warm day, the outdoor temperature may be 27°C (300 K). Use your graph to find the volume of the gas in the balloon at that temperature. Then predict the volume of the gas at 600 K (327°C).

____

____

____

2. At which temperature did the gas molecules in the balloon have the most kinetic energy?

____

____

____

3. At low temperatures, gases liquefy and solidify. What happens to gases at very high temperatures?

____

____

____

EXTENSION AND APPLICATION

1. Liquefying gases is a multi-million dollar industry. What are some commercial uses of liquefied gases? Why is it an advantage to use liquefied gases?
2. How are liquid helium, hydrogen, and oxygen used in the space program?

Teacher Guide

LAB MANUAL

11-2

Charles's Law

In 1787 Jacques Charles, a French scientist, discovered the connection between the volume of a gas and its temperature. In 1848 William Thomson, a British physicist whose title is Lord Kelvin, proposed the absolute temperature scale, which now bears his name.

The behavior of gases, as observed by Charles, can be explained by the kinetic theory. According to the kinetic theory, an ideal gas is described as follows.

1. Gas particles are in random motion.
2. The volume of the gas particle itself is negligible when compared to the space occupied.
3. No forces of attraction or repulsion exist between the gas molecules.
4. The average kinetic energy of the molecules is directly proportional to the Kelvin temperature.

PROCESS SKILLS

1. Students will heat and cool a sample of air and determine its volume.
2. Students will prepare a graph of their data.
3. Students will relate the temperature of a gas to its volume and kinetic energy.

PREPARATION HINTS

- Inflate a balloon beforehand to make sure it is of appropriate size for the glassware.
- Iron rings (for weighting the balloons) may be prepared by bending a strip of steel to the appropriate size.
- Before doing the experiment, have students blow up the balloons to overcome the initial resistance of the rubber to stretching. This will allow for the air in the balloons to expand and contract by small amounts without any appreciable change in pressure. The smallest round balloons available will work best.

MATERIALS

thermometer
1-L beaker
hot plate
iron ring, smaller than the diameter of the beaker
ice
string or thread
small balloon, round
tongs
metric ruler

PROCEDURE HINTS

- Before proceeding, be sure students have read and understood the purpose, procedure and safety precautions for this laboratory activity.
- To save time, have students begin heating a water bath before doing step 1. If possible, have them use a hot plate instead of a burner.
- You may wish to have students measure the volume of the air at a few more temperatures and add their data to the graph.

DATA AND OBSERVATIONS

Table 1

	Temperature	Circumference of balloon
1.	25°C	102 mm
2.	1°C	89 mm
3.	99°C	135 mm

ANALYSIS

Table 2

Circumference (C) of balloon (mm)	Diameter (D) of balloon (mm) $C / \pi = D$	Radius (r) of balloon (mm) $D / 2 = r$	Volume (V) of gas in the balloon (mm^3) $4/3 \pi r^3 = V$
Sample Data: **1.** 102	32.5	16.25	18 000
2. 100	31.8	15.9	16 800
3. 110	35.0	17.5	22 400

Sample Graph

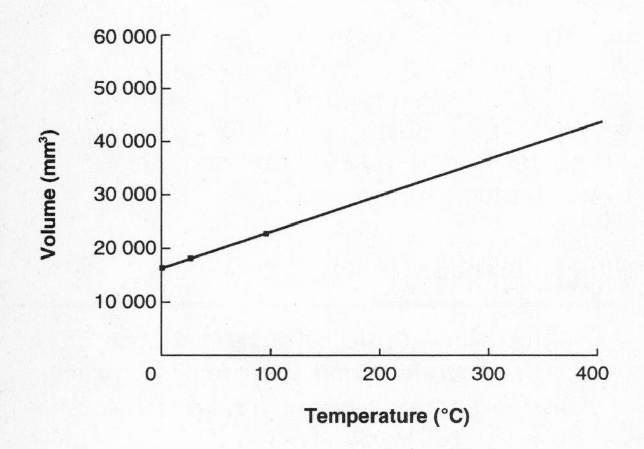

CONCLUSIONS

1. Answers will vary. The volume of the gas at 600 K would be twice its volume at 300 K.
2. The gas molecules had the most kinetic energy at the highest temperature.
3. At very high temperatures, gases start to lose electrons and ionize. Eventually, all matter at high temperature would ionize and then form a plasma. These temperatures approximate the temperature on the surface and interior of the sun. Even though plasma is the most common form of matter in the universe, it is a rare form on Earth.

EXTENSION AND APPLICATION

1. Answers will vary but may include that liquefied gases are used for refrigeration (nitrogen), medical uses (oxygen), beverage carbonation (carbon dioxide) and fuel (propane, natural gas). Liquefied gases allow for the storage of large volumes of gas in small-volume containers.
2. Liquid helium is used as a cooling agent and also in altitude propulsion jets. Liquid hydrogen and oxygen are the fuel and oxidizer of choice, respectively, for booster rockets that propel space vehicles into orbit.

Name

Date Class

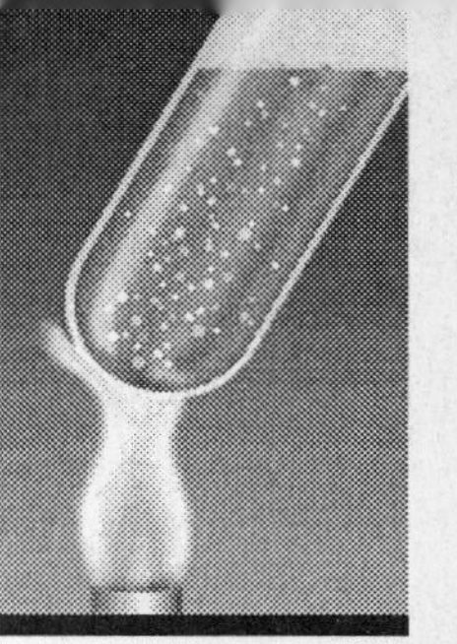

LAB MANUAL

12-1

Determining the Avogadro Constant

One of the most important concepts in the study of chemistry is the mole concept. The mole is a counting unit. One mole of any substance contains 6.02×10^{23} atoms, molecules, or formula units. The value 6.02×10^{23}, called the Avogadro constant, has been determined by precise experiments. This laboratory activity will allow you to explore one way of finding the Avogadro constant. The experiment is not precise and the value it gives is only an approximation. Your value, therefore, may be close to the correct value, or it may be quite different.

At a given temperature, molecules of any gas have the same average kinetic energy. The result is that, at the same temperature and pressure, one mole of gas occupies the same amount of space as any other gas. The Avogadro constant can be determined by dividing the volume occupied by one mole of gas by the volume occupied by one molecule of gas.

The volume occupied by a measured mass of butane gas is found experimentally in this activity. This volume is proportionally converted to the volume that would be occupied by one mole. The volume occupied by one molecule is estimated as the space occupied by an imaginary cylinder containing one moving molecule of the gas. The diameter of the cylinder equals the estimated diameter of a butane molecule, 4.5×10^{-8} cm. The height of the cylinder equals the mean free path of the molecule. The mean free path is the average distance traveled by a molecule before it collides with another molecule and is approximately 1×10^{-4} cm.

In this activity, the butane gas will be collected by the displacement of water. When you collect a gas by displacing water in a container, some of the water evaporates and mixes with the collected gas. The total pressure of the gases is the sum of the vapor pressure of the water and the pressure of the collected gas. The warmer the water, the greater its vapor pressure will be. To account for the presence of the water vapor, you must subtract its vapor pressure from the total pressure of the gases in the container. Obtain the value for water vapor pressure (P_{water}) from the following table.

Vapor Pressure of Water

°C.	kPa	°C.	kPa
0	0.6	22	2.6
5	0.9	23	2.8
10	1.2	24	3.0
15	1.7	25	3.2
16	1.8	26	3.4
17	2.0	27	3.6
18	2.1	28	3.8
19	2.2	29	4.0
20	2.3	30	4.2
21	2.5		

In this laboratory activity, you will measure the volume occupied by a quantity of butane gas and use this information to calculate the Avogadro constant.

Copyright © Glencoe/McGraw-Hill, a division of The McGraw-Hill Companies, Inc.

OBJECTIVES

- **Determine** the molar volume of butane.
- **Calculate** a value for the Avogadro constant.

MATERIALS

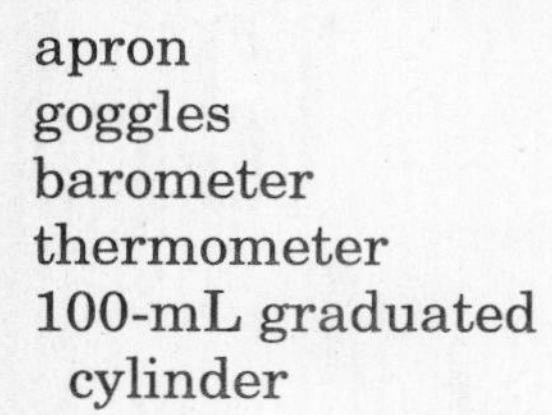

apron	large beaker, trough, or dish pan
goggles	butane lighter
barometer	balance
thermometer	paper towel
100-mL graduated cylinder	

PROCEDURE

1. You will collect a sample of butane gas and determine its molar volume. Before you proceed, think about what properties of the gas you must measure and what equation you must use to convert the volume of butane to a form that you can use in your calculations. Record your thoughts in the Data and Observations section.
2. Find the atmospheric pressure by reading a barometer, or obtain the value from your teacher. Record the value.
3. Determine the mass of a butane lighter to the nearest 0.01 g. Record the mass.
4. Fill the 100-mL graduated cylinder all the way to the top with tap water.
5. Add water to the large beaker, trough, or dish pan until it is half full.
6. Measure the temperature of the water in the beaker or trough and record it.
7. Invert the water-filled graduated cylinder in the trough or beaker so that the top is under water. Hold it in place as shown in Figure A. Make certain there is no air trapped in the cylinder.
8. Hold the butane lighter under the water with the valve in the position shown in Figure A so that you can release gas from the lighter into the cylinder.
9. Press the valve and release gas until the cylinder is filled with a little less than 100 mL of the gas.

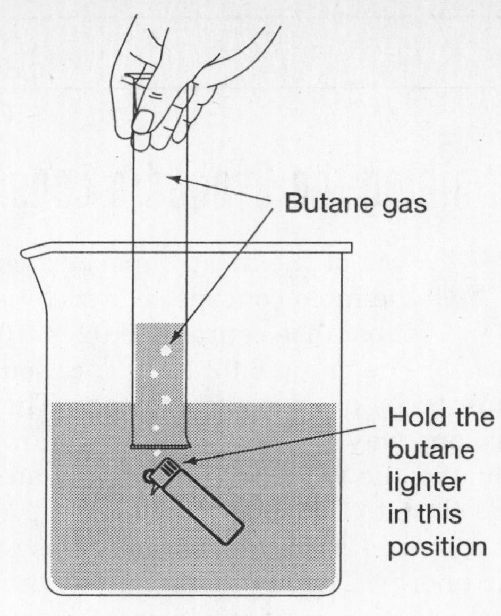

Figure A

10. Remove the lighter from the water.
11. Adjust the cylinder in the trough until the water levels inside and outside the cylinder are equal. This assures that the pressure of the gas inside the cylinder is the same as atmospheric pressure. Hold the cylinder in that position and read the volume of the trapped gas to the nearest milliliter. Record the volume in liters.
12. Release the gas from the graduated cylinder as your teacher instructs.
13. Dry the lighter very thoroughly. Use a paper towel or tissue to get rid of any water trapped in the valve area of the lighter. Do this very carefully; even a little trapped water will affect results.
14. Determine the mass of the dried lighter and record it.
15. Rinse the graduated cylinder, and pour the water in the beaker or trough down the drain.

DATA AND OBSERVATIONS

What properties of the gas must you measure? What equation must you use to convert the volume of butane to a form that you can use in your calculations?

____

____

1. Barometric pressure ________ mm Hg
2. Initial mass of lighter ____________ g

Copyright © Glencoe/McGraw-Hill, a division of The McGraw-Hill Companies, Inc.

3. Temperature of water ____________ °C

4. Volume of gas ____________ L

5. Final mass of lighter ____________ g

ANALYSIS

1. Convert the barometric pressure to kilopascals.

   $$\text{Pressure (kPa)} \times \frac{\text{data 1}}{7.50 \text{ mm Hg/kPa}}$$

   ____

2. Calculate the mass of butane collected.

   Mass = data 2 − data 5

   ____

3. Calculate the number of moles of butane collected.

   $$n = \frac{\text{analysis 2}}{\text{molar mass of butane}}$$

   ____

4. In the table on page 105, find the vapor pressure of water at the water bath temperature.

   ____

5. Determine the pressure of dry butane collected.

   $$P_{\text{dry gas}} = P_{\text{total}} - P_{\text{water}}$$
   $$= (\text{analysis 1}) - (\text{analysis 4})$$

   ____

6. Convert the water temperature to kelvins.

   Temperature (K) = data 3 + 273

   ____

7. Use the combined gas law to calculate the volume that the collected butane would occupy at STP.

   $$V = \frac{\text{data 4} \times \text{analysis 5} \times 273 \text{ K}}{101.3 \text{ kPa} \times \text{analysis 6}}$$

   ____

8. Calculate the volume of one mole of the butane gas collected in this experiment.

   $$\text{Volume of one mole} = \frac{\text{volume of gas collected at STP}}{\text{number of moles of gas collected}}$$
   $$= \frac{\text{analysis 7}}{\text{analysis 3}}$$

   ____

CONCLUSIONS

1. Find the volume of a hypothetical cylinder swept out by one butane molecule. Assume that the molecule has a diameter 4.5×10^{-8} cm and travels the mean free path distance of 1.0×10^{-4} cm/molecule, as shown in Figure B.

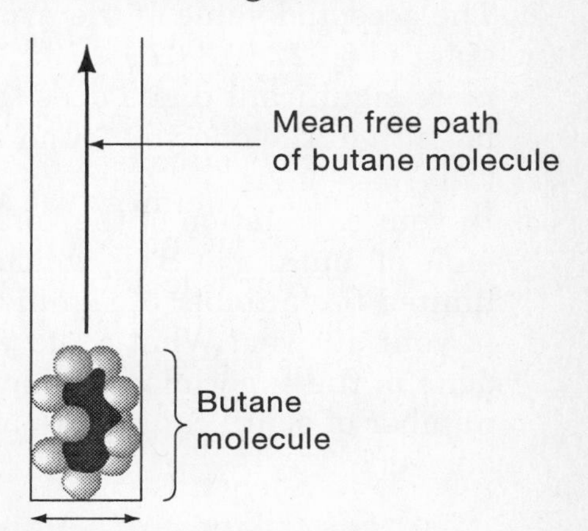

Figure B

$$\text{Volume of cylinder swept by one molecule} = \pi \times (\text{radius})^2 \times \text{height of cylinder}$$
$$= \pi \times \left(\frac{\text{molecular diameter}}{2}\right)^2 \times \text{mean free path}$$

____

2. Find the Avogadro constant.

   Avogadro constant =

   $$\frac{\text{volume of one mole (L/mol)} \times 1000 \text{ cm}^3\text{/L}}{\text{molecule cylinder volume (cm}^3\text{/molecule)}}$$
   $$= \frac{\text{analysis 8}}{\text{conclusions 1}}$$

   ____

Copyright © Glencoe/McGraw-Hill, a division of The McGraw-Hill Companies, Inc.

3. How does your result compare with the correct value of 6.02×10^{23}?

____

____

____

____

____

EXTENSION AND APPLICATION

1. A super computer can count at the rate of one billion (1×10^9) per second. To the nearest million, how many years would it take for this computer to count up to the Avogadro constant?
2. The accepted value of the Avogadro constant is $6.022\ 013\ 67 \times 10^{23}$. How many more significant digits does this value have than the one you found experimentally?
3. In your calculation of the volume of one mole of butane at STP, which quantity limited the number of significant digits in your answer? What could you have done in the experiment to increase the number of significant digits in your answer?

Teacher Guide

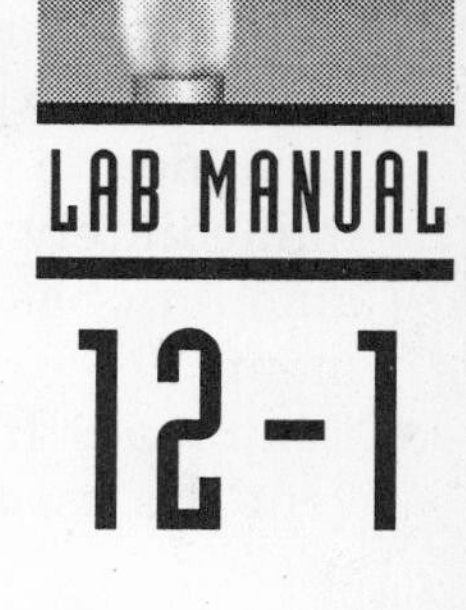

LAB MANUAL

12-1

Determining the Avogadro Constant

This laboratory activity is a mixture of process and theory. Students obtain values for the Avogadro constant by using their own experimental data together with data from other sources. The premise is that the volume occupied by one mole of a gas depends on the total volumes swept out by all the molecules. Students measure the mass and volume of a butane sample. Then, by using the combined gas law, they determine the volume that would be occupied by one mole of the gas under standard conditions.

The volume swept out by each molecule depends on the size of the molecule and the length of the path that it travels. The volume is approximated by assuming that it is the cylinder shown in Figure B. The size of the molecule is estimated, since a butane molecule can be in any orientation as it translates from place to place. The height of the cylinder is taken as the mean free path of a gas molecule.

Misconception: Students may have the idea that the volume occupied by one mole of a gas is always equal to 22.4 L. Point out to them that this volume depends on the temperature and pressure, and that STP are used as universally understood conditions for comparing volumes.

PROCESS SKILLS

1. Students will collect a sample of butane and calculate its volume at standard conditions.
2. Students will calculate the molar volume of butane.
3. Students will determine an approximate value for the Avogadro constant.

PREPARATION HINTS

- Students will need large containers to hold water for their inverted graduated cylinders. Gas collection troughs are ideal for this purpose. If you use 1-L or larger beakers, warn students that these large beakers break easily.
- Any commercial butane cigarette lighter can be used, and lighters can be used again in classes that follow. The lighters may be saved and reused from year to year.

MATERIALS

apron
goggles
barometer
thermometer
100-mL graduated cylinder
large beaker or pneumatic trough
butane lighter
balance
paper towel

PROCEDURE HINTS

- Before proceeding, be sure students have read and understood the purpose, procedure, and safety precautions for this laboratory activity.
- Note that P_{water} is given for only a small range of temperatures. Try to have water in this range.
- Students should be told that with the tools available to them, the procedure can yield only an approximate value for the Avogadro constant. Some students, however, get surprisingly good results.
- A large portion of the lab time will be devoted to calculations and data analysis. Be sure to provide that time.
- **Troubleshooting:** The most common problem in this laboratory procedure is finding a mass that is too low for the collected butane. This problem occurs when students do not dry the butane lighter thoroughly. A droplet of water trapped in

the valve assembly of the lighter has enough mass to distort the results. Be sure to caution students to dry their lighters completely before finding the final mass.

- **Safety Precaution:** No open flames should be permitted in the laboratory while this experiment is going on. It is a good idea to collect the butane lighters immediately after the students collect the amount of gas required for their experiments.
- **Disposal:** If necessary, students should vent the used butane gas in a fume hood.

DATA AND OBSERVATIONS

Students' answers will vary, but they should show some understanding that they need to measure the mass and volume of the gas collected and use the combined gas equation.

1. Barometric pressure	745 mm Hg
2. Initial mass of lighter	17.81 g
3. Temperature of water	23 °C
4. Volume of gas	0.100 L
5. Final mass of lighter	17.55 g

ANALYSIS

1. Pressure (kPa) =
745 mm Hg / (7.50 mm Hg/1 kPa)
= 99.3 kPa

2. Mass = 17.81 g – 17.55 g = 0.26 g

3. n = 0.26 g butane/(58 g butane/mole)
= 0.0045 mol

4. Vapor pressure of water = 2.8 kPa

5. $P_{dry\ gas}$ = 99.3 kPa – 2.8 kPa
= 96.5 kPa

6. Temperature (K) = 23°C + 273 = 296 K

7. $$V = \frac{0.100\ \text{L} \mid 96.5\ \text{kPa} \mid 273\ \text{K}}{\mid 101.3\ \text{kPa} \mid 296\ \text{K}}$$

= 0.0879 L

8. $$\frac{0.0879\ \text{L}}{0.0045\ \text{mol}} = 19.5\ \text{L/mol}$$

CONCLUSIONS

1. Volume of a cylinder swept by one molecule

$= \pi \times (2.25 \times 10^{-8}\ \text{cm})^2 \times (1.0 \times 10^{-4}\ \text{cm})$

$= 1.6 \times 10^{-19}\ \text{cm}^3$

2. Avogadro constant =

$$\frac{(19.5\ \text{L/mol}) \times 1000\ \text{cm}^3/\text{L}}{1.6 \times 10^{-19}\ \text{cm}^3/\text{molecule}}$$

$= 1.2 \times 10^{23}$ molecules/mol

3. Students' answers will vary, depending on their results.

EXTENSION AND APPLICATION

1. It would take 19 million years to count up to 6.02×10^{23}.

2. The accepted value has 7 more significant digits than the students' answers, which should have two significant digits.

3. The number of significant digits was limited to two because the mass of butane was found to 0.01 g. The significant digits in the answer could have been increased to three by massing the butane to 0.001 g.

Copyright © Glencoe/McGraw-Hill, a division of The McGraw-Hill Companies, Inc.

Name

Date Class

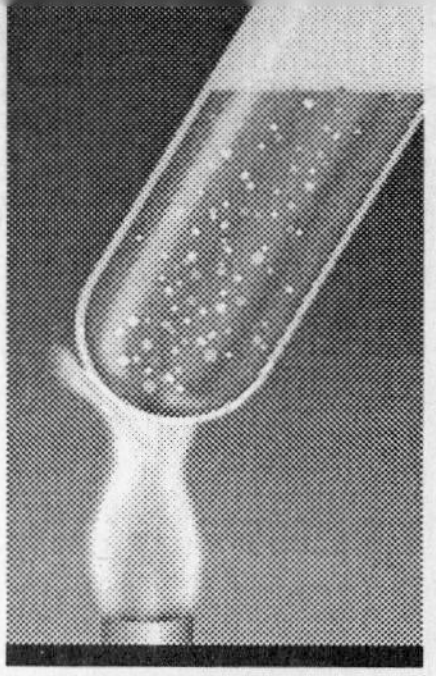

LAB MANUAL

12-2

Stoichiometry of a Chemical Reaction

The following balanced equation shows the reaction between the compounds sodium hydrogen carbonate and hydrochloric acid.

$$NaHCO_3(s) + HCl(aq) \rightarrow NaCl(aq) + CO_2(g) + H_2O(l)$$

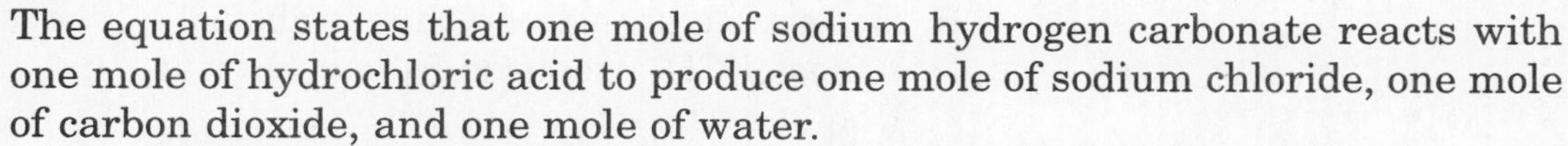

The equation states that one mole of sodium hydrogen carbonate reacts with one mole of hydrochloric acid to produce one mole of sodium chloride, one mole of carbon dioxide, and one mole of water.

In this experiment, you will react samples of three known carbonates and an unknown carbonate with hydrochloric acid. You will determine the stoichiometric relationships among the reactants and products. You will then determine the identity of the unknown carbonate by comparing its experimental results with those of the known carbonate.

OBJECTIVES

- **React** known amounts of carbonates and hydrogen carbonates with acid.
- **Determine** the stoichiometric relationships among the reactants and products.
- **Identify** an unknown carbonate by the amount of product formed.

MATERIALS

24-well microplate
thin-stem pipet
balance
distilled water
paper towels

PROCEDURE

Part 1: Determining Stoichiometric Relationships in a Reaction

1. Obtain a sample of a carbonate, approximately 1.0 g mass. Record the molecular formula that appears on the label of the container. Record all data for this carbonate in the Data and Observations section.
2. Place the microplate on a balance and record the mass to the nearest 0.01 g.
3. Put *all* the carbonate into well A4 of the microplate. Measure and record the mass of the microplate and the sample.
4. Fill a thin-stem pipet with 8*M* HCl solution. **CAUTION:** *Be careful with HCl. It burns skin and clothing. Do not inhale the vapors. If spillage occurs, notify your teacher promptly.*
5. Wipe the outside of the pipet filled with HCl, and stand it stem up in well A3.
6. Measure and record the total mass of the unreacted setup (the microplate, the sample, and the pipet with HCl solution).
7. Take the pipet from well A3 and, drop by drop, add the HCl solution to the sample of carbonate in well A4. Allow the bubbles of gas to escape after each drop before adding more HCl.
8. Continue to add acid solution one drop at a time until the carbonate has dissolved and the solution produces no more bubbles.
9. Return the pipet, stem upward, to well A3 and again mass the tray, the solution, and the thin-stem pipet. Record the mass of the complete reacted setup.
10. Dispose of the reacted chemicals as directed by your teacher.
11. Rinse the microplate with water and dry with a paper towel.
12. Repeat steps 1–11 with the other known carbonates.

Part 2: Identifying an Unknown Carbonate by Stoichiometric Principles

Repeat steps 1–11 with the unknown carbonate. Record your data for the unknown carbonate.

DATA AND OBSERVATIONS

Part 1: Data for known carbonates

	$NaHCO_3$	$CaCO_3$	$KHCO_3$
1. Mass of microplate	___ g	___ g	___ g
2. Mass of the microplate and sample	___ g	___ g	___ g
3. Mass of complete unreacted setup	___ g	___ g	___ g
4. Mass of complete reacted setup	___ g	___ g	___ g
5. Mass of sample	___ g	___ g	___ g
6. Moles of sample (mass/mol mass)	___ moles	___ moles	___ moles
7. Mass of CO_2 lost (data 3 – data 4)	___ g	___ g	___ g
8. Moles of CO_2 lost (data 7/44 g/mole)	___ moles	___ moles	___ moles
9. Mass of CO_2 produced as a percentage of mass of sample $\frac{\text{data 7}}{\text{data 5}} \times 100\%$	___ %	___ %	___ %

Part 2: Data for unknown carbonate

1. Mass of microplate ______ g
2. Mass of the microplate and sample ______ g
3. Mass of complete unreacted setup ______ g
4. Mass of complete reacted setup ______ g
5. Mass of sample ______ g
6. Mass of CO_2 lost (data 3 – data 4) ______ g
7. Mass of CO_2 produced as a percentage of mass of sample (data 6/data 5) 100% ______ %

ANALYSIS

1. Write a balanced equation for the reaction of each known carbonate with HCl.

____

____

____

2. For each known carbonate, calculate the mass of the sample and the number of moles used. Record these values.
3. Calculate and record the mass of the unknown carbonate.
4. The difference between the masses of the unreacted and reacted setups is, in each case, equal to the mass of carbon dioxide given off. For each known and unknown carbonate, determine the mass of CO_2 lost. Record the masses.
5. Determine and record the number of moles of CO_2 produced from each reaction of a known carbonate.
6. Determine the mass of CO_2 produced as a percentage of the mass of each carbonate. Record the percentages.

CONCLUSIONS

1. From the stoichiometric relationships in the reactions in Part 1, determine the number of moles of CO_2 that theoretically would be produced by each reaction.

____

____

____

2. How did the actual number of moles of CO_2 produced by each reaction with a known carbonate compare with the theoretical number of moles?

____

____

3. Because a substance reacts stoichiometrically, the outcome of a reaction can be used to identify an unknown compound. Compare the percentage of CO_2 in the unknown carbonate with the percentages of CO_2 in the three known carbonates. What is the unknown carbonate?

____

EXTENSION AND APPLICATION

Geologists are often interested in the percentage of carbonates in certain rocks and minerals. Find out why.

Teacher Guide

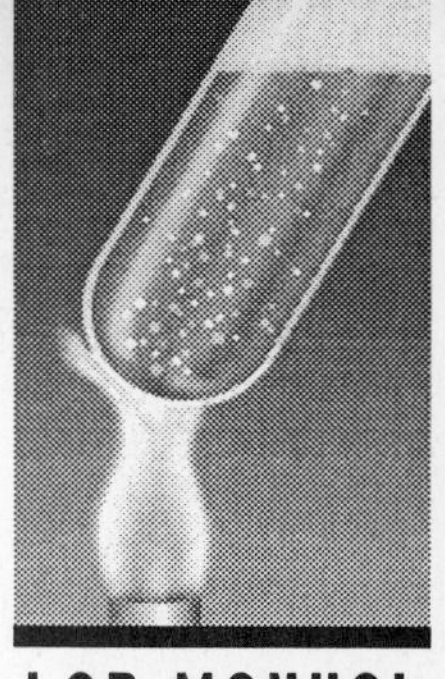

LAB MANUAL

12-2

Stoichiometry of a Chemical Reaction

In a reaction of a carbonate or hydrogen carbonate with a strong acid, a salt of the acid, carbon dioxide, and water are formed. In this experiment, the total mass of the reaction system is taken and the final mass of the reaction system is obtained. The only product that escapes is carbon dioxide. The amount of acid used is in excess; therefore, the reaction goes to completion. The amount of carbon dioxide produced depends on the mass of the carbonate used. The stoichiometric relationships in the reactions can be determined from these masses. Since the percentage of the compound that could generate carbon dioxide is dependent on the total molecular mass, the identity of the unknown compound can be inferred by finding the mass of carbon dioxide produced as a percentage of the mass of compound reacted and comparing this result with similar calculations for the known carbonates.

PROCESS SKILLS

1. Students will react three known carbonates and one unknown carbonate with an acid.
2. Students will measure the mass of one product of the reactions.
3. Students will analyze the stoichiometry of the reaction in terms of moles and percent composition.

PREPARATION HINTS

- Provide small carbonate samples of approximately 1.0 g mass for each group to use in Part 1.
- Provide one of the known carbonates without an identifying label for use as the unknown.
- For a class of 24 students, working in pairs, less than 200 mL of 8*M* HCl solution will be needed for four runs—three knowns and one unknown. The directions for preparing the HCl solution are given for 1 L of solution for convenience.

MATERIALS

The following carbonate and hydrogen carbonates:

$NaHCO_3$ (s)
$CaCO_3$ (s)
$KHCO_3$ (s)

You may substitute other carbonates and hydrogen carbonates as available.

8*M* HCl(aq) (Add 660 mL of concentrated HCl to 340 mL of distilled water. Add the acid to the water, not the reverse. Cool the solution as suggested in Preparation Hints.)

PROCEDURE HINTS

- Before proceeding, be sure students have read and understood the purpose, procedure, and safety precautions for this laboratory activity.
- It is important that students wear goggles and aprons throughout this laboratory activity.
- **Safety Precaution:** Handle the concentrated hydrochloric acid with care. Prepare the 8*M* HCl solution in a fume hood. Place the container with the distilled water in a basin or trough of cool tap water. Be sure to add the acid to the water, not the reverse.
- **Safety Precaution:** The reactions in this activity produce fizzing. Students may be tempted to get too close to the microplates. Remind them to keep faces away from the microplates and to follow all safety rules. Close supervision of students is necessary.
- **Disposal:** Dispose of materials according to instructions in **Disposal** J.

DATA AND OBSERVATIONS

Part 1: Data for known carbonates

	$NaHCO_3$	$CaCO_3$	$KHCO_3$
1. Mass of microplate	48.13 g	48.12 g	48.14 g
2. Mass of the microplate and sample	48.94 g	48.90 g	48.92g
3. Mass of complete unreacted setup	54.18 g	55.01 g	54.26 g
4. Mass of complete reacted setup	53.76g	54.66 g	53.88 g
5. Mass of sample	0.81 g	0.78 g	0.78 g
6. Moles of sample (mass/mol mass)	0.0096 moles	0.0078 moles	0.0078 moles
7. Mass of CO_2 lost (data 3 – data 4)	0.42 g	0.35 g	0.38 g
8. Moles of CO_2 lost (data 7 / 44 g/mole)	0.0095 moles	0.0080 moles	0.0086 moles
9. Mass of CO_2 produced as a percentage of mass of sample $\frac{\text{data 7}}{\text{data 5}} \times 100\%$	52%	45%	49%

Part 2: Data for unknown carbonate

$NaHCO_3$	
1. Mass of microplate	48.16 g
2. Mass of the microplate and sample	48.93 g
3. Mass of complete unreacted setup	55.67 g
4. Mass of complete reacted setup	55.27 g
5. Mass of sample	0.77 g
6. Mass of CO_2 lost (data 3 – data 4)	0.40 g
7. Mass of CO_2 produced as a percentage of mass of sample $\frac{\text{data 6}}{\text{data 5}} \times 100\%$	52%

ANALYSIS

1. $NaHCO_3(s) + HCl(aq) \rightarrow NaCl(aq) + CO_2(g) + H_2O(l)$

$CaCO_3(s) + 2HCl(aq) \rightarrow CaCl_2(aq) + CO_2(g) + H_2O(l)$

$KHCO_3(s) + HCl(aq) \rightarrow KCl(aq) + CO_2(g) + H_2O(l)$

2–6. See Data and Observations.

CONCLUSIONS

1. Each of the reactions with the known carbonates theoretically produce one mole of CO_2 for each mole of carbonate used.
2. For the known carbonates, approximately 0.01 mole of carbonate produced approximately 0.01 mole of CO_2. The 1:1 ratio of reactant to product is similar to the theoretical 1:1 mole ratio for the known carbonates.
3. The percentage of CO_2 in the unknown carbonate is 52%. The percentage of CO_2 in $NaHCO_3$ is also 52%. Therefore, the unknown carbonate is $NaHCO_3$.

EXTENSION AND APPLICATION

Rocks that contain a high percentage of carbonates are sedimentary rock, formed millions of years ago. Geologists know that these rocks often have deposits of oil or natural gas.

Name

Date Class

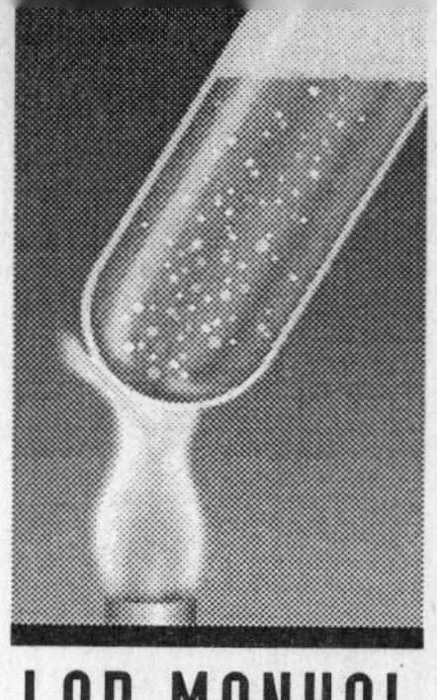

LAB MANUAL

12-3

Molar Volume of a Gas

It is important in chemistry to know how the volume, mass, and number of molecules in a measured amount of gas are related. The work of Amadeo Avogadro explains this relationship. Avogadro's principle states that at equal temperatures and pressures, equal volumes of gases contain equal numbers of particles. Because gases respond to changes in temperature and pressure, the conditions under which the volume of a gas is measured must be standardized. Standard temperature and pressure (STP) is defined as 0°C (273K) and 101.325 kPa (760 mm Hg or 1 atmosphere). At STP, 1 mole of any gas occupies 22.4 liters. This value is called the molar volume of a gas.

In this experiment, you will determine an experimental value for the molar volume of a gas.

OBJECTIVES

- **React** a metal and an acid to produce a gas.
- **Measure** the volume of the gas produced in the reaction.
- **Calculate** the molar volume of the gas.

MATERIALS

apron	metric ruler
goggles	plastic baby bottle
thermometer	400-mL beaker
barometer	graduated cylinder

PROCEDURE

1. Fill a 400-mL beaker about three-fourths full of tap water. Let the water come to room temperature. Check the temperature with the thermometer. Then record the temperature of the water in the Data and Observation section.
2. Record the barometric pressure from a barometer or get the value from your teacher.
3. From your teacher, obtain a piece of magnesium ribbon approximately 7 cm–10 cm long. Record the exact length in both cm and m.
4. Measure 15 mL of 6*M* HCl into a graduated cylinder. **CAUTION:** *Handle the acid solution with care. HCl is highly corrosive and burns the skin and damages clothing. Notify your teacher immediately of any acid spills.*
5. Slowly pour the acid into the baby bottle, letting the acid run down the side of the bottle.
6. Carefully add tap water from the beaker to the bottle. Again, add the water slowly, pouring it down the side of the bottle so that the water and acid mix as little as possible. Add water until you fill the bottle completely.
7. Add water to the beaker until it is one-half full.
8. Place the strip of magnesium ribbon into the nipple of the baby bottle. Being careful not to dislodge the magnesium ribbon, screw the top of the bottle in place.
9. Invert the bottle into the beaker of water.
10. Eventually the magnesium metal will begin to react with the hydrochloric acid. From time to time, tap the side of the baby bottle to dislodge bubbles of hydrogen gas that form.
11. When no more gas bubbles are generated, you know that the reaction is complete. Keeping the bottle inverted, read the volume of hydrogen directly from the bottle. Record the volume.
12. Empty the contents of the baby bottle and the beaker into the collection tub as directed by your teacher. Rinse the bottle and beaker and return them to your teacher. **CAUTION:** *The beaker contains hydrochloric acid. Handle the solution with care.*

DATA AND OBSERVATIONS

1. Temperature of water in the beaker ________ °C
2. Barometric pressure ________ kPa
3. Length of Mg ribbon (in cm) ________ cm
4. Length of Mg ribbon (in m) ________ m
5. Mass of 1 meter of Mg ribbon (obtain value from teacher) ________ g/m
6. Volume of H_2 collected ________ mL

ANALYSIS

1. Convert the water temperature to kelvins.

   Temperature (K) = data 1 + 273

   ____

2. Determine the mass of the piece of Mg ribbon.

   Mass = data 4 × data 5

   ____

3. Find the number of moles of Mg in your sample.

   $$\text{Moles} = \frac{\text{analysis 2}}{\text{molar mass of Mg}}$$

   ____

4. Calculate the pressure of dry H_2 as explained in Lab 12.1 on page 105.

Vapor Pressure of Water

°C	kPa	°C	kPa
15	1.7	22	2.6
16	1.8	23	2.8
17	2.0	24	3.0
18	2.1	25	3.2
19	2.2	26	3.4
20	2.3	27	3.6
21	2.5	28	3.8

$$\underset{(\text{dry } H_2)}{\text{Pressure}} = \underset{(\text{total})}{\text{pressure}} - \underset{(\text{water vapor})}{\text{pressure}}$$

____

5. Convert the volume of H_2 from milliliters (mL) to liters (L).

   Volume (L) = data 6/1000 mL/L

   ____

6. Calculate the volume of H_2 at STP.

   $$\text{Volume} = \frac{\text{Data 2} \times \text{Analysis 5} \times 273 \text{ K}}{\text{Analysis 1} \times 101.3 \text{ kPa}}$$

   ____

7. Write a balanced equation for the reaction of magnesium metal with hydrochloric acid.

   ____

8. Use the number of moles of magnesium to determine the number of moles of hydrogen gas produced by the reaction.

   ____

9. Using the volume of H_2 at STP that you calculated in Analysis 6, and the number of moles of H_2 that you determined in Analysis 8, calculate the experimental value for one mole of H_2 gas at STP.

   $$V_{1\text{ mol}} = \frac{\text{Analysis 6}}{\text{Analysis 8}}$$

CONCLUSIONS

The accepted value for the molar volume of a gas at STP is 22.4 L. Calculate your percentage error.

$$\% \text{ error} = \frac{\text{expected} - \text{observed}}{\text{expected}} \times 100\%$$

____

EXTENSION AND APPLICATION

What would your results have been if the magnesium metal had been coated with a heavy layer of oxide?

Teacher Guide

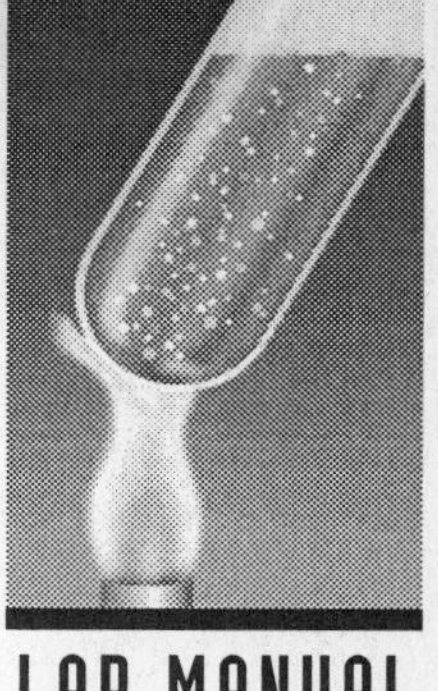

LAB MANUAL

12-3

Molar Volume of a Gas

This experiment is a quantitative way to relate the number of moles determined by mass to the number of moles of a gas actually produced in a chemical reaction. The number of moles of hydrogen produced depends on the number of moles of magnesium used.

PROCESS SKILLS

1. Students will measure the volume of the gas produced by the reaction of magnesium and hydrochloric acid.
2. Students will determine the relationship between the volume of the gas and the number of moles.

PREPARATION HINTS

- Store magnesium ribbon in a sealed plastic bag to prevent the reaction of the ribbon with water and oxygen in the air.
- Measure the mass of 1 m of the magnesium ribbon. Write the mass on the chalkboard for students to use in their calculations.
- When you prepare the 6*M* HCl, be sure to add the acid to the water and not the reverse.
- The baby bottle has several advantages over either a eudiometer or a buret. The bottle is safer, cheaper, and more convenient for students to use. Since the bottle is calibrated, students will be able to estimate the number of milliliters of hydrogen to the nearest 5 mL.

MATERIALS

Mg ribbon (cut to approximately 7 cm–10 cm lengths)

6M HCl (200 mL concentrated HCl to 200 mL water; allow to cool to room temperature.)

apron	metric ruler
goggles	plastic baby bottle
thermometer	400-mL beaker
barometer	graduated cylinder

PROCEDURE HINTS

- Before proceeding, be sure students have read and understood the purpose, procedure, and safety precautions for this laboratory activity.
- Students should make careful readings of the volume of gas collected.
- Be sure students wear safety goggles and lab aprons.
- **Safety Precaution:** Have a bottle of $NaHCO_3$ on hand to neutralize HCl spills.
- **Safety Precaution:** Be sure to remind students that, after the experiment, the bath water will contain a significant amount of hydrochloric acid. Caution them to handle this solution with care.
- **Safety Precaution:** Students should not have free access to the magnesium ribbon. Magnesium ribbon can be set on fire with a match. Magnesium fires burn intensely and cannot be put out with water or CO_2 fire extinguishers. Have sand available as an extinguishing agent.
- **Disposal:** Have a collection tub prepared for the bottles and beakers. Dispose of liquids according to instructions in **Disposal** C.

DATA AND OBSERVATIONS

1.	Temperature of water in the beaker	22°C
2.	Barometric pressure	755 mm Hg
3.	Length of Mg ribbon	5 cm
4.	Length of Mg in meters	0.05 m
5.	Mass of 1 meter of Mg ribbon	0.94 g/m
6.	Volume of H_2 collected	45 mL

ANALYSIS

1. $22°C + 273 = 295\ K$
2. The mass of the Mg is calculated as follows:
   data 4 × data 5 = 0.05 m × 0.94 g/m
   = 0.047 g Mg
3. The number of moles of Mg is found as follows:

$$\frac{\text{Analysis 2}}{\text{molar mass Mg}} = 0.047\ g/24\ g/mole$$

$$= 0.002\ mole$$

4. To convert the barometric pressure from millimeters of mercury (mm Hg) to kiloPascals (kPa):
   Data 2 × 0.133 kPa/mm Hg = 100.4 kPa
5. Pressure (dry H_2) = pressure (total) – pressure (water vapor)
   = analysis 4 – vapor pressure of water at data 1 temperature
   = 100.4 kPa – 2.6 kPa = 97.8 kPa
6. To convert the volume of H_2 from milliliters (mL) to liters (L):
   Volume (L) = data 6/1000 mL/L
   = 45 mL/1000 mL/L
   = 0.045 L
7. To calculate the volume of H_2 at STP:

$$\text{Volume} = \frac{\text{Analysis 4} \times \text{Analysis 6} \times 273\ K}{\text{Analysis 1} \times 101.3\ kPa}$$

$$= \frac{100.4\ kPa \times 0.045L \times 273\ K}{295K \times 101.3\ kPa}$$

$$= .04\ L$$

8. $Mg + 2HCl \rightarrow MgCl_2 + H_2$
9. Since the stoichiometry of the equation is 1 mole of Mg to 1 mole of H_2, you would expect to produce 0.0020 mole of H_2 when 0.0020 mole of Mg was used.
10.

$$n = \frac{\text{pressure} \times \text{volume}}{\text{gas constant} \times \text{temperature}}$$

$$n = \frac{\text{analysis 5} \times \text{analysis 6}}{8.31 \times \text{analysis 1}}$$

$$= \frac{97.5\ kPa \times 0.045\ L}{8.31\ L \bullet kPa/mol\ K \times 295\ K}$$

$$= 0.0018\ mol$$

CONCLUSIONS

$$\%\ \text{error} = \frac{\text{expected} - \text{observed}}{\text{expected}} \times 100\%$$

$$= \frac{0.0020 - 0.0018}{0.0020} \times 100\%$$

$$= 10\%$$

EXTENSION AND APPLICATION

If some of the sample of magnesium had been converted to magnesium oxide, the mass of magnesium available for the reaction would have been lower and the amount of hydrogen collected would have been smaller.

Copyright © Glencoe/McGraw-Hill, a division of The McGraw-Hill Companies, Inc.

Name

Date Class

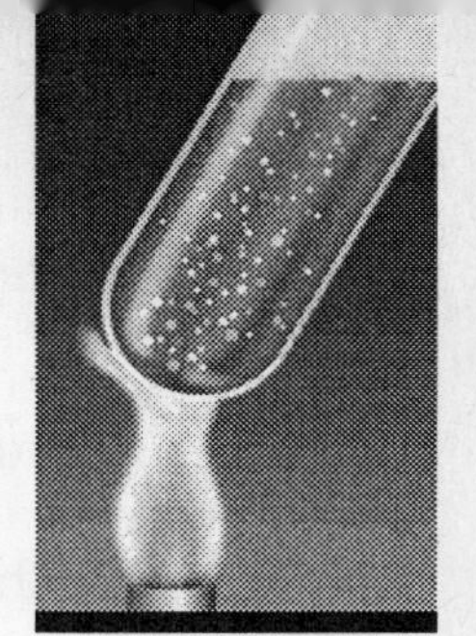

LAB MANUAL

13-1

Relating Solubility and Temperature

All substances dissolve to a greater or lesser degree in water. The amount of a given substance that dissolves in a given amount of water depends on the temperature of the water. The solubility of gases decreases with an increase in water temperature, but the solubility of solids usually increases with an increase in water temperature. There are exceptions, however. The solubility of cesium chloride, for example, decreases when water temperature increases. Also, temperature has little effect on the solubility of sodium chloride in water.

The ability of a solid to dissolve in water can be predicted by determining the enthalpy of solution of the solid. If heat is absorbed when a solute dissolves, then the enthalpy of solution (ΔH) is a positive value and solubility will be aided by an increase in temperature. If the enthalpy of solution of a solute is negative, a decrease in the temperature of a solution will aid solubility. Because most substances dissolve to a greater degree in water that is higher in temperature, the heat of solution for most substances is positive, and the process of solution is endothermic.

In this experiment, you will examine the relationship between the solubility of a salt and the temperature of the solution.

OBJECTIVES

- **Measure** the mass of a solute.
- **Construct** a graph of solubility vs. temperature.
- **Compare** the solubility of various solids.

MATERIALS

apron	10-mL graduated cylinder
goggles	50-mL graduated cylinder
aluminum foil	hot plate
scissors	thermometer
metric ruler	large test tube
marking pencil	distilled water
balance	forceps
glass stirring rod	
400-mL beaker	

PROCEDURE

1. Record under Data and Observations the name of the solid you have been assigned.
2. Cut aluminum foil into five 12-cm squares. Fold up the sides of the foil to form five dishes like the one shown in the diagram.

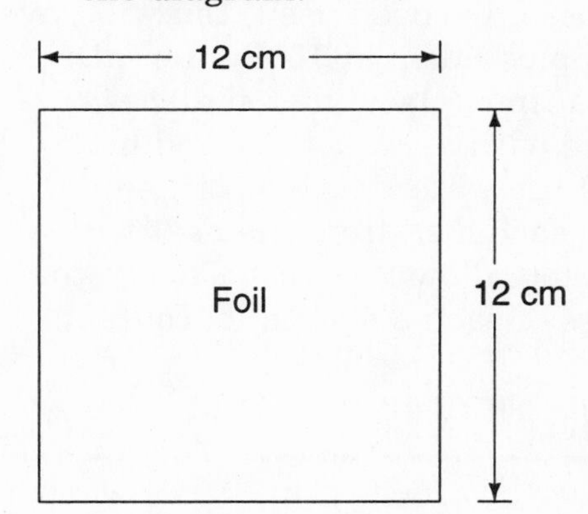

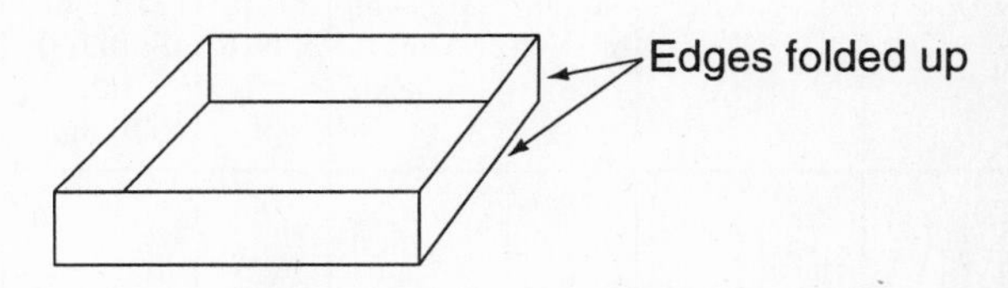

3. With the marking pencil, label the dishes 1 to 5.
4. Determine the mass of each dish. Record the masses in the table under Data and Observation.
5. Half fill a 400-mL beaker with water and begin heating the water to a temperature 10°C above room temperature.
6. Use the larger graduated cylinder to

measure and pour 25 mL of distilled water into the test tube.

7. Measure 20 g of solid and add it to the test tube. Stir gently with a glass rod until as much solid as possible has dissolved. Leave the rod in the test tube.
8. Measure the temperature of the solution in the test tube. Record the temperature in the table in the Dish #1 row.
9. Pour 5 mL of the solution into foil dish #1, using the smaller graduated cylinder.
10. Place the test tube containing the solution and the stirring rod in the beaker of water. Place the small graduated cylinder in the beaker as well.
11. When the water has reached 10°C above room temperature, stir and pour 5 mL of solution into the graduated cylinder in the beaker. Then transfer this sample into foil dish #2.
12. Increase the temperature of the water bath another 10°C. (Add small amounts of the same solid to the test tube to ensure that the solution remains saturated.)
13. Repeat steps 10 through 12 until all the foil dishes have been used. You will have five samples taken at 10°C intervals.
14. One at a time, place the foil dishes on a hot plate with a low setting and heat until all the water has evaporated.
15. Remove each dish from the heat with forceps and allow it to cool. Then record the mass of each dish and its contents.

DATA AND OBSERVATIONS

Solid ______________________

Dish #	Temperature °C	Mass of Dish (g)	Mass of Dish and Salt (g)	Mass of Salt (g)	Mass of Salt/100 mL (g/100mL)
1					
2					
3					
4					
5					

ANALYSIS

1. Calculate the mass of solid in each dish. Record these masses in the table.
2. Calculate the mass of salt in each dish per 100 mL of solution. (Multiply the number of grams of solid you recorded for each dish by 20.) Record these figures in the table.
3. Make a graph of your data. Place the temperature of the solution on the x-axis and the grams of solid (solute) per 100 mL on the y-axis.
4. Compare your graph with a graph of the solubility of your salt in a reference book. How did your results compare?

____

____

____

____

CONCLUSIONS

1. What is the relationship between the solubility of the salt and the temperature of the solution?

____

____

____

2. Compare your graph with one prepared by a group that used a different salt. Which salt was more soluble at room temperature ($\approx$20°C)?

____

____

EXTENSION AND APPLICATION

What is supersaturation? How does it differ from saturation?

Teacher Guide

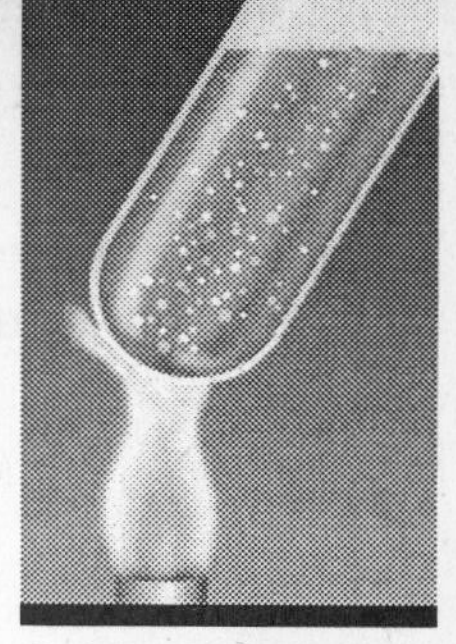

LAB MANUAL

13-1

Relating Solubility and Temperature

This laboratory activity can serve as an introduction to the concept of solubility. The procedure encourages students to work independently, but also requires them to share information in order to arrive at a common graph of solubilities versus temperature.

Misconception: Students are often surprised to learn that some solids dissolve more readily in cool water than in hot water.

PROCESS SKILLS

1. Students will prepare a saturated solution of a salt in water.
2. Students will measure the mass of salt dissolved at different temperatures.
3. Students will graph their results and compare the graphs of solubilities with published graphs of solubilities for the various salts.

PREPARATION HINTS

- Use only the solids suggested. Do not substitute organic solvents or solids. Organic compounds are highly flammable or explosive and may have toxic vapors.
- A lab burner and ring stand set-up can be used in place of a hot plate.

MATERIALS

samples of solid KCl, NaCl, $CaCl_2$

apron
goggles
aluminum foil
scissors
metric ruler
marking pencil
balance
glass stirring rod
400-mL beaker
10-mL graduated cylinder
50-mL graduated cylinder
hot plate (or burner with ring stand, ring and wire gauze)
thermometer
18 × 150 mm test tube, heat-resistant
distilled water
forceps

PROCEDURE HINTS

- Before proceeding, be sure students have read and understood the purpose, procedure, and safety precautions for this laboratory activity.
- The graduated cylinder remains in the water bath to prevent heat loss while the solution is measured out. Students should measure and pour out the solution as quickly as possible. Make sure they do not transfer any solid from the bottom of the test tube to the foil dishes.
- It may be helpful to have each student group keep a water bath at a specific temperature. Then students can move from bath to bath with their graduated cylinders, tubes of solution, and foil dishes.
- The solid that remains after students have completed the experiment may be recovered and used again.
- Comparison of solubilities is possible if different lab sections use different salts.

DATA AND OBSERVATIONS

Solid: KCl

Dish #	Temperature °C	Mass of Dish (g)	Mass of Dish and Salt (g)	Mass of Salt (g)	Mass of Salt/100 mL (g/100ml)
1	22	1.23	3.01	1.78	35.6
2	32	1.26	3.05	1.79	35.8
3	42	1.32	3.18	1.86	37.2
4	52	1.40	3.51	2.11	42.2
5	62	1.29	3.52	2.23	44.6

ANALYSIS

1–2. See sample data in the table.

3.

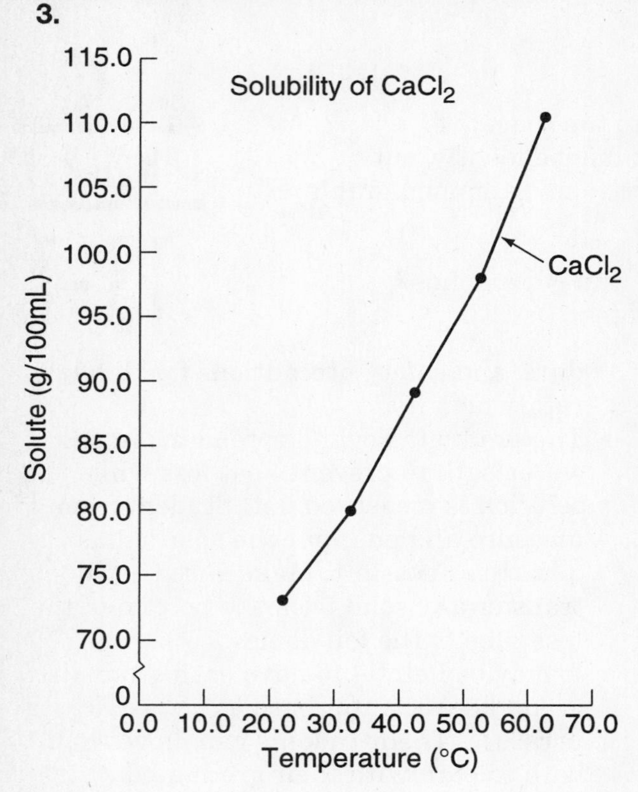

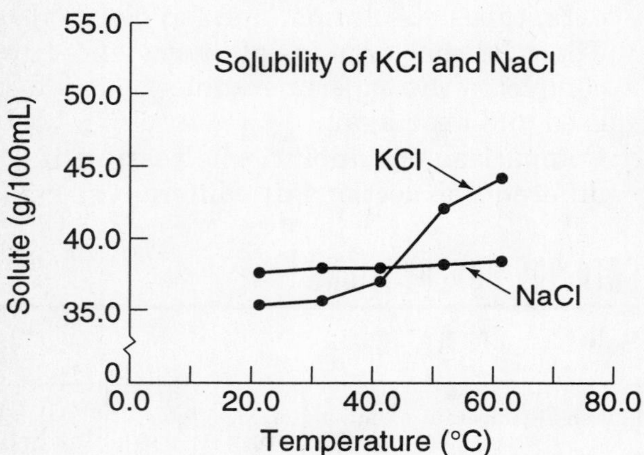

4. Answers will vary.

CONCLUSIONS

1. In general, the higher the temperature of the solution, the higher the solubility of the salt.
2. The most soluble salt was calcium chloride. The least soluble was potassium chloride.

EXTENSION AND APPLICATION

Supersaturation is a condition in which a solution contains more solute than it normally would hold at a particular temperature. A supersaturated solution can be prepared by slowly cooling a saturated solution. A single crystal of solute or even a dust particle can trigger the formation of crystals in the solution until the solution reaches the normal saturation point for that temperature.

Name

Date Class

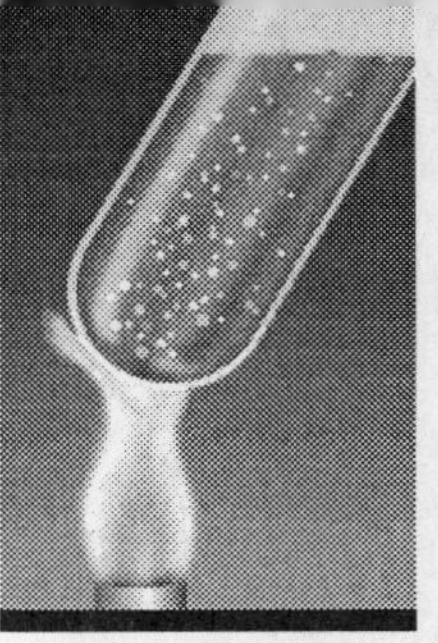

LAB MANUAL

13-2

The Effect of a Solute on Freezing Point

Pure water freezes at 0°C at standard atmospheric pressure. At this point, the vapor pressures of liquid water and solid water are the same. If there is a nonvolatile compound dissolved in the water, the solution will not freeze until the temperature is lower than 0°C. At a lower temperature, the vapor pressure of the solid will equal the lowered vapor pressure of the solution. Freezing point lowering, boiling point elevation, and osmotic pressure are all colligative properties, meaning that they are dependent only on the concentration of solute particles, not on their nature.

How far the freezing point of any solution is lowered depends on what the solvent is and on the ratio of solute particles to solvent particles. Therefore, in studying freezing point depression, the concentration of the solution is always expressed as the unit of molality, which is the number of moles of solute per kilogram of solvent. In this experiment, you will observe the effect of a solute on the freezing point of water.

OBJECTIVES

- **Observe** the freezing point depression of a solution.
- **Determine** the effect of solute concentration on the freezing point of a solvent.
- **Determine** the value of the molal freezing point constant for water.

MATERIALS

apron
goggles
notebook paper
shaved or crushed ice
rubber band
ketchup cups (2) with lid (1)
thermometer
paper punch
balance

PROCEDURE

1. **Hypothesize** what will happen to the freezing point of water as the concentration of sodium chloride added to the water is increased.
2. Prepare five samples of NaCl, each with a mass of 0.25 g, and place the samples on separate pieces of paper.
3. Construct a plastic calorimeter. Put a rubber band around one ketchup cup and then place this cup inside a second ketchup cup. With a paper punch make a hole in the lid of a ketchup cup.
4. Measure and record the mass of the empty calorimeter and its lid under Data and Observations.
5. Prepare a data table like the one in the Data and Observations section. Your table should have columns for Readings 1 through 6.
6. From this point on, you must work quickly. Read steps 7–12 so that you will be prepared for action.
7. Fill the calorimeter with crushed ice and replace the lid. Remove some ice if the top does not fit snugly.
8. Measure the mass of the calorimeter with its lid and the ice. Record this mass.
9. Insert the thermometer through the hole in the calorimeter top. Measure the temperature of the crushed ice to the nearest 0.1°C and record it in your data table under Reading 1.
10. Open the calorimeter and add one of the prepared NaCl samples. Replace the cover and thermometer. Swirl the calorimeter to mix the contents. Continue until the NaCl is completely dissolved. Remove the cover very briefly to check.
11. Record in your data table the lowest temperature of the mixture and the total mass of NaCl in the mixture for this reading (Reading 2).

12. Repeat steps 10 and 11 with each of the remaining NaCl samples. Observe the ice as more NaCl is added.

DATA AND OBSERVATIONS

Hypothesis: ______________________

____

____

1. Mass of calorimeter (with lid) ______ g
2. Mass of calorimeter and ice ______ g

	Reading 1	Reading 2	Reading 3
NaCl mass (grams)			
Temperature (°C)			
Moles of NaCl			
Moles of ions			
Moles of ions/kilogram of ice			

ANALYSIS

1. For each reading, convert grams of NaCl to the number of moles of NaCl and record the results in your data table.
2. For each reading, calculate the number of moles of ions (moles NaCl × 2 ions/mole). Record the results.
3. Calculate the mass of ice you started with, in grams, and convert to kilograms.

   ______________________ kg
4. Complete your data table for moles of ions per kilogram of ice in each reading.
5. Construct a graph of your results. Plot moles of ions per kilogram of ice on the *x*-axis and the temperature of the mixture on the *y*-axis. Draw the best-fitting straight line through your points. Find the slope of the straight line. What are the units of the slope of the line?

   ____

CONCLUSIONS

1. What happened to the temperature of the mixture as NaCl was added? Was your hypothesis correct? Explain.

   ____

2. What happened to the amount of ice remaining as NaCl was added?

   ____

3. The absolute value of the slope of the graph you plotted is the molal freezing point constant. Compare your result with the actual value of 1.853 C°/molal solution.

   ____

EXTENSION AND APPLICATION

1. Explain why molality instead of molarity is used in studying colligative properties.
2. What would happen if you repeated the experiment with ethyl alcohol instead of NaCl? What would happen if you used calcium chloride, $CaCl_2$? Why is NaCl not used as an antifreeze in automobile radiators?

Teacher Guide

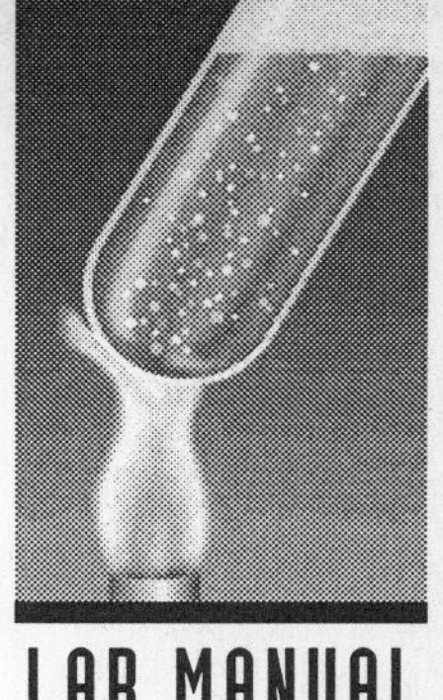

LAB MANUAL

13-2

The Effect of a Solute on Freezing Point

The addition of any soluble substance to a liquid lowers the freezing point of the liquid. Freezing point depression is a colligative property, meaning that it is dependent only on the concentration of solute particles, whether ions or molecules, and not on their nature. The action of antifreeze in automobile radiators utilizes this principle.

The constant that defines the amount of change in the freezing point of a pure liquid is called the molal freezing point constant (K_{fp}). The value of this constant is different for each solvent. For water, the molal freezing point constant is 1.853 C°/molal solution.

PROCESS SKILLS

1. Students will observe the change in temperature of an ice-water mixture as a solute is added.
2. Students will plot experimental data and find the slope of a straight line.
3. Students will determine the molal freezing point constant for water.

PREPARATION HINTS

- Do not use ice cubes. Shaved or crushed ice is required for this activity.
- Ketchup cups, also called soufflé cups, are available with lids from restaurant supply stores.

MATERIALS

NaCl	ketchup cups (2) with lid (1)
apron	thermometer (showing 0.1C°)
goggles	paper punch
notebook paper	balance
shaved or crushed ice	
rubber band	

PROCEDURE HINTS

- Before proceeding, be sure students have read and understood the purpose, procedure, and safety precautions for this laboratory activity.
- **Troubleshooting:** Warn students not to overload the calorimeter with ice. The top must fit securely on the calorimeter.
- **Troubleshooting:** Students must work relatively quickly with the ice/calorimeter combination. If they work too slowly so that no ice remains in the calorimeter, the addition of more NaCl will not give the expected decrease in temperature.
- **Troubleshooting:** Be sure that students use the best-fitting straight line to connect the points of the graph.
- **Troubleshooting:** Be sure students read thermometers to the nearest 0.1C°.

DATA AND OBSERVATIONS

Students' hypotheses will vary, but should indicate that the freezing point will continue to decrease as the concentration of NaCl increases.

1. Mass of calorimeter (with lid) 2.34 g
2. Mass of calorimeter and ice 17.44 g

Students' tables should resemble the table on the next page.

ANALYSIS

1. See the table on the next page.
2. See the table.
3. Mass of ice (at start) in kg: 0.0151 kg
4. See the table on the next page.

	Reading 1	Reading 2	Reading 3	Reading 4	Reading 5	Reading 6
NaCl mass (grams)	0	0.25	0.50	0.75	1.00	1.25
Temperature (°C)	–1	–2.2	–3.0	–4.2	–5.0	–6.1
Moles of NaCl	0	0.0043	0.0085	0.013	0.017	0.021
Moles of ions	0	0.0086	0.017	0.026	0.034	0.042
Moles of ions/kilogram of ice	0	0.57	1.1	1.7	2.3	2.8

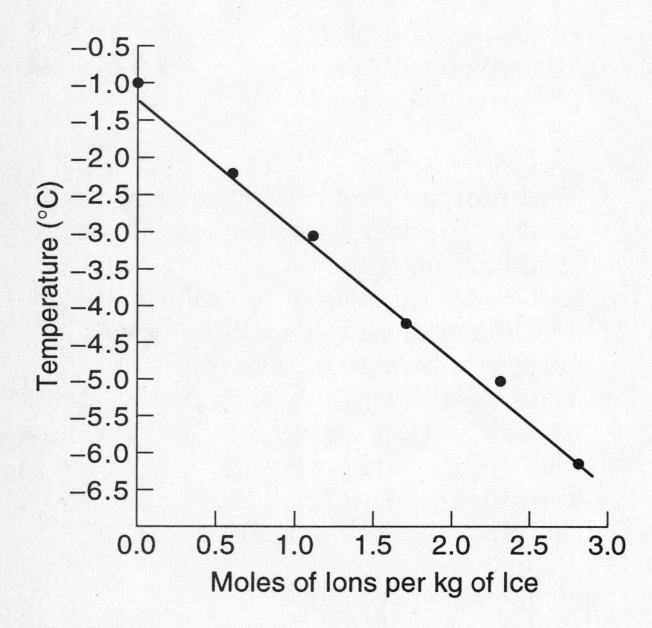

5. The slope of the straight line in the sample plot is –1.82. The units for the slope are C°/moles of ions per kilogram.

CONCLUSIONS

1. As NaCl was added, the temperature of the mixture decreased. Students' hypotheses may vary.
2. As more NaCl was added, more ice melted.
3. Students' answers will vary depending on the accuracy of their results.

EXTENSION AND APPLICATION

1. Freezing point depression depends on the ratio of solute ions or molecules to solvent molecules, which is also given by the ratio of moles of solute to moles of solvent. Molality gives the ratio of the number of moles of solute per kg of solvent, which is a fixed number of moles of solvent. Molarity gives the moles of solute in a fixed volume of solution, but the number of solvent molecules is not known and can vary with concentration.
2. With ethyl alcohol, the freezing point depression would be one-half as much per mole of solute, because there is only one mole of ethyl alcohol particles per mole of solute instead of two moles of particles per mole of NaCl. With $CaCl_2$, the freezing point depression would be 1.5 times larger per mole of solute since there are three moles of particles per mole of $CaCl_2$ instead of two moles of particles per mole of NaCl. If NaCl were used in antifreeze, the radiators would corrode.

Name

Date Class

Acidic and Basic Anhydrides

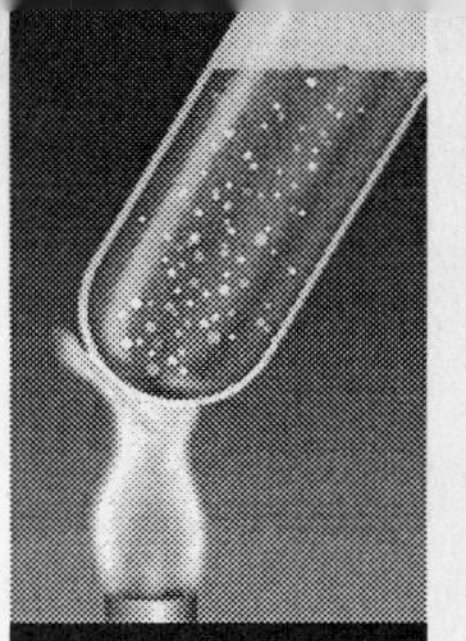

LAB MANUAL

14-1

When dissolved in water, the oxide of a metal forms a basic solution. The oxide of a nonmetal when dissolved in water forms an acidic solution. Unless water is added to them, the oxides of both metals and nonmetals are anhydrides, that is, compounds without water. Because a metallic oxide in water forms a basic solution, a metallic oxide is called a basic anhydride. The purpose of this activity is to form acidic and basic solutions by using an acidic anhydride and a basic anhydride.

OBJECTIVES

- **Prepare** a basic solution using a basic anhydride.
- **Prepare** an acidic solution using an acidic anhydride.
- **React** the acidic and basic solutions to form a salt.

MATERIALS

thin-stem pipet	tap water
microtip pipets	scissors
24-well microplate	toothpick
24-well template	clear tape
marble chips	chemical scoop

Use the thin-stem micropipet for CO_2 generator and microtip pipets when drops are added.

PROCEDURE

1. Form a **hypothesis** about whether the solutions formed by dissolving calcium oxide in tap water and a carbon oxide in tap water will be acidic or basic. Write your hypothesis under Data and Observations.
2. Use scissors to make a slit in the bulb of a thin-stem pipet, as shown in Figure A.

Thin-stem pipet

Cut here

Figure A

3. Push a small piece of marble through the slit into the bulb.
4. Seal the slit with a piece of clear tape.

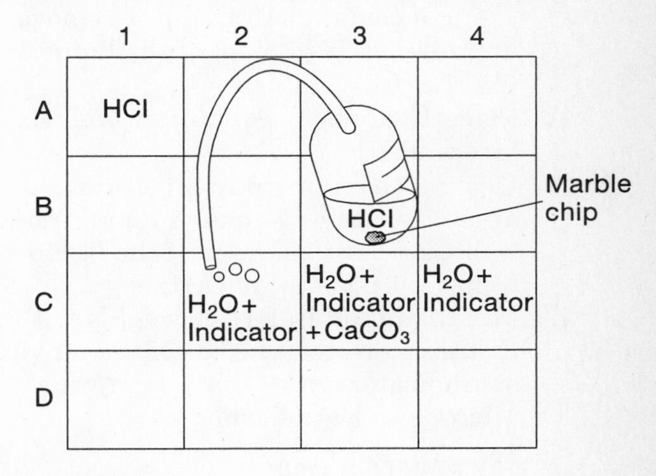

Figure B

5. Label a 24-well template and a Microplate Data Form as shown in Figure B. Place the microplate on the template with the numbered columns away from you and the lettered rows to the left.
6. Using a clean microtip pipet, place ½-pipet of tap water in wells C2, C3, and C4 of the microplate.
7. Add 2 drops of universal indicator to wells C2, C3, and C4. Note the color of the universal indicator and the approximate pH in your Microplate Data Form.
8. Using a clean microtip pipet, place ½-pipet of HCl in well A1 of the microplate. **CAUTION:** *HCl is corrosive. It burns skin and clothing.*
9. Squeeze the bulb of the thin-stem pipet containing the piece of marble to eject

the air from the pipet. Continue to hold the bulb so that air cannot enter the pipet.

10. Invert the thin-stem pipet and place the stem into well A1. Release the bulb of the pipet and draw up approximately ¼-pipet of HCl into the bulb.
11. Place the bulb of the thin-stem pipet in well B3 and direct the stem into well C2.
12. Allow the gas produced to bubble through well C2. Note the color of the universal indicator-water solution in your Microplate Data Form.
13. Using a chemical scoop, place a few crystals of calcium oxide into well C3. Note the color of the universal indicator-water solution in your Microplate Data Form.
14. Allow the solid to settle out in well C3. Using a clean microtip pipet, remove approximately ⅛-pipet of the liquid from well C3.
15. Place 10 drops of liquid from well C3 into well D1.
16. Using a clean microtip pipet, remove approx-imately ⅛-pipet of liquid from well C2. Place 10 drops of the liquid from well C2 into well D1.
17. Mix the liquid in D1 thoroughly with a toothpick. Note the color of the universal indicator-water solution in your Microplate Data Form.

DATA AND OBSERVATIONS

Hypothesis: ______________________

____

1. What color was the universal indicator in tap water alone?

____

2. What was the color of the universal indicator when the gas was bubbled through the water in well C2?

____

3. What was the color of the universal indicator when calcium oxide was added to the water in well C3?

____

4. What happened to the universal indicator when the two solutions were mixed in well D1?

____

ANALYSIS

1. Balance this equation, which describes the reaction in the micropipet containing the marble chip.

$$\_\_\_ CaCO_3 + \_\_\_ HCl \rightarrow \_\_\_ CO_2 + \_\_\_ H_2O + \_\_\_ CaCl_2$$

2. Balance this equation, which describes what happened to the water in well C2.

$$\_\_\_ CO_2 + \_\_\_ H_2O \rightarrow \_\_\_ H_2CO_3$$

3. Balance this equation, which describes what happened when calcium oxide was added to the water in well C3.

$$\_\_\_ CaO + \_\_\_ H_2O \rightarrow \_\_\_ Ca(OH)_2$$

4. Balance this equation, which describes the reaction in well D1.

$$\_\_\_ Ca(OH)_2 + \_\_\_ H_2CO_3 \rightarrow \_\_\_ CaCO_3 + \_\_\_ H_2O$$

CONCLUSIONS

1. Is CO_2 an acidic or a basic anhydride?

____

2. Is CaO an acidic or a basic anhydride?

____

3. a. Is the product of the chemicals from wells C2 and C3 acidic, basic, or neutral?

____

b. Explain your answer.

____

____

EXTENSION AND APPLICATION

1. When a salt is dissolved in water, the salt solution almost never has a neutral pH. Why?
2. How might the pH of a solution be changed by a salt?

Teacher Guide

LAB MANUAL

14-1

Acidic and Basic Anhydrides

When the oxide of a nonmetal or a metal is placed in water, the water reacts with the oxide to form an acidic solution or a basic solution, respectively.

Misconceptions: Students often assume that salt solutions always have a neutral pH. Students seldom realize that water can react with a salt to form an acid and a base in solution. Students often interpret an indicator color change to be no reaction. Point out that a reaction to form an acid or a base has taken place.

PROCESS SKILLS

1. Students will form an acidic solution and a basic solution from an acid anhydride and a basic anhydride.
2. Students will observe the effect of the anhydride in water on the pH of the solution.
3. Students will perform a neutralization reaction in which a salt and water are formed.

PREPARATION HINTS

- 2*M* HCl is prepared by dissolving 20 mL of concentrated HCl in 100 mL of distilled water.
- Check your stock of marble chips to be certain they are of an appropriate size for the microplate.
- For each group, make a copy of the 24-well template on p. T6.

MATERIALS

universal indicator	24-well template
calcium oxide (CaO)	tap water
2*M* HCl	scissors
marble chips ($CaCO_3$)	toothpick
thin-stem pipet	clear tape
microtip pipets	chemical scoop
24-well microplate	

Use the thin-stem micropipet for CO_2 generator and microtip pipets when drops are added.

PROCEDURE HINTS

- Before proceeding, be sure students have read and understood the purpose, procedure, and safety precautions for this laboratory activity.
- Small pieces of marble are desirable. Do not use powdered calcium carbonate in the micropipet.
- The solution of carbonic acid will not stay acidic for a long time. Advise students to use the solution from well C2 as soon as possible.
- The water in well C4 is used for purposes of comparison. Because carbon dioxide in the air can change the pH and the indicator color in a short time, do not save the solution in well C4 for another class.
- **Disposal:** Dispose of excess HCl according to instructions in **Disposal** D, the marble chips according to **Disposal** F, and the microplate materials according to **Disposal** J.

DATA AND OBSERVATIONS

Students' hypotheses will vary. A correct hypothesis is that calcium oxide will form a basic solution and carbon oxide will form an acidic solution.

1. The universal indicator in water alone should show a color indicating a relatively neutral pH. The color of the indicator is usually green or yellow-green depending on the pH of the local water supply.

Copyright © Glencoe/McGraw-Hill, a division of The McGraw-Hill Companies, Inc.

2. When carbon dioxide is bubbled through water, the weak acid, carbonic acid is formed and the universal indicator turns yellow or red.
3. When calcium oxide is added to water, the weak base calcium hydroxide is formed and the universal indicator turns blue.
4. When the carbonic acid solution from well C2 is mixed with the calcium hydroxide solution from well C3, the acid and the base react to form a salt and water. A green or yellow-green color indicates a neutral solution.

	1	2	3	4
C		Before CO_2 green pH-7 / After CO_2 yellow pH<7	Before CaO green pH-7 / After CaO blue pH>7	Before CONTROL pH-7 / After CONTROL pH=7
D	Mixture of C_2 and C_3			

ANALYSIS

1. $CaCO_3 + 2HCl \rightarrow CO_2 + H_2O + CaCl_2$
2. $CO_2 + H_2O \rightarrow H_2CO_3$
3. $CaO + H_2O \rightarrow Ca(OH)_2$
4. $Ca(OH)_2 + H_2CO_3 \rightarrow CaCO_3 + 2H_2O$

CONCLUSIONS

1. CO_2 is an acidic anhydride.
2. CaO is a basic anhydride.
3. **a.** The product from reacting the contents of wells C2 and C3 is neutral.
   b. An acidic solution neutralizes a basic solution:

   acid + base → salt + water

EXTENSION AND APPLICATION

1. When a salt dissolves in water, the ions that make up the salt react with water molecules in a process called hydrolysis.
2. The ions in the salt might alter the balance of H^+ and OH^- in the solution.

Copyright © Glencoe/McGraw-Hill, a division of The McGraw-Hill

Name

Date Class

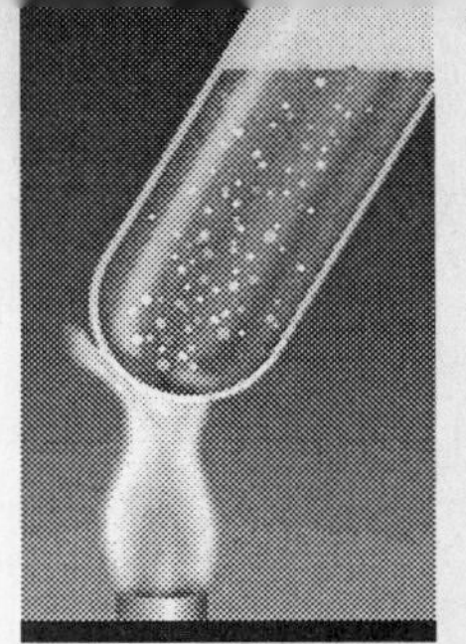

Using Indicators to Determine pH

Many kinds of solutions must be tested to determine whether they are acidic, basic, or neutral. The pH scale represents the hydronium ion concentration of a solution and is used to indicate how acidic or basic a solution is. The normal pH range is from the highly acidic pH 1 to the neutral pH 7 to the highly basic pH 14. The most accurate method of pH determination is by use of an electronic instrument known as a pH meter. For simpler though less accurate pH determination, indicators are commonly used. pH indicators are organic compounds that change color with changing pH. Some indicators are mixtures of compounds, each of which changes color at a different pH. Indicators are chosen for specific applications according to the pH range in which they show a color change. In this experiment, you will observe the variation in color with pH of several indicators.

OBJECTIVES

- **Prepare** solutions of varying pH.
- **Observe** the color changes of pH indicators in these solutions.
- **Determine** the pH range of each indicator.

MATERIALS

apron
goggles
96-well microplate
96-well template
microtip pipets (3)
100-mL beaker
distilled water

PROCEDURE

1. Label a 96-well template and a Microplate Data Form as described under Data and Observations. Place a microplate on the template with the numbered columns away from you and the lettered rows to the left.
2. Fill a 100-mL beaker half-full with distilled water.
3. Place 9 drops of distilled water in each of wells 2 through 11 in rows A through D.
4. Add 10 drops of 0.1*M* HCl to each of wells A1, B1, C1, and D1.
5. Using a clean microtip pipet, add 10 drops of 0.01*M* NaOH to each of the wells A12, B12, C12, and D12.
6. Remove 1 drop of acid solution from well A1 and add it to well A2 with your acid pipet. Mix the contents of well A2 by drawing up the contents of the well into the pipet and returning the liquid immediately to well A2.
7. Remove 1 drop of solution from well A2 and place it in well A3. Mix well A3 in the same way that you did well A2.
8. Repeat the dilution and mixing process of steps 6 and 7 through well A6, *but do not transfer any solution to well A7.*
9. Repeat the acid dilution process used in row A for the wells in rows B, C, and D. Do not transfer any solution to well 7 in any row.
10. Use a clean pipet to remove 1 drop of solution from well A12 and place it in well A11. Mix as in step 6.
11. Repeat the dilution and mixing process of step 10 from well A12 back through well A8, *but do not add any solution to well A7.*
12. Repeat the base dilution process used in row A for the wells in rows B, C, and D. The rows of solutions now have H^+ concentrations that vary by a factor of 10. Each row of wells approximates the pH scale from 1 to 12. Well 4 of each row contains a solution that is about pH 4, well 9 of each row contains a solution that is about pH 9, and so on.
13. Add 1 drop of universal indicator to each well in row A.

Copyright © Glencoe/McGraw-Hill, a division of The McGraw-Hill Companies, Inc.

14. Add 1 drop of bromothymol blue to each well in row B.
15. Add 1 drop of phenolphthalein to each well in row C.
16. Finally, add 1 drop of thymol blue to each well in row D.
17. Observe the wells in each row. Record your observations of colors on your Microplate Data Form. Discard the solutions in your microplate according to your teacher's instructions and rinse the microplate.

DATA AND OBSERVATIONS

Label the rows of your Microplate Data Form as follows.
Row A: Universal indicator
Row B: Bromothymol blue
Row C: Phenolphthalein
Row D: Thymol blue

ANALYSIS

1. For each indicator, list the color changes and the pH range or ranges at which they occur.

   Universal indicator:

   Bromothymol blue:

   Phenolphthalein:

   Thymol blue:

2. Which indicators changed color at more than one pH?

CONCLUSIONS

1. Which of the indicators tested is a good acid pH range indicator? Why?
2. Which of the indicators tested is a good basic pH range indicator? Why?
3. Which of the indicators tested is the best choice when the acidity or basicity of a solution is not known? Why?
4. For what reason do you think universal indicator is called *universal*?

EXTENSION AND APPLICATION

1. Suggest a reason for the common use of bromothymol blue to test the pH of water in swimming pools.
2. Phenolphthalein is not a good indicator to use in testing the pH of water in a fish tank. Suggest a reason why.
3. What characteristics are necessary for a universal indicator? Based on your observations in this experiment, suggest a composition for a universal indicator.

Copyright © Glencoe/McGraw-Hill, a division of The McGraw-Hill Companies, Inc.

Teacher Guide

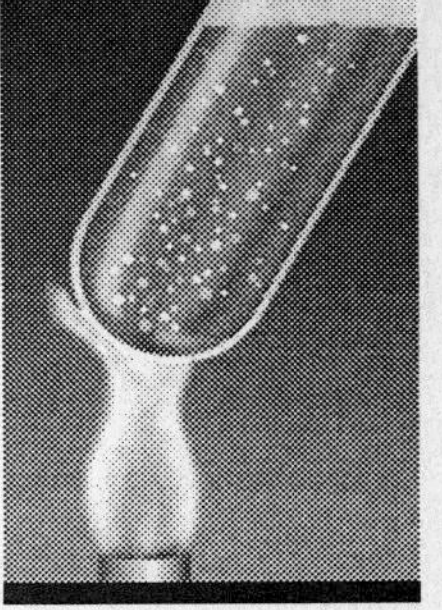

LAB MANUAL

14-2

Using Indicators to Determine pH

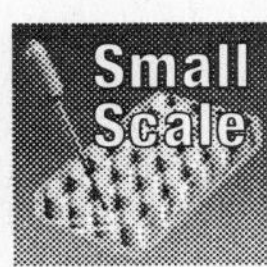

The relative acid or base concentration of a solution can be tested in several ways. The advantages and disadvantages of each method should be emphasized in preparation for this experiment. Students should infer from the experiment that a universal indicator is a mixture of indicators chosen so that there are color changes over the entire pH range.

Misconceptions: Students do not tie indicators to everyday applications outside of the laboratory. They do not realize that testing the pH of solutions, such as swimming-pool water, blood plasma, and urine, is often done by using indicator papers.

PROCESS SKILLS

1. Students will prepare series dilutions of acids and bases to cover the entire pH range.
2. Students will observe the color changes of several indicators.
3. Students will determine the pH ranges in which the indicators are useful.

PREPARATION HINTS

- Use pipets to transfer small quantities of HCl and NaOH solutions to student supply bottles.
- Phenolphthalein is a powerful laxative. Do not allow solid phenolphthalein to come into contact with skin. Wear rubber or plastic gloves when preparing phenolphthalein solution.
- For each group, make a copy of the 96-well template on page T8.

MATERIALS

0.1*M* HCl (Dilute 8.6 mL of concentrated HCl to 1 L of solution, or dilute 1*M* HCl 10:90 to make 0.1 L.)

0.01*M* NaOH (Dissolve 0.4 g of sodium hydroxide pellets in enough water to make 1 L of solution, or dilute 0.1*M* NaOH 10:90 with distilled water to make 0.1 L.)

2% phenolphthalein (Dissolve 2 g of solid phenolphthalein in a mixture of 50 mL of ethyl alcohol and 50 mL of distilled water.)

bromothymol blue (Dissolve 0.1 g of bromothymol blue in a solution made by combining 16 mL of 0.01*M* NaOH and 234 mL of distilled water. Note: If your bromothymol blue is in the form of sodium salt, then dissolve 0.1 g in 250 mL of distilled water.)

thymol blue (Dissolve 0.1 g of thymol blue in a solution made by combining 21.5 mL of 0.01*M* NaOH and 228.5 mL of distilled water. Note: If your thymol blue is in the form of sodium salt, then dissolve 0.1 g in 250 mL of distilled water.)

universal indicator (commercial preparation)

apron
goggles
96-well microplate
96-well template
microtip pipets (3)
100-mL beaker
distilled water

PROCEDURE HINTS

- Before proceeding, be sure students have read and understood the purpose, procedure, and safety precautions for this laboratory activity.
- Phenolphthalein, bromothymol blue, and thymol blue can be combined to make a universal indicator. Students enjoy making their own indicator and should be encouraged to do so.
- The indicator solutions can be saved in the wells for several weeks as reference for future laboratory experiments. Run a continuous strip of clear tape over the top of a row of wells. Run your finger over the top of the wells to seal in the liquid.

DATA AND OBSERVATIONS

		1	2	3	4	5	6	7	8	9	10	11	12
Universal indicator	A	red	red	orange	yellow	yellow	green	green	green	blue/ green	blue	← violet →	
Bromothymol blue	B	←		yellow		→	green	green	green	←		blue	→
Phenolphthalein	C	←			colorless			→	←	slight pink	→	pink	pink
Thymol blue	D	←	red	→	←		yellow		→	←		blue	→

- **Disposal:** All solutions should be collected in a wash basin or trough. Dispose of the solutions according to the instructions in **Disposal** J.

ANALYSIS

1. Universal indicator: Several color changes, usually red, to orange, to yellow in the pH 1–5 range; green at pH 6–8; blue and violet at pH 9–12.
   Bromothymol blue: yellow to green at pH 6; green to blue at pH 9.
   Phenolphthalein: colorless to pink at pH 8–10.
   Thymol blue: red to yellow at pH 1–3; yellow to blue at about pH 9.
2. Universal indicator gives several color changes across the entire pH range. Bromothymol blue changes from yellow in acidic solution, to green in neutral solution, and then to blue in basic solution. Thymol blue changes from red in acidic solution, to yellow in neutral solution, and then to blue in basic solution.

CONCLUSIONS

1. Thymol blue is a good indicator for the acid pH range because it changes color in this range.
2. Phenolphthalein is a good indicator for the basic pH range because it changes color only in this range.
3. Universal indicator is the best choice to determine the pH of a solution of unknown pH because it has distinctive colors for several different pHs.
4. The word *universal* indicates a broad range of applications and shows that this indicator can be used at any pH.

EXTENSION AND APPLICATION

1. Bromothymol blue is the best indicator for swimming-pool water because it can indicate a neutral solution. Swimming-pool water must have a near-neutral pH to be safe for human use.
2. Phenolphthalein shows a color change only in basic solutions of pH 9 or higher. Most likely, fish will not stay healthy in such solutions but require neutral or slightly acidic solutions.
3. Because most indicators change color at only one pH or a narrow pH range, a universal indicator must be a mixture of indicators that show different colors over the entire pH range. A mixture of the three indicators in this experiment could be used as a universal indicator with color changes at acidic, neutral, and basic pH. Note, however, that bromothymol blue exhibits two color changes. The second is at pH 8–9.5. Thymol blue also exhibits two color changes.

Name

Date Class

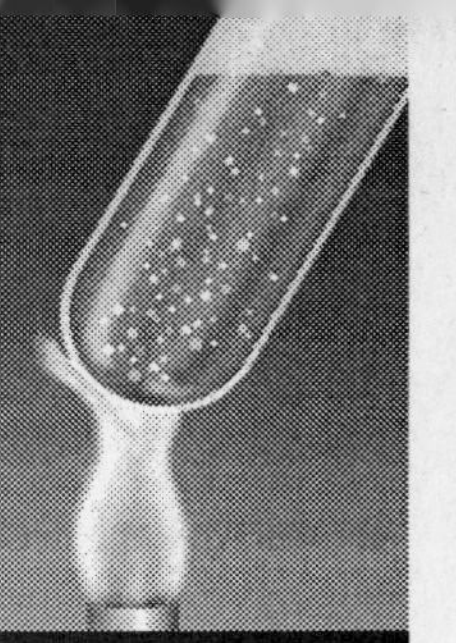

LAB MANUAL

15-1

Hydrolysis of Salts

Many of the reactions you have carried out have been in a water solution. While water is a diluting agent, it is also a part of the chemistry itself. When salts dissolve in water, you might expect the solution to be neutral. Many salt solutions, such as those of $NaNO_3$ and KCl, are neutral. Other salts, when dissolved in water, form solutions that are either acidic or basic.

When salts dissolve in water, they dissociate to form metal and nonmetal ions. The metal and nonmetal ions can interact with water molecules to form an acidic or basic solution in a reaction called hydrolysis.

For example, the salt sodium acetate, $NaC_2H_3O_2$, dissociates in water according to the equation

$$NaC_2H_3O_2 \rightarrow Na^+ + C_2H_3O_2^-$$

In this example, the Na^+ ions are from a strong base, NaOH, so they will not react with the water. The acetate ions, $C_2H_3O_2^-$, are from a weak acid, acetic acid, and will react with the water to form acetic acid and OH^- ions. The presence of OH^- ions produces a basic solution.

In general, a salt made from a strong base, such as KOH and NaOH, and a weak acid dissolves in water to form a basic solution. A salt made from a strong acid, such as HCl, HNO_3, and H_2SO_4, and a weak base dissolves in water to form an acidic solution. Salts made from strong acids and strong bases, as well as salts made from weak acids and weak bases, dissolve in water to form neutral solutions.

In this experiment, you will dissolve many salts in water and determine in each case whether an acidic, basic, or neutral solution is formed.

OBJECTIVES

- **Determine** the pH of a series of salt solutions.
- **Deduce** the relative strengths of the acid and base when each salt dissolves in water.
- **Write** a balanced equation for each hydrolysis reaction.

MATERIALS

apron
goggles
small test tubes (6)
solid rubber stoppers to fit test tubes (6)
test-tube rack
marking pencil
spatula
10-mL graduated cylinder
thin-stem pipet
universal indicator
distilled water
chart of pH and indicator colors
paper towels

PROCEDURE

1. The table under Data and Observations lists the formulas of the twelve salts you will be using. Using your knowledge of hydrolysis, form a **hypothesis** about which of the salt solutions you will prepare will be acidic, which will be basic, and which, if any, will be neutral. Record your hypothesis under Data and Observations. Use your hypothesis to make the predictions called for in the second column of the table. Write your predictions in the table.
2. Set up a test-tube rack with six test tubes in a row.
3. Use the marking pencil to number the test tubes to correspond to the numbers of the first six salts in the table.
4. Use the spatula to place a small amount of the first six salts in the corresponding test tubes 1–6. Put one salt in each tube.

Clean the spatula with a paper towel between each use. **CAUTION:** *$Pb(NO_3)_2$ is toxic*.

5. To each of these test tubes, add 10 mL of distilled water.
6. Close each of the tubes with a rubber stopper, and shake each tube until the solid salt is dissolved completely.
7. Remove the rubber stoppers. Then use the thin-stem pipet to add 3 drops of universal indicator to each test tube.
8. Use the chart of pH and indicator colors to determine, and record in the table, whether each of the solutions in test tubes 1–6 is acidic, basic, or neutral. Then record the approximate pH of each solution.
9. Discard the solutions as directed by your teacher. Rinse the test tubes and stoppers with distilled water and discard the rinse water in the same way as the solutions.
10. Repeat steps 2–9 with the last six solutions.
11. Some of the 12 salts you tested will undergo hydrolysis to form weak acids or weak bases. Ohter salts will not undergo hydrolysis. Determine which salts will hydrolyze, complete the last two columns in the table by writing in the corect column the formula for the species produced.

DATA AND OBSERVATIONS

Hypothesis: ______________________________

____

ANALYSIS

1. Write a net ionic equation to show the species formed for each salt that undergoes a hydrolysis reaction. Use the last equation in the introduction as a model. Write *No reaction* if the salt does not hydrolyze.

   1. ______________________________
   2. ______________________________
   3. ______________________________
   4. ______________________________
   5. ______________________________
   6. ______________________________
   7. ______________________________
   8. ______________________________
   9. ______________________________
   10. ______________________________
   11. ______________________________
   12. ______________________________

Reactions of Salt Solutions

Salt	Prediction (acidic, basic, or neutral)	Test result (acidic, basic, or neutral)	pH of solution	Weak acid formed	Weak base formed
1. $NaCl$					
2. Na_2CO_3					
3. $NaHCO_3$					
4. $NaC_2H_3O_2$					
5. $Al_2(SO_4)_3$					
6. $ZnSO_4$					
7. Na_3PO_4					
8. NaH_2PO_4					
9. NH_4Cl					
10. $Pb(NO_3)_2$					
11. KBr					
12. $MnSO_4$					

2. Which salt solutions, if any, do not undergo a hydrolysis reaction? Explain.

CONCLUSIONS

1. Did your results confirm your hypothesis? What was your percentage of accuracy in predicting whether each solution was acidic, basic, or neutral? What information would you have needed to make more successful predictions?

2. What is the role of water in the process of hydrolysis?

EXTENSION AND APPLICATION

For each of the first six reactions, what specific indicator could you have used to show a color change at the pH you determined? You can choose the indicators from those listed in table 9 on page xxvi of this laboratory manual.

Teacher Guide

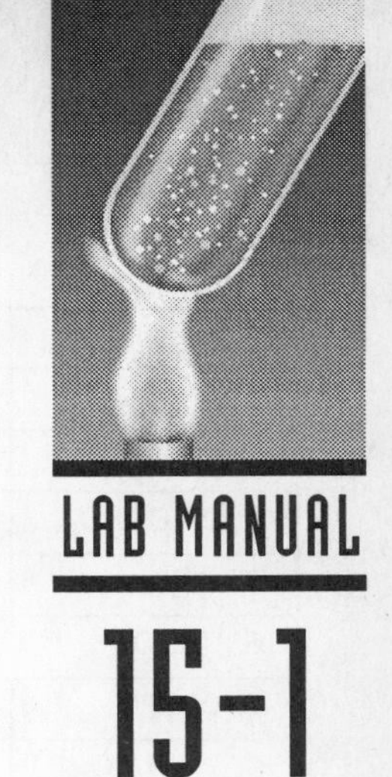

LAB MANUAL

15-1

Hydrolysis of Salts

When equivalent amounts of acid and base react, a salt solution is the only product. However, when salts dissociate in water, the products—a metal ion and a nonmetal ion—interact with water to form solutions that usually are not neutral. In fact, a neutral salt solution is a rarity, not the rule. In solution, the water can act as a source of H_3O^+ and OH^- ions. The properties of the acid and the base give the salt its properties in aqueous solutions. Of course, the relative strengths of the acids and bases will affect the pH of the solution.

Misconception: Most students think that all acids are strong and that all acids can cause chemical burns. The characteristics of strong bases are rarely part of their experiences. Yet, strong bases can be just as harmful to skin as strong acids. Also, students usually think that all salts are neutral and that combining equal amounts of an acid and a base results in a neutral solution.

PROCESS SKILLS

1. Students will use universal indicator to determine the pH of twelve salt solutions.
2. Students will infer the strengths of the species formed in each solution from the pH they determined.
3. Students will write a balanced chemical equation for each hydrolysis (and nonhydrolysis) reaction.

PREPARATION HINTS

- An adequate supply of distilled water is essential for this laboratory.
- You may wish to provide the smallest available bottles of salts.

MATERIALS

universal indicator (commercial preparation, obtainable from suppliers)
solid salts: $NaCl$, Na_2CO_3, $NaHCO_3$, $NaC_2H_3O_2$, $Al_2(SO_4)_3$, $ZnSO_4$, Na_3PO_4, NaH_2PO_4, NH_4Cl, $Pb(NO_3)_2$, KBr, $MnSO_4$

apron
goggles
13 × 100-mm test tubes (6)
spatula
10-mL graduated cylinder
thin-stem pipet
solid rubber stoppers to fit test tubes (6)
test-tube rack
marking pencil
distilled water
chart of pH and indicator colors
paper towels

PROCEDURE HINTS

- Before proceeding, be sure students have read and understood the purpose, procedure, and safety precautions for this laboratory activity.
- Students should wear their goggles and aprons throughout this experiment.
- The process of managing a large number of salt samples can be a challenge. The best method depends on your particular lab situation. One way to handle this problem is to have students take their test tubes in the test-tube racks to the supply of sample salts and obtain the six salts they will use at one time. This method cuts down on the movement in the room and helps to avoid accidents and spillage.
- Students should be instructed to take small samples of salts.
- Make sure students understand the need for accuracy in determining pH with the universal indicator.
- **Safety Precautions:** $MnSO_4$ may irritate body tissues. $Pb(NO_3)_2$ is a strong oxidant. There is significant fire risk if $Pb(NO_3)_2$ comes in contact with organic materials.
- **Disposal:** The lead and manganese solutions should be disposed of according to instructions for heavy metals. Students may rinse the other solutions down the drain using plenty of water.

Reactions of Salt Solutions

Salt	Prediction (acidic, basic, or neutral)	Test result (acidic, basic, or neutral)	pH of solution	Weak acid formed	Weak base formed
1. NaCl	Predictions	N	7		
2. Na_2CO_3	will vary.	B	10	H_2CO_3	
3. $NaHCO_3$		B	9	H_2CO_3	
4. $NaC_2H_3O_2$		B	8	$HC_2H_3O_2$	
5. $Al_2(SO_4)_3$		A	5		$Al(OH)_3$
6. $ZnSO_4$		A	4		$Zn(OH)_2$
7. Na_3PO_4		B	9	H_3PO_4	
8. NaH_2PO_4		B	8	H_3PO_4	
9. NH_4Cl		A	5		NH_3
10. $Pb(NO_3)_2$		A	4		$Pb(OH)_2$
11. KBr		N	7		
12. $MnSO_4$		A	4		$Mn(OH)_2$

DATA AND OBSERVATIONS

Hypothesis: Hypotheses will vary, but should indicate that a solution of a salt formed from a strong base and weak acid will be basic, one formed from a strong acid and weak base will be acidic, and one formed from both a strong acid and a strong base might be neutral.

ANALYSIS

1. The following equations are numbered the same as the salts in the table.

1. no reaction
2. $CO_3^{2-} + 2H_2O \rightarrow H_2CO_3 + 2OH^-$
3. $HCO_3^- + H_2O \rightarrow H_2CO_3 + OH^-$
4. $C_2H_3O_2^- + H_2O \rightarrow HC_2H_3O_2 + OH^-$
5. $Al^{3+} + 3H_2O \rightarrow Al(OH)_3 + 3H^+$
   (most likely answer)
   $Al^{3+} + H_2O \rightarrow AlOH^{2+} + H^+$
   (better answer)
6. $Zn^{2+} + 2H_2O \rightarrow Zn(OH)_2 + 2H^+$
   (most likely answer)
   $Zn^{2+} + H_2O \rightarrow ZnOH^+ + H^+$
   (better answer)
7. $PO_4^{3-} + 3H_2O \rightarrow H_3PO_4 + 3OH^-$
   (most likely answer)
   $PO_4^{3-} + H_2O \rightarrow HPO_4^{2-} + OH^-$
   (better answer)
8. $H_2PO^{4-} + H_2O \rightarrow H_3PO_4 + OH^-$
9. $NH^{4+} + H_2O \rightarrow NH_3 + H_3O^+$
10. $Pb^{2+} + 2H_2O \rightarrow Pb(OH)_2 + 2H^+$
   (most likely answer)
   $Pb^{2+} + H_2O \rightarrow PbOH^+ + H^+$
   (better answer)
11. no reaction
12. $Mn^{2+} + 2H_2O \rightarrow Mn(OH)_2 + 2H^+$
   (most likely answer)
   $Mn^{2+} + H_2O \rightarrow MnOH^+ + H^+$
   (better answer)
   (Note: All H^+'s could also be shown as H_3O^+'s by adding one more water molecule to the reactant side of the equations.)

2. NaCl and KBr do not hydrolyze. Na^+, Cl^-, K^+, and Br^-, are from strong electrolytes. Their solutions are neutral.

CONCLUSIONS

1. Answers will vary, depending on students' hypotheses and predictions. Knowledge of the strengths of the acids and the bases from which the salts formed would have helped students.

2. It is a source of H_3O^+ and OH^- ions.

EXTENSION AND APPLICATION

Reaction	pH	Indicator
1	7	bromothymol blue
2	10	alizarin yellow R
3	9	phenolphthalein
4	8	phenol red
5	5	2, 5-dinitrophenol
6	4	2, 5-dinitrophenol

Name

Date Class

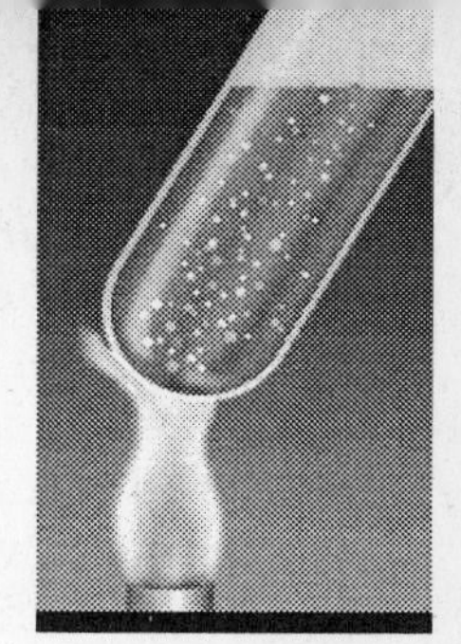

Acid/Base Titration

Chemists use many techniques to determine the way in which substances react. If a reaction involving two solutions forms a solid precipitate, the solid can be separated from the liquid and its mass determined. But suppose two solutions are mixed and the temperature increases, indicating a chemical reaction has occurred and no precipitate is formed. How can the mass of the products of the reaction be determined while they are still in solution?

If such a product is an acid or a base, the concentration can be measured by titration. In a titration, measured amounts of acid and base combine to produce a salt and water. If the salt formed is soluble in water, no precipitate is formed. A titration is based on the molar relationship between H_3O^+ and OH^- ions reflected in the balanced equation of an acid–base reaction. When the endpoint of the titration is reached, the number of H_3O^+ ions is equivalent to the number of OH^- ions. To determine the endpoint of the titration, a small amount of indicator is added to the reaction mixture.

In this laboratory, you will prepare a standardized base solution by titrating NaOH with an oxalic acid solution. You will then titrate an unknown acid with the standardized NaOH solution to determine the molecular mass of the acid.

OBJECTIVES

- **Determine** by titration the molarity of a sodium hydroxide solution.
- **Neutralize** a known mass of an unknown monoprotic (one H^+) acid with the standardized sodium hydroxide solution.
- **Calculate** the molecular mass of the unknown monoprotic acid.

MATERIALS

apron
goggles
50-mL buret
buret or utility clamp
ring stand
50-mL beaker
400-mL beakers (2)
125-mL Erlenmeyer flask
500-mL graduated cylinder
50-mL graduated cylinder
stirring rod
distilled water
balance
sheet of white paper
paper towel

PROCEDURE

Part 1: Standardization of a Base Solution

1. Set up the buret using the buret clamp and ring stand as shown in the diagram.

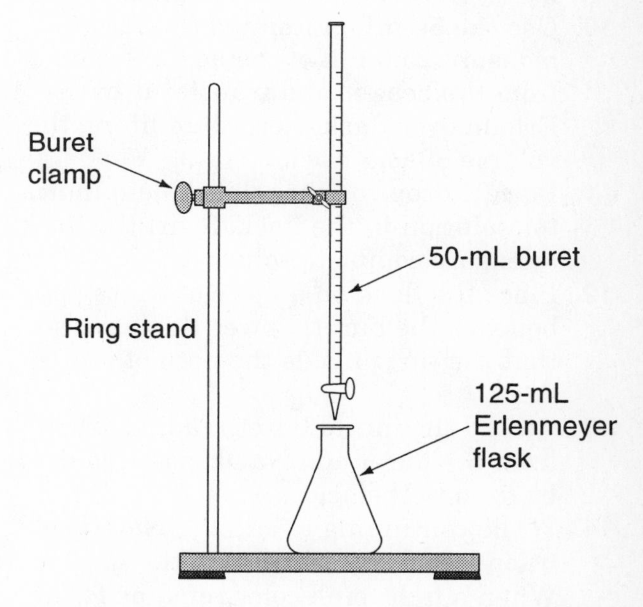

2. Label a 400-mL beaker *NaOH*. Measure the mass of the beaker. Record the mass under Data and Observations. Use the Trial 1 column.
3. Add approximately 2 g of sodium hydroxide, NaOH, pellets to the beaker. Record the exact mass of the beaker and the sodium hydroxide. **CAUTION:** *Sodium hydroxide is highly corrosive. Do not handle it with your hands. If the base comes in contact with your skin or clothing,*

wash the area with plenty of water. Report immediately all spills of sodium hydroxide solid or liquid to your teacher.

4. Add 250 mL of distilled water to the beaker and stir until the solid is dissolved completely. Record, in liters, the volume of water used.
5. Carefully fill the buret with the solution in the beaker. Fill the tip of the buret by opening the stopcock and releasing a small amount of the base into the 50-mL beaker. Discard this base.
6. Label the other 400-mL beaker $H_2C_2O_4 \cdot 2H_2O$ (oxalic acid dihydrate).
7. Determine and record the mass of the beaker.
8. Add about 1 g of oxalic acid crystals to the beaker. Record the exact mass of the beaker and the solid oxalic acid. **CAUTION:** *Oxalic acid is toxic as well as a tissue irritant.*
9. Add 250 mL of distilled water to the beaker. Stir with a clean stirring rod until the solid has dissolved. Record, in liters, the volume of water used.
10. Use the 50-mL graduated cylinder to measure 25.0 mL of the acid solution from the beaker and transfer it to the Erlenmeyer flask. Record, in liters, the volume of acid used.
11. Place 4 drops of phenolphthalein indicator solution in the flask. Swirl the flask to ensure complete solution.
12. Place the flask on a piece of white paper beneath the buret. Lower the buret so that the tip is inside the neck of the flask.
13. Record the initial buret reading. Then from the buret add NaOH solution drop by drop to the acid.
14. While continuing to add the NaOH solution, swirl the contents of the flask. When a light pink color remains for at least 30 seconds, record the final buret reading and determine the number of mL of NaOH solution used to neutralize the acid.
15. Convert the number of mL to L, and record the volume.
16. Discard the solution in the flask according to the directions of your teacher. Rinse the flask with distilled water and discard the rinse water in the same way as the solution.
17. Repeat the titration using additional amounts of the same acid and base solutions. Record your results in the Trial 2 column.

Part 2: Molecular Mass of an Unknown Acid

1. Your teacher will provide you with an unknown acid in a sample vial. Record the mass of the vial and acid under Data and Observations, Part 2. Use the Trial 1 column.
2. Pour the solid acid into the clean Erlenmeyer flask.
3. Measure and add 5 mL of distilled water to the sample vial. Close and shake the vial to dissolve all the acid crystals.
4. Pour the liquid from the vial into the Erlenmeyer flask.
5. Dry the inside of the sample vial and its stopper with a piece of paper towel. Find the mass of the closed vial and record the mass.
6. Determine the mass of the unknown acid and record it.
7. Add 45 mL of distilled water and 4 drops of phenolphthalein indicator solution to the acid in the Erlenmeyer flask. Record in liters the total volume of water used, including the water used to rinse the vial.
8. Refill your buret to the zero mark with NaOH solution. Record the initial buret reading.
9. Start to add NaOH solution to the acid.
10. Add NaOH to the flask drop by drop. As you are adding the NaOH, swirl the contents of the flask. When a light pink color remains for at least 30 seconds, record the final buret reading and determine the number of mL of NaOH solution used. Convert the number of mL to L, and record the volume.
11. Discard the solution in the flask according to the directions of your teacher. Rinse the flask with distilled water and discard the rinse water in the same way as the solution.
12. Repeat the titration with a second sample of the same unknown acid. Record your results in the Trial 2 column.

DATA AND OBSERVATIONS

Part 1: Standardizing NaOH

	Trial 1	Trial 2
1. Mass of beaker labeled *NaOH*	___ g	
2. Mass of beaker and NaOH	___ g	
3. Mass of NaOH	___ g	
4. Volume of water used	___ L	
5. Mass of beaker labeled $H_2C_2O_4 \cdot 2H_2O$	___ g	
6. Mass of beaker and $H_2C_2O_4 \cdot 2H_2O$	___ g	
7. Mass of $H_2C_2O_4 \cdot 2H_2O$	___ g	
8. Volume of water used	___ L	
9. Volume of $H_2C_2O_4 \cdot 2H_2O$ used	___ L	___ L
10. Initial buret reading	___ mL	___ mL
11. Final buret reading	___ mL	___ mL
12. Volume of NaOH solution used to neutralize $H_2C_2O_4 \cdot 2H_2O$	___ L	___ L

Part 2: Unknown Acid

	Trial 1	Trial 2
1. Mass of vial and acid	___ g	___ g
2. Mass of vial alone	___ g	___ g
3. Mass of acid	___ g	___ g
4. Volume of water used	___ L	___ L
5. Initial buret reading	___ mL	___ mL
6. Final buret reading	___ mL	___ mL
7. Volume of NaOH solution used to neutralize acid	___ L	___ L

ANALYSIS

Part 1: Standardizing NaOH

1. Calculate the molarity of the NaOH solution.
   a. Balance the following equation for the neutralization reaction of NaOH with oxalic acid.

$$NaOH + H_2C_2O_4 \rightarrow Na_2C_2O_4 + H_2O$$

____

   b. Determine the molarity of oxalic acid used.

$$M = \frac{\text{mass of acid/molecular mass of } H_2C_2O_4 \cdot 2H_2O}{\text{volume of water used}}$$

____

   c. Calculate the number of moles of $H_2C_2O_4 \cdot 2H_2O$ used in the titration.

$$\text{moles} = \text{molarity of } H_2C_2O_4 \cdot 2H_2O \times \text{volume of acid solution used}$$

$$= \text{analysis b} \times \text{data 9}$$

____

   d. Look at the balanced equation. Determine the number of moles of NaOH used to neutralize the acid.

____

   e. Determine the molarity of the base solution.

$$M = \frac{\text{moles of NaOH used}}{\text{volume of base used}} = \frac{\text{analysis d}}{\text{data 12}}$$

____

Part 2: Unknown Acid

1. Calculate the molecular mass of the unknown acid.
   a. Balance the equation for the neutralization reaction of NaOH with the unknown monoprotic acid (HX).

$$NaOH + HX \rightarrow NaX + H_2O$$

____

   b. Determine the number of moles of NaOH used to neutralize the unknown acid (HX).

moles of NaOH = molarity of NaOH × volume of NaOH

moles of NaOH = analysis e (Part 1) × data 7 (Part 2)

____

c. Look at the balanced equation. Determine the number of moles of acid used in relation to the number of moles of base.

____

d. Determine the molecular mass of the unknown acid.

$$\text{molecular mass of HX} = \frac{\text{mass of acid}}{\text{mol of acid}} = \frac{\text{data 3 (Part 2)}}{\text{analysis c (Part 2)}}$$

____

CONCLUSIONS

1. Consider the equation in analysis 1a (Part 2). What is the number of moles of unknown acid used in the titration? Hint: Because the acid is monoprotic, the number of moles of acid in the titration is equal to the number of moles of base used to neutralize it.

____

____

____

____

____

____

2. The amount of water added to the Erlenmeyer flask in Part 2 could have been 10 mL more or less, but the results would have been the same. Why?

____

____

____

____

____

____

____

____

____

EXTENSION AND APPLICATION

Quality control laboratories use titration as an easy and effective means of determining the concentration of acids and bases. What commercial products could use titration as a method of determining the quality of a product?

Teacher Guide

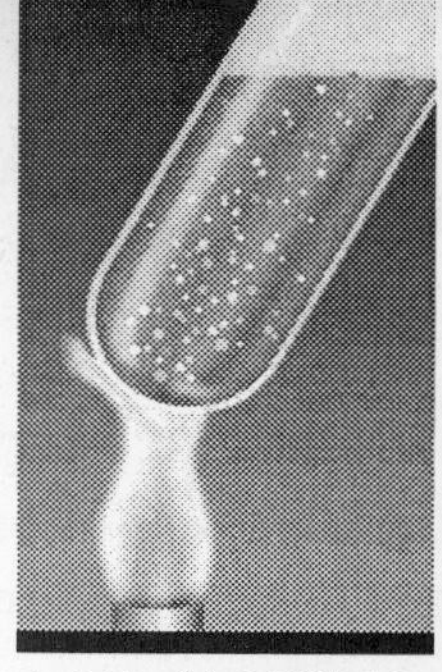

LAB MANUAL

15-2

Acid/Base Titration

Acid-base indicators act by establishing an equilibrium between their molecular and ionized forms. In a titration, an indicator added to the reaction mixture reaches an equilibrium when there is neither excess acid nor excess base. In the process of reaching an equilibrium, the indicator changes color. This change in color tells the chemist that excess acid or base has been neutralized—the endpoint of the titration has been reached. During a titration, it is necessary to add enough base or acid to neutralize the reaction and just shift the equilibrium of the indicator. The indicator itself does absorb or release H_3O^+ ions, but the indicator is in such low concentration that the small amount of ions it adds or releases is of no consequence.

PROCESS SKILLS

1. Students will perform an acid/base titration to determine the molarity of a solution of NaOH.
2. Students will calculate the molecular mass of the unknown acid, using their series of measurements.

PREPARATION HINTS

Use a fresh supply of cream of tartar, KHC4H4O6 for the unknown acid. Cream of tartar will not be suitable if the sample is old or has been exposed to moisture. You will need to provide each student group with two samples of cream of tartar if two trials are to be run. Samples should be about 0.5 to 1.0 g.

MATERIALS

solid sodium hydroxide, NaOH(pellets)
solid oxalic acid, $H_2C_2O_4 \cdot 2H_2O$
unknown solid acid: potassium hydrogen bitartrate, $KHC_4H_4O_6$, cream of tartar
phenolphthalein indicator solution

apron	50-mL beaker
goggles	400-mL beakers (2)
50-mL buret	125-mL Erlenmeyer flask
buret or utility clamp	500-mL graduated cylinder
ring stand	balance
stirring rod	sheet of white paper
50-mL graduated cylinder	paper towel
distilled water	

PROCEDURE HINTS

- Before proceeding, be sure students have read and understood the purpose, procedure, and safety precautions for this laboratory activity.
- Students must wear goggles and aprons at all times during this laboratory activity.
- If students have not had experience using burets, they will need help in the proper technique for titration.
- Although the glassware used is calibrated in mL, the liquid measurements should be recorded in L for ease of calculations when data are analyzed.
- Citric acid (a triprotic acid) may be substituted, if necessary, for oxalic acid.
- Solid monoprotic acids, for example sodium or potassium hydrogen sulfate ($NaHSO_4$ or $KHSO_4$), may be substituted for the cream of tartar.
- **Safety Precaution:** Solid NaOH is extremely caustic. Do not permit students to have access to large quantities of sodium hydroxide. Students must be careful not to spill sodium hydroxide. This solid is very hygroscopic. A pellet of sodium hydroxide on a lab table or balance will rapidly absorb moisture and dissolve during a class period. This concentrated solution could easily be mistaken for a drop of water.
- **Safety Precaution:** Oxalic acid is poisonous. It should be handled carefully. Students who may be allergic to oxalic acid should not be exposed to this compound.

- **Disposal:** All solutions should be collected and disposed of according to the instructions in **Disposal** C.

DATA AND OBSERVATIONS

Part 1: Standardizing NaOH

	Trial 1	Trial 2
1. Mass of beaker labeled *NaOH*	68.10 g	
2. Mass of beaker and NaOH	70.46 g	
3. Mass of NaOH	2.36 g	
4. Volume of water used	0.25 L	
5. Mass of beaker labeled $H_2C_2O_4 \cdot 2H_2O$	68.20 g	
6. Mass of beaker and $H_2C_2O_4 \cdot 2H_2O$	68.99 g	
7. Mass of $H_2C_2O_4 \cdot 2H_2O$	0.79 g	
8. Volume of water used	0.25 L	
9. Volume of $H_2C_2O_4 \cdot 2H_2O$ used	0.025 L	0.025 L
10. Initial buret reading	0.0 mL	0.0 mL
11. Final buret reading	5.0 mL	7.0 mL
12. Volume of NaOH solution used to neutralize $H_2C_2O_4 \cdot 2H_2O$	0.005 L	0.007 L

Part 2: Unknown Acid

	Trial 1	Trial 2
1. Mass of vial and acid	10.55 g	11.19 g
2. Mass of vial alone	9.96 g	10.40 g
3. Mass of acid	0.59 g	0.79 g
4. Volume of water used	0.050 L	0.050 L
5. Initial buret reading	0.0 mL	0.0 mL
6. Final buret reading	13.0 mL	17.0 mL
7. Volume of NaOH solution used to neutralize acid	0.013 L	0.017 L

ANALYSIS

Part 1: Standardizing NaOH

Values from Trial 1 are used in these calculations.

1. Molarity of NaOH solution:

a. $2NaOH + H_2C_2O_4 \rightarrow Na_2C_2O_4 + 2H_2O$

b. Molarity of oxalic acid used:

$$M = \frac{\text{mass of acid/ molecular mass of } H_2C_2O_4 \cdot 2H_2O}{\text{volume of water used}}$$

0.79 g/126 g/mole /0.25 L =
0.0063 moles /0.25 L = 0.025M

c. Number of moles of $H_2C_2O_4 \cdot 2H_2O$ used:
moles =
molarity of $H_2C_2O_4 \cdot 2H_2O \times$ volume of acid solution used

0.025 moles/L × 0.025 L = 0.0006 moles

d. Number of moles of NaOH used: (Two moles of NaOH are required for one mole of oxalic acid.)

0.0006 moles acid × 2 base groups/mole =
0.0012 moles NaOH

e. Molarity of base solution:

M = moles of NaOH used/volume of base used

0.0012 moles NaOH/0.005 L = 0.24M

Part 2: Unknown Acid

1. Molecular mass of the unknown acid (HX):

a. $NaOH + HX \rightarrow NaX + H_2O$

b. Number of moles of NaOH used to neutralize the unknown acid (HX):

moles = molarity of NaOH × volume of NaOH used

0.24 moles/L × 0.013 L = 0.0031 moles

c. 0.0031 mole of acid reacts with 0.0031 mole of NaOH.

d. Molecular mass of unknown acid:

mass of acid/moles of acid =
0.59 g/0.0031 mole = 190 g/mole

CONCLUSIONS

1. The number of moles of acid must equal the number of moles of base because the number of H_3O^+ and OH^- ions is in a 1:1 ratio. So 0.0031 moles of NaOH were required to neutralize 0.0031 moles of the solid unknown acid.
2. The water was used only as a solvent for the dissociation of the acid to give H_3O^+ ions. The water would not increase or decrease the number of H_3O^+ ions from the solid acid.

EXTENSION AND APPLICATION

Any commercial product that has an acid or base content, such as household ammonia, may be analyzed by means of titration.

Name

Date Class

Oxidation/Reduction of Vanadium

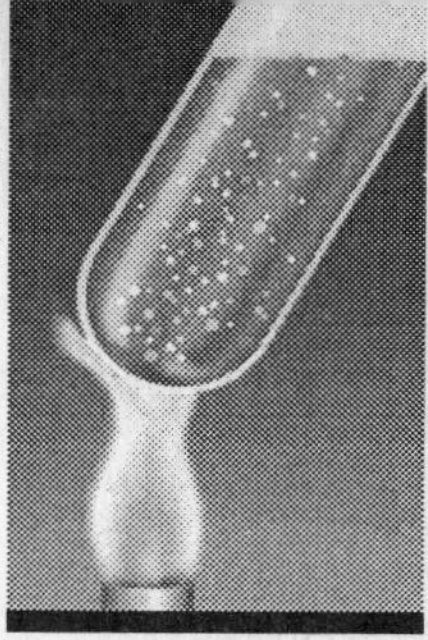

LAB MANUAL

16-1

Vanadium, element 23 on the periodic table, is one of the transition elements of Period 4. One characteristic of transition elements is that they have variable oxidation numbers. This means that elements 21 through 29 and the elements directly below them in the periodic table have several stable oxidation numbers. For example, the vanadium atom in an ionic form can have oxidation numbers of 5+, 4+, 3+, and 2+. The purpose of this activity is to prepare ions of an element that have different oxidation numbers.

OBJECTIVES

- **Prepare** ions of the same element that have different oxidation numbers.
- **Compare** some properties of vanadium ions with different oxidation numbers.

MATERIALS

24-well microplate
24-well template
microtip pipets (2)
thin-stem pipet
scissors

PROCEDURE

Part 1: Reduction of Vanadium

1. Form a **hypothesis** about whether vanadium ions with different oxidation numbers will have different physical and chemical properties. Write your hypothesis under Data and Observations.
2. Place a 24-well microplate on a template with the numbered columns away from you and the lettered rows to the left.
3. Make a microscoop by cutting the bulb of a thin-stem pipet, as shown in Figure A.

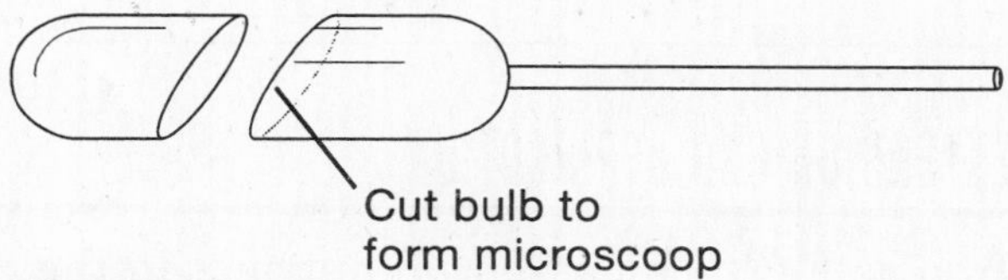

Figure A

4. Place a half-microscoop of zinc metal into well A1.
5. Using a microtip pipet, place ½-pipet of NH_4VO_3 in well A2. Ammonium vanadate contains vanadium with a 5+ oxidation number. Well A2 will serve as a control well. Use the diagram of your microplate in Figure B as a guide to the steps that follow. **CAUTION:** *Ammonium vanadate is toxic.*

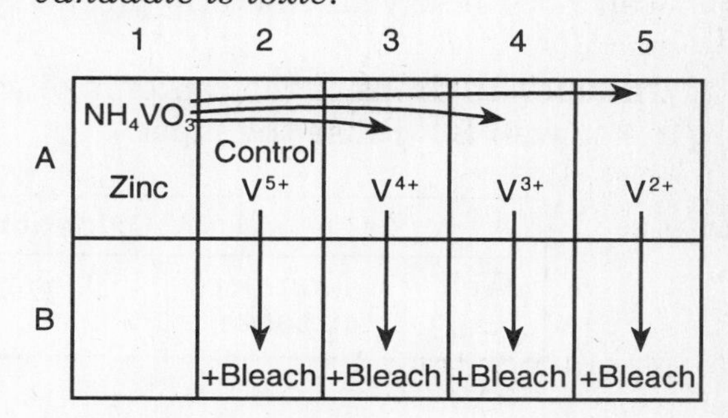

Figure B

6. Fill the microtip pipet three-quarters full with NH_4VO_3 solution. Place the NH_4VO_3 in well A1 with the zinc metal.
7. Draw the liquid in well A1 back up into the microtip pipet. Then return the liquid to well A1. This is one pass. Repeat this process for as many passes as necessary until the solution turns color. The solution now contains vanadium as V^{4+}. Record in the data table the number of times you must return the V^{5+} solution to the zinc for a color change.
8. Place 20 drops of the V^{4+} solution into well A3.
9. Place the rest of the V^{4+} solution remaining in the pipet back into well A1 with the zinc metal.
10. Draw the liquid in well A1 back up into the pipet. Then return the liquid to well

A1. Repeat passes until the solution turns color. The solution of vanadium in A1 is now in the V^{3+} state. Record the number of passes in your data table.

11. Place 20 drops of the V^{3+} solution into well A4.
12. Place the rest of the V^{3+} solution remaining in the pipet back into well A1 with the zinc metal.
13. Draw the liquid in well A1 back up into the pipet. Return the liquid to well A1. Repeat passes until the solution turns color. The solution of vanadium in A1 is now in the V^{2+} state. Record the number of times you must return the solution to the zinc for a color change.
14. Place 20 drops of the V^{2+} solution in well A5. Return any remaining solution to well A1.

Part 2: Oxidation of Vanadium

1. Rinse your microtip pipet and transfer 10 drops of the contents of well A5 to well B5. Rinse your microtip pipet again.
2. Transfer 10 drops of the contents of well A4 to well B4. Rinse the pipet.
3. Transfer 10 drops of the contents of well A3 to well B3. Rinse the pipet.
4. Transfer 10 drops of the contents of well A2 to well B2.
5. Fill a clean microtip pipet with diluted household bleach.
6. Drop by drop, add the diluted bleach to well B5 until a color change occurs. Record your observations and the number of drops required in the column labeled *Color with bleach* in the data table.
7. Following the procedure in step 6, add diluted bleach to wells B4, B3, and B2. Record your observations.

DATA AND OBSERVATIONS

Hypothesis: ______________________

____

____

Oxidation/Reduction Reactions

Well	Number of passes	Color	Oxidation state	Color with bleach
A2				
A3				
A4				
A5				

ANALYSIS

1. How many changes in oxidation number did vanadium go through in Parts 1 and 2?

____

____

2. Which change in oxidation number required the greatest number of passes through the zinc metal?

____

____

CONCLUSIONS

If V2+ is allowed to stand for any length of time, it reverts to V5+. What factors do you think will slow this process?

____

EXTENSION AND APPLICATION

Iron as Fe^{2+} is an essential nutrient in the human diet, but iron as Fe^{3+} has no value as a nutrient. Some iron-enriched cereals contain iron filings. What is the advantage of using iron in this form?

Teacher Guide

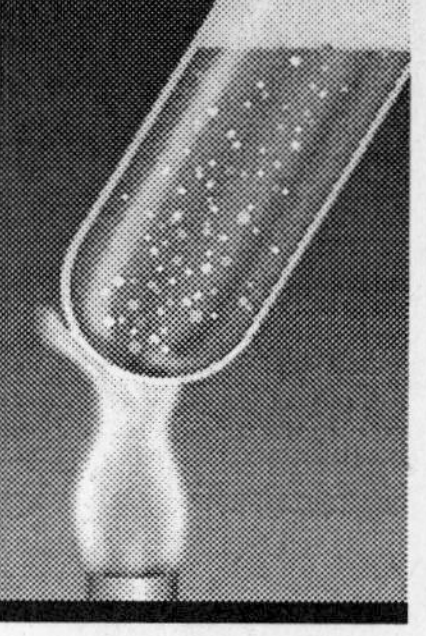

LAB MANUAL

16-1

Oxidation/Reduction of Vanadium

Students will cause the same element to undergo six changes in oxidation number. The reaction responsible for the reduction of V^{n+} is

$$V^{n+} + Zn^0 \rightarrow Zn^{2+} + V^{(n+)-1}$$

The reaction responsible for the oxidation of vanadium ions is

$$V^{n+} + 2OCl^- + 2H^+ \rightarrow V^{n+(+1)} + 2Cl^0 + 2OH^-$$

PROCESS SKILLS

1. Students will observe the properties of an element with four different oxidation numbers.
2. Students will produce ions of the same element that have different oxidation numbers.

PREPARATION HINTS

- Prepare 0.1*M* NH_4VO_3 by dissolving 0.4 g of NaOH in 50 mL of distilled water. Stirring constantly, add 1.0 g of NH_4VO_3. When the NH_4VO_3 has dissolved completely, add 1 mL of concentrated H_2SO_4. Then add distilled water to make a total volume of 100 mL.
- Ammonium vanadate is sometimes marketed under the name ammonium metavanadate.
- 40-mesh zinc metal provides the best combination of surface area and size.
- Prepare diluted bleach by adding 50 mL of commercially prepared household bleach to 50 mL of distilled water.
- For each group, make a copy of the 24-well template on p. T6.

MATERIALS

zinc granules
0.1*M* NH_4VO_3
household bleach (diluted 1:1)
24-well template
24-well microplate
microtip pipets (2)
thin-stem pipet
scissors

PROCEDURE HINTS

- Before proceeding, be sure students have read and understood the purpose, procedure, and safety precautions for this laboratory activity.
- It is important that students use a microtip pipet to dispense the vanadium solution over the zinc metal.
- It is essential that no zinc metal be carried over to any well from well A1.
- The first color change that students will notice is from yellow to green. This is not a change in the oxidation state of vanadium. The color results from a mixture of both yellow V^{5+} ions and blue V^{4+} ions.
- **Safety Precaution:** Ammonium vanadate (NH_4VO_3) is toxic and gives off poisonous ammonia gas if heated. Observe proper precautions.
- **Disposal:** Dispose of materials according to instructions in **Disposal** J.

DATA AND OBSERVATIONS

Students' hypotheses will vary but should indicate that vanadium ions with different oxidation numbers have different properties.

The numbers of passes between pipet and well will differ because color changes are gradual, and it is a matter of judgment as to when the color changes are complete.

Students will observe these colors:

Yellow	V^{5+}
Blue	V^{4+}
Green	V^{3+}
Violet	V^{2+}

ANALYSIS

1. Vanadium goes through three changes in oxidation number from V^{5+} to V^{2+} with zinc and three more changes from V^{2+} to V^{5+} with bleach.
2. The change from V^{3+} to V^{2+} required the greatest number of passes through the zinc metal. Reduction becomes more difficult as more electrons are added to the metal ion.

CONCLUSIONS

The process can be slowed by limiting the exposure of V^{2+} to oxygen in the air.

EXTENSION AND APPLICATION

Iron filings are used in enriched products because the metal is less easily oxidized than the Fe^{2+} ion. If iron were added as Fe^{2+} (as iron(II) sulfate, for example), the compound would be easily oxidized by oxygen in the air to Fe^{3+}. The metal Fe is converted to Fe^{2+} by the hydrochloric acid in the stomach.

Copyright © Glencoe/McGraw-Hill, a division of The McGraw-Hill Companies, Inc.

Name

Date Class

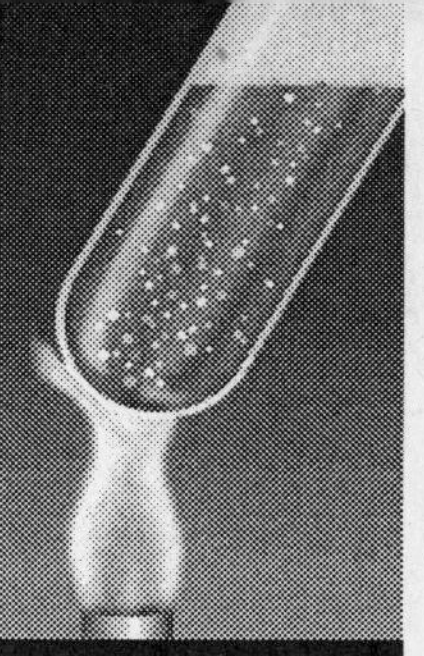

Corrosion as an Electrochemical Process

The most common electrochemical reaction does not, as you might think, occur in batteries. It is corrosion, which can occur in metal objects exposed to the environment. The best known kind of corrosion is the formation of rust on iron. The two half-reactions in rusting are oxidation of the iron

$$Fe \rightarrow Fe^{2+} + 2e^-$$

and reduction of oxygen from the air in the presence of water.

$$O_2 + 2H_2O + 4e^- \rightarrow 4OH^-$$

The red solid that we call rust results when Fe^{2+} is further oxidized by air to Fe^{3+} and precipitates as the hydrated oxide with a variable amount of water, $Fe_2O_3 \cdot nH_2O$.

Rusting occurs when a piece of iron is part of a naturally occurring redox reaction. The electrons released in the oxidation of iron travel through the iron to the site of the reduction of oxygen dissolved in water. Another metal in contact with iron can either increase or decrease iron corrosion, depending on whether the metal is less or more active than iron. In this experiment, you will observe the effects of distilled water, salt solution, copper, and magnesium on the corrosion of iron.

OBJECTIVES

- **Demonstrate** the corrosion of iron.
- **Observe** the results of corrosion.
- **Investigate** protection of iron from corrosion by another metal and by a protective coating.

MATERIALS

apron
goggles
iron nails about 3 cm long (12)
thin-stem pipet
nail polish or model paint (any color)
24-well microplate
24-well template
distilled water

PROCEDURE

1. Before you begin, form a **hypothesis** about the effects of wrapping an iron nail in copper and in magnesium on corrosion of the nail. Will corrosion be increased or decreased by these metals? Record your hypothesis under Data and Observations.
2. Place a 24-well microplate on a template with the lettered rows at the left and the numbered columns away from you.
3. Obtain 12 common nails.
4. Paint the bottom half of 6 nails with nail polish or model paint. Allow the coating to dry.
5. While waiting for the polish or paint to dry, wrap a small piece of magnesium ribbon around three of the other nails.
6. Wrap a small piece of copper wire around the three remaining nails.
7. Place the nails with the copper wires in wells B1, B2, and B3 of a microplate.
8. Place the nails with the magnesium ribbon in wells B4, B5, and B6.
9. Place three of the painted nails in wells A1, A2, and A3 with the painted half down.
10. Place three of the painted nails in wells A4, A5, and A6 with the painted half up.
11. Half-fill each of wells A1, B1, A4, and B4 with sodium chloride solution.
12. Half-fill each of wells A2, B2, A5, and B5 with distilled water.
13. Do not add any solution or water to wells A3, B3, A6, or B6.

14. Label the Microplate Data Form to show what you placed in each well.
15. Leave the microplate overnight in a warm area of the classroom or laboratory.
16. The next day, carefully examine the solutions, the nails, any solids present, and the wires. Record on the Microplate Data Form your observations for each well and compare what has happened in the different wells. Record your comparisons.
17. Discard the nails and solutions as instructed by your teacher. Rinse your microplate.

DATA AND OBSERVATIONS

Hypothesis: ______________________

____

____

Record all your observations in the Microplate Data Form.

ANALYSIS

1. Due to formation of Fe^{3+} by oxidation, corrosion will produce one or more of the following: red-brown rust on the nail, a yellowish to red-brown precipitate, a yellow to orange color in the solution. In which wells in row A did iron corrosion occur? Compare the extent of corrosion in these wells.

____

____

____

2. In which wells in row B did iron corrosion occur? Compare the extent of iron corrosion in these wells. Also compare the extent of corrosion in these wells with the corrosion in the wells of row A. Was there any evidence of corrosion of copper or magnesium?

____

____

____

3. Was there a difference in the amount of corrosion in NaCl solution and in distilled water? Explain.

____

____

____

CONCLUSIONS

1. What was the difference in the painted and unpainted ends of the nails exposed to distilled water and sodium chloride in row A? Did paint protect the nails from corrosion? Was there any difference in the extent of corrosion in salt water compared with distilled water?

____

____

2. What was the effect of contact with copper on the corrosion of iron?

____

____

3. What was the effect of contact with magnesium on the corrosion of iron?

____

____

4. Was your hypothesis about the effects of copper and magnesium on iron corrosion correct? Explain.

____

____

EXTENSION AND APPLICATION

1. Predict what would be observed if this experiment were conducted with zinc instead of copper or magnesium.
2. Using reference books, investigate anodized aluminum. Explain the process of anodizing and the result of the process.

Teacher Guide

Corrosion as an Electrochemical Process

LAB MANUAL

16-2

By definition, metal corrosion is the result of metal combining with other elements in the environment to form oxides or salts. The term *corrosion* is generally reserved for processes that destroy objects exposed to the environment. The formation of rust on a car, for example, makes that part of the car brittle, and it eventually crumbles away. Iron corrosion is an electrochemical process.

Misconception: Students rarely make the connection between the textbook description of the physical and chemical processes of corrosion and the real world. Remind them that any of the changes they see in this experiment can happen to iron in the environment. Note that plumbers, electricians, carpenters, and home do-it-yourselfers should be careful about putting different metals in contact with each other.

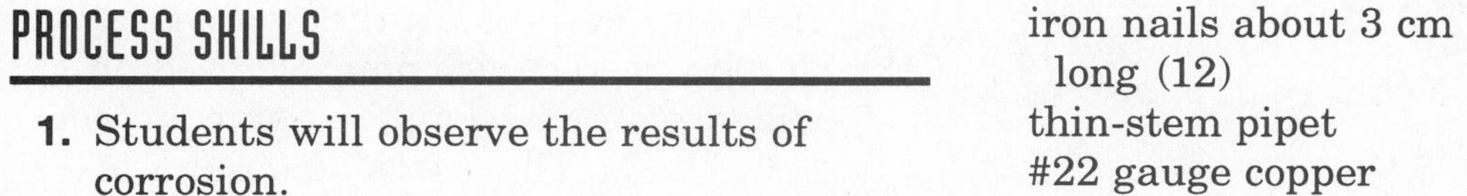

PROCESS SKILLS

1. Students will observe the results of corrosion.
2. Students will demonstrate the effects of copper and magnesium in contact with iron.
3. Students will demonstrate the effect of a protective coating on the iron.

PREPARATION HINTS

- Common iron nails should be used for this experiment. Do not use galvanized nails. Galvanized nails have a coating of zinc, which will interfere with the experiment. The common nails should be washed in detergent to remove the usual coating of oil.
- Students may wish to try the experiment with other nails as an extra assignment. In this case, galvanized nails are appropriate. Students can use copper, aluminum, and coated nails also.
- For each group, make a copy of the 24-well template in p. T6.

MATERIALS

1*M* NaCl (Add 58.5 g of NaCl to enough distilled water to give 1 L of solution.)

apron
goggles
magnesium ribbon
nail polish or model paint (any color)
iron nails about 3 cm long (12)
thin-stem pipet
#22 gauge copper wire
24-well microplate
24-well template
distilled water

PROCEDURE HINTS

- Before proceeding, be sure students have read and understood the purpose, procedure, and safety precautions for this laboratory activity.
- **Troubleshooting:** It is not possible to predict which of the various hydrated iron ions and iron oxides will be observed. Therefore, students should not be surprised or troubled if their precipitates and solutions vary somewhat in color.
- **Safety Precaution:** Magnesium ribbon is extremely flammable. Keep it away from open flames.
- **Disposal:** Dispose of the metals according to the instructions in **Disposal** A. Dispose of the solutions according to the instructions in **Disposal** C.

DATA AND OBSERVATIONS

Students' hypotheses will vary, but they may be able to recognize that copper is less active than iron and should accelerate iron corrosion, whereas magnesium is more active than iron and should slow down or prevent iron corrosion.

Sample observations:

A1 Solution has slight orange color. Tip of nail shows slight rust. Rest of nail is unchanged.

A2 Solution has slight orange color. Slight rust in shaft of nail where it has been machined.

A3 No change.

A4 Large amount of yellow precipitate. Nail is discolored and rusted below the solution line. No change on painted end.

A5 Large amount of yellow precipitate. Nail is discolored and rusted below the solution line. No change on painted end.

A6 No change.

B1 Largest amount of precipitate, yellow-orange, in clumps at bottom of well. End of nail is very rusty. No change in the appearance of the copper.

B2 Large amount of yellow-orange precipitate in clumps at bottom of well. Less rust on nail than in B1 but more than in wells A4 and A5. No change in appearance of copper.

B3 No change.

B4 No precipitate or color in solution. No rust on nail. Magnesium has a white powdery coating and there is some white coating on the nail under the magnesium.

B5 No precipitate or color in solution. No rust on nail. Slight white powdery coating on magnesium and on nail under magnesium.

B6 No change.

ANALYSIS

1. In row A, there was no corrosion of the nails exposed only to air in wells A3 and A6. Slight corrosion occurred in wells A1 and A2. Presumably, some spots on the part of the nails that were below the level of the water or salt solution were not covered with paint; had the paint completely covered the immersed part of the nail, there would have been no evidence of corrosion. The most corrosion occurred in A4 and A5. In these wells, the nail ends that were in the water or salt solution were not protected by paint.
2. In row B, there was no corrosion on the nails in wells B3 and B6, which were exposed only to air. Iron corrosion occurred in wells B1 and B2, where the nails were wrapped with copper. There was more precipitate in the salt water than in any of the wells in row A. The copper wire was not corroded in B1 and B2. There was no iron corrosion in B4 and B5, where the nails were wrapped with magnesium ribbon. Magnesium corrosion was shown by the white coating on the magnesium and the nails.
3. In the A row, there did not seem to be a difference in the amount of corrosion in the wells containing sodium chloride solution and in the distilled water wells, but in the B row, there was more corrosion in the wells with sodium chloride than in the wells with distilled water. Sodium chloride is an electrolyte, which made movement of electrons easier.

CONCLUSIONS

1. The parts of the nails that were completely covered with paint showed no evidence of corrosion; the paint protected the nails. There was no observable difference in the extent of corrosion in salt water compared with distilled water for the painted or unpainted nails.
2. Copper increased the extent of corrosion of iron in salt water, as shown by the larger amount of precipitate on the copper-wrapped nail in well B1 compared with the nails in wells A4 and A5, which had no copper.
3. The corrosion of iron was prevented in the presence of magnesium. Instead, the magnesium corroded.
4. Answers will vary, depending on students' hypotheses.

EXTENSION AND APPLICATION

1. Zinc, like magnesium, is a more active metal than iron. Therefore, it should act to protect iron, just as magnesium does. The zinc would be oxidized and the iron would not corrode.
2. In the process of anodizing, a metal such as aluminum is made the anode in an electrochemical cell. The "anodized" metal gets a coating of oxide, which adheres firmly to the metal surface and protects it from corrosion. The surface coating on anodized aluminum can be dyed to produce a smooth, colored surface with a metallic sheen.

Copyright © Glencoe/McGraw-Hill, a division of The McGraw-Hill Companies, Inc.

Name

Date Class

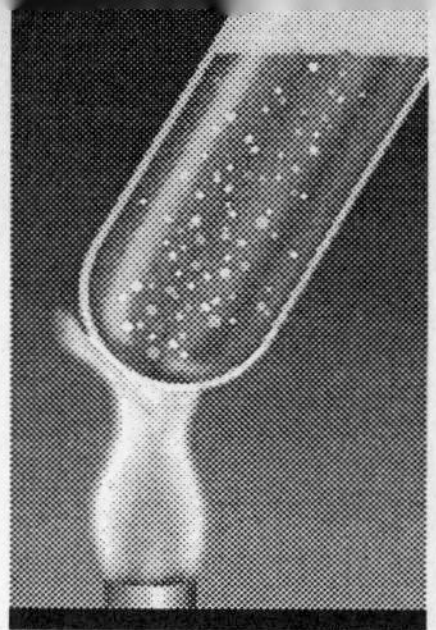

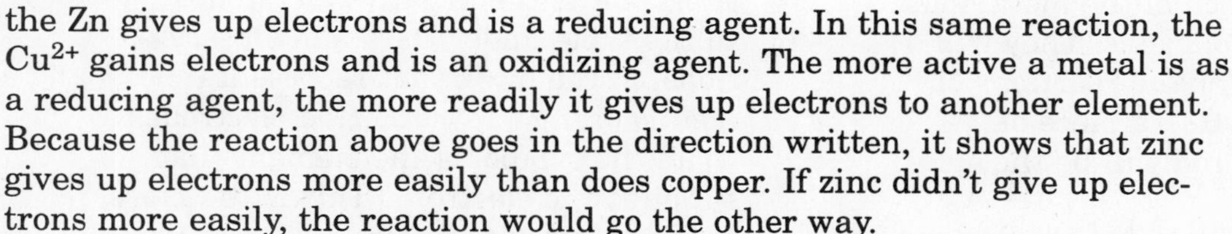

Comparing the Abilities of Metals to Give Up Electrons

In the reaction between zinc metal and copper(II) ions in aqueous solution,

$$Zn(s) + Cu^{2+} \rightarrow Zn^{2+} + Cu(s)$$

the Zn gives up electrons and is a reducing agent. In this same reaction, the Cu^{2+} gains electrons and is an oxidizing agent. The more active a metal is as a reducing agent, the more readily it gives up electrons to another element. Because the reaction above goes in the direction written, it shows that zinc gives up electrons more easily than does copper. If zinc didn't give up electrons more easily, the reaction would go the other way.

A list of metals arranged in order of activity as reducing agents, from the most active at the top to the least active at the bottom, is known as an activity series. A table of standard reduction potentials is also an activity series, although it includes other reducing and oxidizing agents in addition to metals and their ions. Activity series are used to predict the directions in which reactions will proceed. In this laboratory activity, you will test nine common metals in reactions like the one between zinc and copper ions and determine an activity series for this group of metals.

OBJECTIVES

- **Set up** a group of microgalvanic cells.
- **Infer** the direction of electron flow in the galvanic cells.
- **List** the metals studied and hydrogen in an activity series.

MATERIALS

apron
goggles
24-well microplate
24-well template
thin-stem pipets (10)
small beaker
sandpaper or laboratory file
galvanometer
filter paper
wire leads with clips at both ends (2)
paper towels
scissors
forceps

PROCEDURE

1. Two pairs of the metals included in this experiment are magnesium and silver, and nickel and copper. Using Table 17.1 on page 602 of your textbook, form a **hypothesis** about which metal in each pair will be the more active metal and will therefore give up electrons to the ion of the other metal. Write your hypothesis under Data and Observations.
2. Label a 24-well template as shown in Figure A. Place your microplate on the template so that the numbered columns are away from you and the lettered rows are at the left.
3. **CAUTION:** *Many of the chemicals you will use are toxic. Follow proper chemical hygiene procedures. Wash your hands thoroughly after completing this laboratory activity.* Fill a thin-stem pipet approximately half-full with $Cu(NO_3)_2$ solution.
4. Place the $Cu(NO_3)_2$ solution in well A1.
5. Using a clean thin-stem pipet, place an approximately equal volume of $Pb(NO_3)_2$ solution in well A2.
6. Repeat step 5 for each of the remaining solutions in the order shown in the microplate diagram in Figure A. Use wells A3 through A5 and then wells B1 through B5.

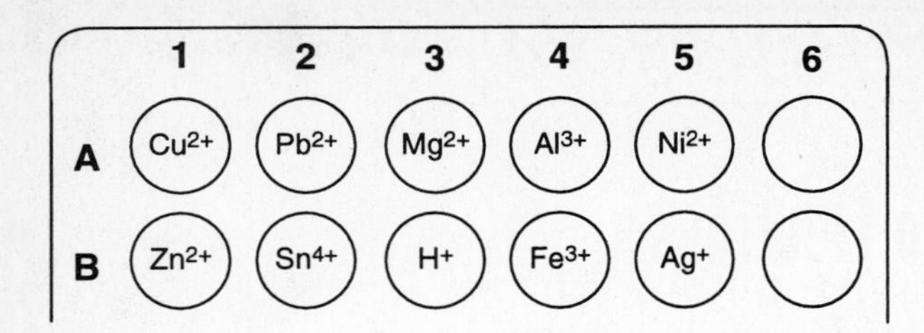

Figure A

7. Obtain small wires or strips of metals corresponding to each of the metal ions in the microplate solutions. They will act as electrodes. For the reactions of hydrogen ion (well B3), a piece of graphite or carbon rod will be the electrode.
8. Clean each piece of metal with sandpaper or a file.
9. Place each piece of metal in the well containing the ion solution of the same metal, for example, copper metal in the Cu^{2+} solution in well A1.
10. Cut a piece of filter paper into the shape shown in Figure B by cutting along the solid lines. Each of the "fingers" should be of a size to fit into a well.

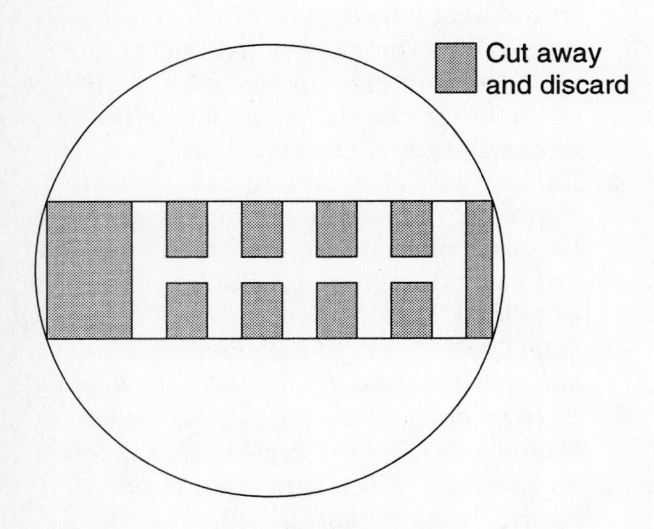

Figure B

11. Place this piece of cut filter paper (which will be the salt bridge) into a small beaker containing a few milliliters of 1*M* KNO_3. Allow the paper to become soaked with the solution.
12. Use forceps to remove the filter paper from the beaker. Bend down the filter paper fingers and arrange the paper so that a finger dips into each of the wells containing an ion solution.
13. Using the clips, attach two separate wires to the terminals of a galvanometer or very low range voltmeter.
14. Using the clips, connect the other ends of the galvanometer wires to the pieces of metal in two different wells (the electrodes).
15. Watch the needle of the meter. The needle will show if current is flowing. In Figure C, electrode 2 is supplying electrons—it is the negative electrode.
16. In the table under Data and Observations, in the block that represents the metals of the two connected electrodes, write the symbol of the element that supplied the electrons. This is the more active metal of the pair.

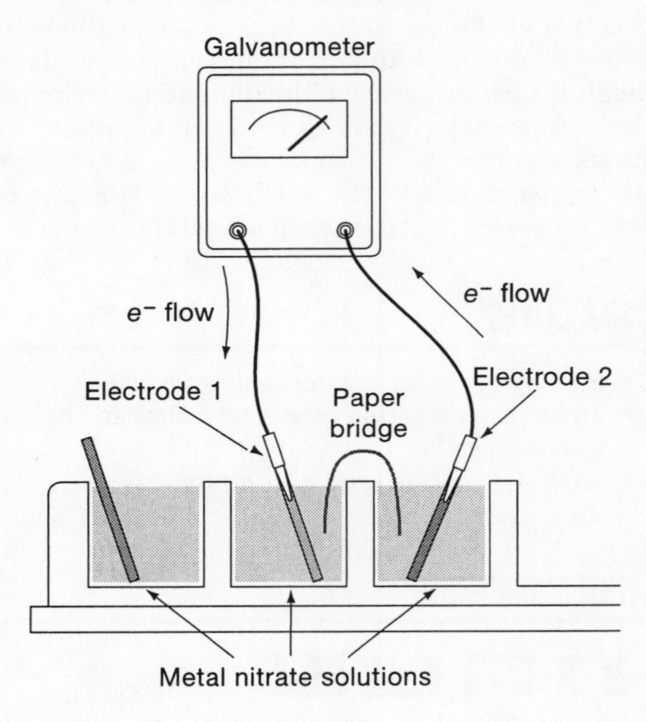

Figure C

17. Repeat this process for each of the 45 possible combinations of electrodes. For each one, write the symbol of the more active element in the table.
18. Return each of the ten solutions in the microplate wells to the original containers, using a clean thin-stem pipet for each solution. Do not mix up the solutions. Discard the piece of filter paper in the trash.
19. Rinse and dry the microplate.

Copyright © Glencoe/McGraw-Hill, a division of The McGraw-Hill Companies, Inc.

DATA AND OBSERVATIONS

Hypothesis: ____________________

____

____

Record your observations in the table below.

Entries are needed only to the right of the Xs in each row. Enter the symbol of the metal that supplies electrons in each pair. For example, when the Cu and Pb electrodes are connected, Pb supplies electrons.

	Cu	Pb	Mg	Al	Ni	Zn	Sn	H	Fe	Ag
Cu	X	Pb								
Pb	X	X								
Mg	X	X	X							
Al	X	X	X	X						
Ni	X	X	X	X	X					
Zn	X	X	X	X	X	X				
Sn	X	X	X	X	X	X	X			
H	X	X	X	X	X	X	X	X		
Fe	X	X	X	X	X	X	X	X	X	
Ag	X	X	X	X	X	X	X	X	X	X

ANALYSIS

1. In the table, count the number of times each electrode gave up electrons to the ions in solution. Record the results.

Cu ________ Zn ________

Pb ________ Sn ________

Mg ________ H ________

Al ________ Fe ________

Ni ________ Ag ________

2. Arrange the electrodes in an activity series in the order of the number of times each acted as a reducing agent by giving up electrons. Arrange the electrodes from most active to least active.

__________ __________

__________ __________

__________ __________

__________ __________

__________ __________

CONCLUSIONS

1. Which was the most active metal in this experiment?

2. Which was the least active metal in this experiment?

3. Write the activity series for the species tested according to their order in Table 17.1 on page 602 of your textbook. (Where there is a choice between two possible reductions of the same metal, use the value for reduction of the 2+ ion to the free element.)

4. Compare your experimental results with the activity series. How are they similar? How are they different?

5. Suggest why the experimental order of activity is different from that in Table 17.1.

6. Did your hypothesis, based on the activity series, correctly predict the experimental results? Explain.

EXTENSION AND APPLICATION

Make a list of your daily activities that depend upon electrochemical reactions.

Teacher Guide

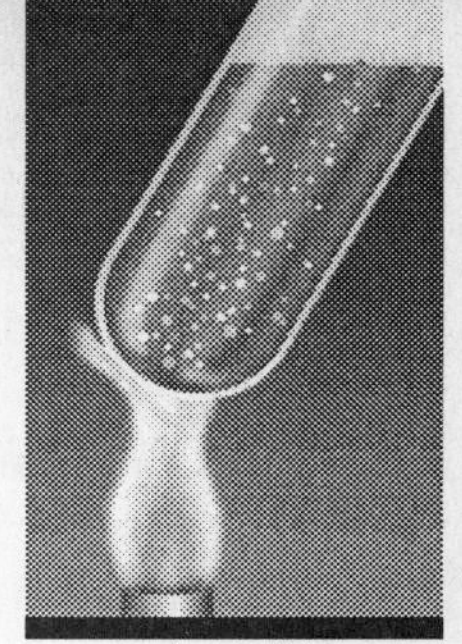

Comparing the Abilities of Metals to Give Up Electrons

LAB MANUAL

17-1

A table of standard reduction potentials is a list of elements in the order of their ability to act as reducing agents. Lithium, the strongest elemental reducing agent, has the largest negative standard reduction potential and gives up electrons most easily of all the metals. Lithium can be expected to act as a reducing agent in combination with all potential oxidizing agents below it in the table. In this laboratory activity, students are able to compare the activities of nine metals and a hydrogen electrode in galvanic cells and then develop their own activity series.

Misconceptions: Students have little or no concept of the differences in reactivity of metals unless the differences are tied to spectacular reactions such as that of sodium with water. Students have little experience with the reactivity differences that lead to common phenomena such as corrosion.

PROCESS SKILLS

1. Students will set up solutions, metals, and a salt bridge to create microgalvanic cells.
2. Students will observe the direction of electron flow in 45 different galvanic cells.
3. Students will list metals and hydrogen in an activity series.

PREPARATION HINTS

- In preparing the salt solutions, be sure to check the hydration of your stock salts and adjust the masses needed, if necessary. **CAUTION:** *Nitrates are strong oxidants. Avoid contact with organic materials.*
- All metal nitrate solutions can be prepared weeks or even months in advance.
- The tin(IV) sulfate solution must be prepared immediately before the laboratory period. Do not use tin(IV) sulfate solution if it has been stored for more than two days.
- Microplates with large wells, either combination plates with 12 large and 48 small wells or those with 24 large wells, are essential for this experiment. Do not try to use microplates with small wells; the spacing of the wells, electrodes, and salt bridge is very difficult to work with.
- Thin-stem pipets, not microtip pipets, should be used.
- Provide a series of small, well-labeled stock bottles for the solutions, which can be collected and reused.
- For each group, make a copy of the 24-well template on p. T6.

MATERIALS

metal wires or strips: Cu, Pb, Mg, Al, Ni, Zn, Sn, Fe, Ag

graphite rods

1*M* KNO_3 solution (Add 101 g of KNO_3 to enough distilled water to give 1 L of solution.)

0.1*M* solutions:

$Cu(NO_3)_2$ (Add 29.6 g of $Cu(NO_3)_2 \cdot 6H_2O$ to enough distilled water to give 1 L of solution.)

$Pb(NO_3)_2$ (Add 33.1 g of $Pb(NO_3)_2$ to enough distilled water to give 1 L of solution.)

$Mg(NO_3)_2$ (Add 25.6 g of $Mg(NO_3)_2 \cdot 6H_2O$ to enough distilled water to give 1 L of solution.)

$Al(NO_3)_3$ (Add 21.3 g of $Al(NO_3)_3 \cdot 6H_2O$ to enough distilled water to give 1 L of solution.)

$Ni(NO_3)_2$ (Add 29.0 g of $Ni(NO_3)_2 \cdot 6H_2O$ to enough distilled water to give 1 L of solution.)

$Zn(NO_3)_2$ (Add 29.7 g of $Zn(NO_3)_2 \cdot 6H_2O$ to enough distilled water to give 1 L of solution.)

$Sn(SO_4)_2$ (Add 34.6 g of $Sn(SO_4)_2 \cdot 2H_2O$ to enough distilled water to give 1 L of solution.)

HNO_3 (Add 18.9 mL of concentrated HNO_3 to enough distilled water to give 1 L of solution.)

$Fe(NO_3)_3$ (Add 40.4 g of $Fe(NO_3)_3 \cdot 9H_2O$ to enough distilled water to give 1 L of solution.)

$AgNO_3$ (Add 17.0 g of $AgNO_3$ to enough distilled water to give 1 L of solution.)

apron
goggles
24-well microplate
24-well template
thin-stem pipets (10)
small beaker
sandpaper or laboratory file
galvanometer or low-voltage voltmeter
filter paper
wire leads with clips at both ends (2)
paper towels
scissors
forceps

Copyright © Glencoe/McGraw-Hill, a division of The McGraw-Hill Companies, Inc.

PROCEDURE HINTS

- Before proceeding, be sure students have read and understood the purpose, procedure, and safety precautions for this laboratory activity.
- It is not necessary to have each well filled to the top with solution. Remind the students that the volume of solution and the amount of metal in the electrode have no effect on the activity of the metal.
- Students should proceed in the order given in the procedure. It is not a good idea to have them perform the laboratory activity as a series of individual experiments. The filter paper salt bridge allows students to take all readings in less than ten minutes.
- All metal nitrate solutions can be returned to the supply bottles and used again.
- Pieces of metal can be dried, kept in a plastic bag, and used again.
- **Troubleshooting:** The major problem in this experiment is failure to clean the metal electrodes sufficiently, especially the more active metals such as magnesium and aluminum. These active metals react readily with oxygen in the air. Remind students to clean the electrodes thoroughly. The potential differences in the small wells leave little margin for error.
- **Disposal:** The piece of filter paper soaked in KNO_3 can be discarded in the trash.

DATA AND OBSERVATIONS

Students' hypotheses will vary, but students should be able to predict, based on Table 17.1 in the text, that magnesium is more active than copper and will give up electrons to copper ions, and that nickel is more active than silver and will give up electrons to silver ions.

ANALYSIS

1. Number of times each electrode gave up electrons to the ions in solution: Cu 1 Pb 2 Mg 9 Al 8 Ni 6 Zn 7 Sn 4 H 3 Fe 5 Ag 0
2. Sample activity series:
   Mg Al Zn Ni Fe Sn H Pb Cu Ag

CONCLUSIONS

1. Magnesium was the most active metal.
2. Silver was the least active metal.
3. The activity series based on standard reduction potentials is
   Mg Al Zn Fe Ni Sn Pb H Cu Ag
4. Students' answers will vary.
5. The order might be different for many reasons: the experiment was not done under standard-state conditions; the metals were not sufficiently clean; differences in the voltages were too small to be detected; whether Fe is oxidized to Fe^{2+} or to Fe^{3+} and whether Sn is oxidized to Sn^{2+} or all the way to Sn^{4+}.
6. Answers will vary depending on students' hypotheses.

EXTENSION AND APPLICATION

Students' lists will vary, but should include activities that depend on batteries.

	Cu	Pb	Mg	Al	Ni	Zn	Sn	H	Fe	Ag
Cu	X	Pb	Mg	Al	Ni	Zn	Sn	H	Fe	Cu
Pb	X	X	Mg	Al	Ni	Zn	Sn	H	Fe	Cu
Mg	X	X	X	Mg	Mg	Mg	Mg	Mg	Mg	Mg
Al	X	X	X	X	Al	Al	Al	Al	Al	Al
Ni	X	X	X	X	X	Zn	Ni	Ni	Ni	Ni
Zn	X	X	X	X	X	X	Zn	Zn	Zn	Zn
Sn	X	X	X	X	X	X	X	Sn	Fe	Sn
H	X	X	X	X	X	X	X	X	Fe	H
Fe	X	X	X	X	X	X	X	X	X	Fe
Ag	X	X	X	X	X	X	X	X	X	X

Name

Date Class

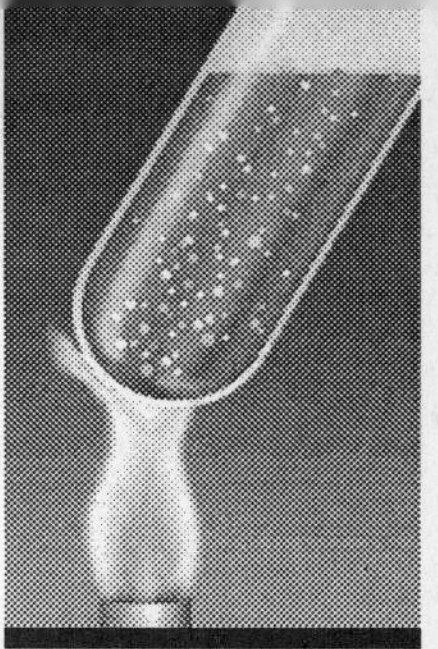

LAB MANUAL

17-2

The Six-Cent Battery

A battery is a device that converts chemical energy into electrical energy. Batteries are a convenient source of portable electrical power. Take a moment to think of the number of devices you use every day that use battery power.

A battery is a series of electrochemical cells connected together. An electrochemical cell consists of two different metals separated by a conducting solution (an electrolyte). The current produced by the difference in electrical potential can be used to do work. The first battery was constructed by Alessandro Volta. Volta used a stack of alternating metal disks separated by paper saturated with an electrolyte. In this activity, you will construct and test electrochemical cells and use them to make a simple battery.

OBJECTIVES

- **Construct** a working model of an electrochemical cell.
- **Evaluate** a series of metal pairs as sources of electrical energy.
- **Construct** a working model of a battery.

MATERIALS

pennies (4)
nickels (4)
dime
quarter
filter paper
scissors
thin-stem pipet
small rubber bands (4)
voltmeter (low range) or galvanometer
connecting wires

PROCEDURE

Part 1: Making an Electrochemical Cell

1. Use scissors to cut filter paper into 24 circles slightly larger than a quarter.
2. Use a pipet to soak the circles with a solution of sodium chloride.
3. Sandwich two soaked filter-paper circles between a penny and a nickel, as shown in Figure A.
4. Secure the coins and filter-paper circles with a rubber band.

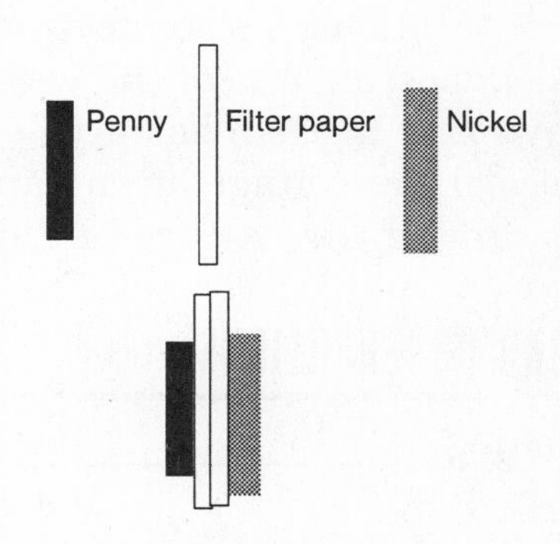

Figure A

5. Hold the cell on edge and carefully touch a voltmeter or galvanometer wire to each coin.
6. Record the voltage or galvanometer reading of the cell in the table under Data and Observations.
7. Repeat the procedure above using all other possible combinations of coins. Record each voltage or galvanometer reading in the data table.

Part 2: Making a Battery

1. Make a hypothesis about the size of the voltage that would be produced if four penny-nickel electrochemical cells were connected in series. Write your hypothesis under Data and Observations.
2. Make four separate penny-nickel cells as you did in Part 1, step 3.

Copyright © Glencoe/McGraw-Hill, a division of The McGraw-Hill Companies, Inc.

3. Place two soaked pieces of filter paper on the penny end of the first cell and set it on the nickel end of the second cell.
4. Repeat step 3 until you have all four cells connected in a series. See Figure B.

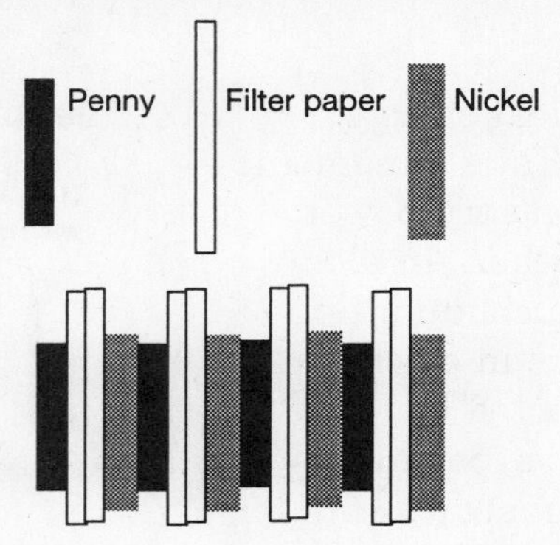

Figure B

5. Hold the coins on edge and carefully touch the meter's connecting wires to the opposite ends of the assembly. You have now made a battery.
6. Record the voltage or galvanometer reading of the battery in the data table.

DATA AND OBSERVATIONS

Hypothesis: ______________________________

____

____

Voltages of Cells				
	Penny	Nickel	Dime	Quarter
Penny	X			
Nickel	X	X		
Dime	X	X	X	
Quarter	X	X	X	X
Voltage of battery ______V				

ANALYSIS

1. Which combination of coins produced the highest voltage?

____

____

2. Which combination of coins produced the lowest voltage?

____

____

CONCLUSIONS

1. Compare the voltage of the four-cell battery with that of a single cell consisting of the same metals. What is your observation about the voltages?
2. Was your hypothesis correct? Why or why not?

____

____

____

EXTENSION AND APPLICATION

1. Have you ever accidentally chewed on a piece of aluminum foil from a gum wrapper? If you have fillings in your teeth, you may have experienced an annoying sensation. What was the sensation, and what caused it?
2. Providing portable electrical power is an engineering challenge. Look in library references for methods by which portable electrical energy is produced.

Copyright © Glencoe/McGraw-Hill, a division of The McGraw-Hill Companies, Inc.

Name Date Class

Copyright © Glencoe/McGraw-Hill, a division of The McGraw-Hill Companies, Inc.

Teacher Guide

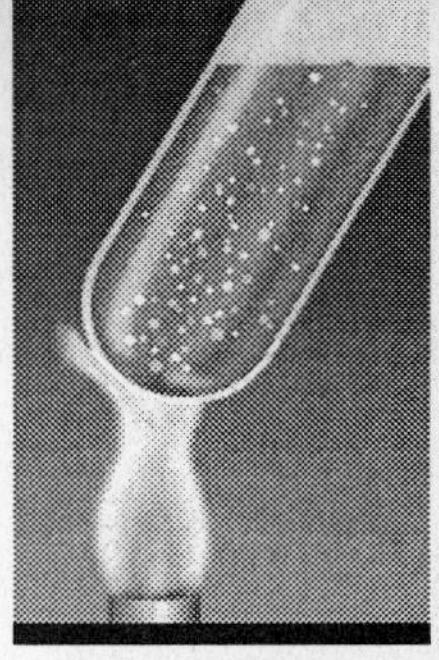

LAB MANUAL

17-2

The Six-Cent Battery

Students are usually more interested in chemistry if it involves common things. Zinc, copper, and magnesium strips can be used in the laboratory instead of coins, but students will probably be more motivated if coins are used.

Misconception: Students often think a dry cell is dry. Tell them that an electrolyte is present in paste form inside the cell.

PROCESS SKILLS

1. Students will make dry cells using different combinations of coins.
2. Students will test the voltage of each dry cell and decide which combination of coins results in the highest voltage.
3. Students will make a battery and test its voltage.

PREPARATION HINTS

- Before beginning the laboratory, it may help to remove excess dirt and corrosion from the coins with steel wool.
- Paper towels may be used in place of filter paper. Direct students to place several folds between the coins.

MATERIALS

1*M* NaCl solution (58.5 g of NaCl dissolved in enough distilled water to make 1 L of solution)
pennies (4)
nickels (4)
dime
quarter
filter paper
scissors
thin-stem pipet
small rubber bands (4)
voltmeter (low range) or galvanometer
connecting wires

PROCEDURE HINTS

- Before proceeding, be sure students have read and understood the purpose, procedure, and safety precautions for this laboratory activity.
- **Disposal:** Dispose of the filter paper according to the instructions in **Disposal** A.

DATA AND OBSERVATIONS

Students' hypotheses will vary but should indicate that the voltage of the electrochemical cells connected in a series will be greater than that of an individual electrochemical cell.

Voltages of Cells				
	Penny	Nickel	Dime	Quarter
Penny	X	0.04	0.05	0.06
Nickel	X	X	0.01	–0.01
Dime	X	X	X	–0.02
Quarter	X	X	X	X
Voltage of battery	0.16 V			

ANALYSIS

1. The combination of a penny and a quarter produced the highest voltage.
2. The nickel/dime and nickel/quarter combinations produced the lowest voltage.

CONCLUSIONS

The battery's voltage should be four times the voltage of a single cell.

EXTENSION AND APPLICATION

1. Fillings are an amalgam of silver and mercury. The silver in the filling and the aluminum in the gum wrapper, with saliva as the electrolyte, form a galvanic cell in the mouth and produce a small voltage.
2. Students may find information about fuel cells, generators, alternators, and solar cells.

Name

Date Class

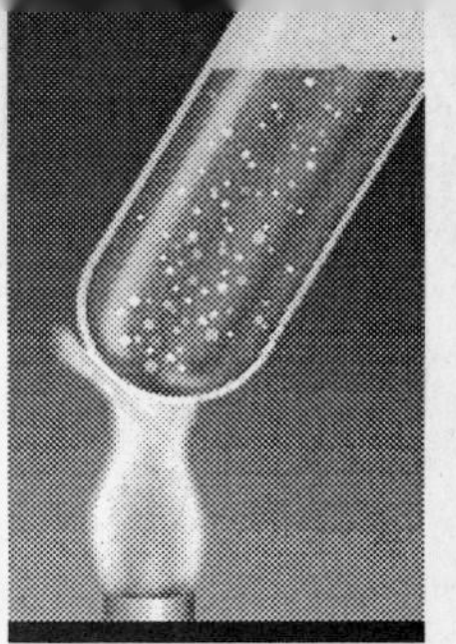

LAB MANUAL

18-1

Examples of Organic Reactions

Organic chemistry is the study of the chemistry of the covalent bonds of carbon. There are some similarities between organic and inorganic reactions. However, many organic reactions are slower and require more care in the control of factors such as temperature and concentration of reactants.

The simplest organic compounds are hydrocarbons—compounds of carbon and hydrogen. There are millions of compounds that contain only carbon and hydrogen. Carbon, unlike most other elements, forms rings and chains in its compounds. Most of these compounds have isomers. Isomers are compounds that have the same molecular formula but differ in structure.

Many carbon compounds contain additional elements, such as oxygen and nitrogen. These elements often form parts of the molecule that have higher reactivity than the rest of the molecule. These more active groups are called functional groups. Functional groups replace some of the hydrogen atoms in the basic organic molecule.

In this experiment, you will prepare ethyne, an unsaturated hydrocarbon, by the reaction of calcium carbide with water. You will then test for differences in reactions between a saturated, or single-bond, hydrocarbon and an unsaturated, or multiple-bond, hydrocarbon.

OBJECTIVES

- **Prepare** a simple hydrocarbon.
- **Compare and contrast** the difference in reactivity between two simple hydrocarbons.
- **Relate** the difference in reactivity between two hydrocarbons to the single and multiple bonds in the compounds.

MATERIALS

apron
goggles
test-tube racks (2)
large heat-resistant test tubes (9)
large beaker
test-tube clamp
ring stand
2-hole rubber stopper
solid rubber stoppers (8)
droppers (2)
forceps
rubber tubing
wood splints
glycerol
10-mL graduated cylinder
sand
marking pencil

PROCEDURE

Part 1: Saturated Hydrocarbon

1. Form a **hypothesis** to account for differences in reactivity of a saturated hydrocarbon—methane or propane—and an unsaturated hydrocarbon—ethyne. Write your hypothesis under Data and Observations.
2. Fill four test tubes with water and invert them in a large beaker approximately half full of tap water. You will collect a gas in these test tubes.
3. Place one end of a rubber tube on the end of a natural gas (methane) jet or the end of a propane tank at your teacher's desk.
4. Place the other end of the tube at the mouth of one of the inverted test tubes.
5. Open the valve of the gas supply and allow the gas to bubble into the test tube. When the tube is filled with gas, discard it because the gas will be mixed with air.
6. Fill the remaining three test tubes in the same way. As each tube is filled, stopper it and place it in a test-tube rack.

Copyright © Glencoe/McGraw-Hill, a division of The McGraw-Hill Companies, Inc.

7. Place the second test tube, stoppered end up, in a test-tube clamp, as shown in Figure A.

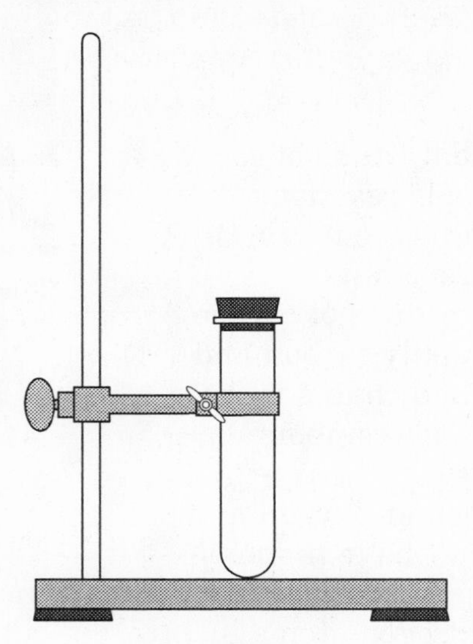

Figure A

8. Light a wood splint, remove the stopper from the test tube in the clamp, and place the burning splint at the mouth of the test tube. Record your observations in the table under Data and Observations.
9. Open one of the other test tubes and place 3 mL of potassium permanganate in the tube. Record your observations in the table.
10. Add 3 mL of household bleach to the fourth test tube. Stopper the tube. Record your observations.
11. Remove the stopper from the test tube with bleach and add 1 mL of hydrochloric acid to the tube to produce chlorine gas. Stopper the test tube. Record your observations. **CAUTION:** *Hydrochloric acid is corrosive. Chlorine gas is poisonous. Do not breathe any gas.*
12. With a marker or labels, label each test tube with the contents and set the rack aside.

Part 2: Unsaturated Hydrocarbon

1. Fill four test tubes with water and invert them in a large beaker approximately half full of tap water.
2. Add sand to another test tube until the tube is about three-quarters full.
3. Use forceps to place two or three pieces of calcium carbide on the surface of the sand in the tube.
4. Lubricate the end of a glass dropper with glycerol and insert it into one hole of a two-hole stopper. Remove the bulb from the dropper.
5. Fill a second dropper with water. Being careful not to release any water, insert this dropper into the other hole of the stopper.
6. Put the stopper assembly into the test tube of sand and calcium carbide. Use a test-tube clamp to attach the test tube to a ring stand.
7. Attach one end of a rubber tube to the end of the glass tube from the dropper.
8. Place the other end of the rubber tube at the mouth of one of the test tubes inverted in the beaker of water, as shown in Figure B.

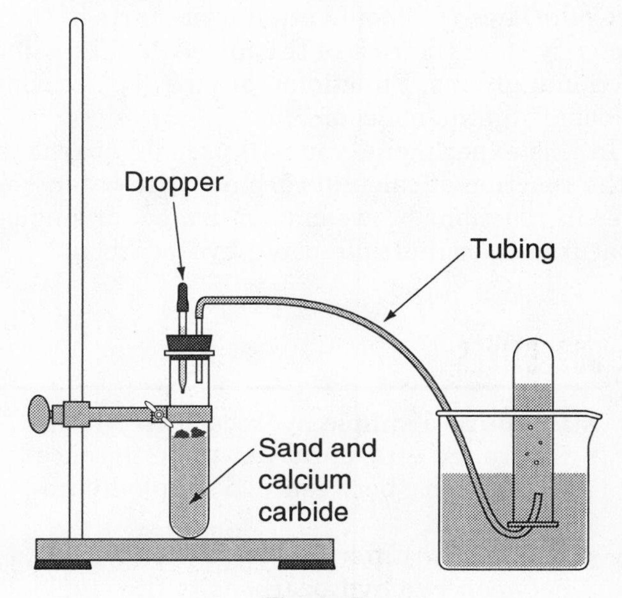

Figure B

9. Add a *few* drops of water to the calcium carbide by squeezing the dropper inserted in the stopper. **CAUTION:** *Water reacts with calcium carbide to form calcium hydroxide and ethyne. Calcium hydroxide is a strong base.*
10. When the inverted test tube is filled with gas, discard it because the gas will be mixed with air.
11. Repeat the process to fill the remaining three tubes with gas.
12. Repeat steps 7-12 of the procedure in Part 1, using the tubes of ethyne gas you have just generated.
13. Dispose of the contents of all the tubes in Part 1 and Part 2 according to your teacher's directions.

DATA AND OBSERVATIONS

Hypothesis: ______________________________

____

____

Reactions of Hydrocarbons		
	Observations	
Reaction	Saturated Hydrocarbon methane or propane CH_4 C_3H_8	Unsaturated Hydrocarbon ethyne C_2H_2
Combustion		
Potassium permanganate $KMnO_4$		
Bleach		
Chlorine gas Cl_2		

ANALYSIS

1. Write the equation for each of the reactions that occurred when the saturated hydrocarbon methane or propane was used in the procedures listed in the table. If no reaction occurred, write *NR*.

   Combustion

   ____

   Reaction with $KMnO_4$

   ____

   Reaction with bleach, NaClO

   ____

   Reaction with Cl_2

   ____

2. Write the equation for each of the reactions that occurred when the unsaturated hydrocarbon ethyne was used in the procedures listed in the table. If no reaction occurred, write *NR*.

   Combustion

   ____

   Reaction with $KMnO_4$

   ____

   Reaction with bleach, NaClO

   ____

   Reaction with Cl_2

   ____

CONCLUSIONS

1. How could you test for the products of combustion for each gaseous hydrocarbon used in this experiment?

   ____

   ____

   ____

   ____

   ____

2. Suppose you received unmarked samples of an unsaturated hydrocarbon and a saturated hydrocarbon. What steps could you use to determine which is the saturated and which is the unsaturated hydrocarbon?

   ____

   ____

   ____

   ____

   ____

3. What difference can be observed between the way methane or propane burns and the way ethyne burns? What would be necessary to make these gases burn in the same way?

4. Which of the hydrocarbons in this experiment was more reactive? How can you account for the differences in reactivity between the compounds?

5. Does your hypothesis accounting for the differences in reactivity of saturated and unsaturated hydrocarbons seem correct at this time? If not, modify your hypothesis.

EXTENSION AND APPLICATION

Compounds that have many double or triple bonds are said to be polyunsaturated. Draw or build molecular models of related compounds that are saturated, unsaturated, or polyunsaturated. What are the differences in the physical and chemical properties of these molecules?

Teacher Guide

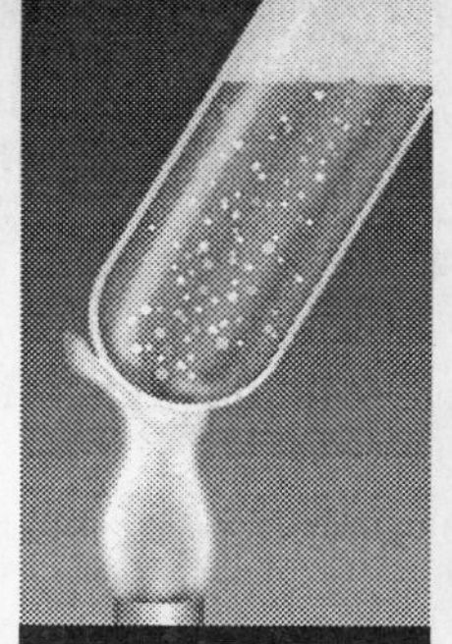

LAB MANUAL

18-1

Examples of Organic Reactions

Organic reactions can be an excellent way to supplement and apply some concepts learned in inorganic chemistry. The reaction of hydrocarbons with oxygen illustrates the processes of oxidation and reduction. The low ionization characteristics of organic acids exemplify the difference between an ionic and a covalent bond. The shape and polarity of isomers of organic compounds show how the physical properties of a compound can be predicted by the structure of a molecule.

PROCESS SKILLS

1. Students will prepare ethyne, an unsaturated hydrocarbon.
2. Students will distinguish between a saturated hydrocarbon and an unsaturated hydrocarbon by carrying out the same four procedures with each hydrocarbon.
3. Students will analyze the relationship between saturated hydrocarbons and single bonds and between unsaturated hydrocarbons and multiple bonds.

PREPARATION HINTS

- The quantities of organic chemicals to which students have access should be kept to a minimum. At no time during this experiment should stock bottles of organic chemicals be available to students.
- When preparing the hydrochloric acid solution, place the container of distilled water in a trough of cool tap water.

MATERIALS

methane or propane gas
0.001*M* potassium permanganate (Dissolve 0.16 g of the salt in 1 L of distilled water.)
household bleach (5% sodium hypochlorite in water)
6.0*M* HCl (Add 500 mL of concentrated hydrochloric acid to 500 mL of distilled water. Add the acid to the water, not the reverse.)

calcium carbide
apron
goggles
test-tube racks (2)
large beaker
test-tube clamp
ring stand
2-hole rubber stopper
18 mm × 150 mm heat-resistant test tubes (9)
forceps
rubber tubing
wood splints
glycerol
solid rubber stoppers (8)
droppers (2)
10-mL graduated cylinder
sand
marking pencil

PROCEDURE HINTS

- Before proceeding, be sure students have read and understood the purpose, procedure, and safety precautions for this lab.
- The reaction of bleach and hydrochloric acid produces chlorine gas. Only small quantities of chlorine gas should be produced in a test tube. **CAUTION:** *Chlorine gas is poisonous. Do not allow students to breathe this gas.*
- If a fume hood is available, conduct the reaction of chlorine with ethyne and the combustion of ethyne in the hood. In any case, make sure the room is very well ventilated. The combustion of ethyne in air produces soot.
- **Safety Precaution:** Except for combustion procedures, do not allow open flames.
- **Disposal:** You will want to dispose of the test tubes as follows:
  - **a.** If the yellow color of chlorine gas persists in any test tubes, the gas should be vented in a fume hood.
  - **b.** If any calcium carbide remains in the test tubes with sand, dry the test tubes thoroughly in a fume hood. The calcium carbide can be saved for future use in this lab, but it should not be returned to a stock bottle.

Copyright © Glencoe/McGraw-Hill, a division of The McGraw-Hill Companies, Inc.

c. The remaining test tubes should be washed routinely.

DATA AND OBSERVATIONS

Students' hypotheses should indicate that the differences in bonds of these compounds account for the differences in reactivity. Unsaturated hydrocarbons are much more reactive.

Reactions of Hydrocarbons

Reaction	Observations	
	Saturated Hydrocarbon methane or propane CH_4 C_3H_8	Unsaturated Hydrocarbon ethyne C_2H_2
Combustion	Flame almost invisible	Smoky flame (Note: The smoke and soot are due to incomplete oxidation of gas.)
Potassium permanganate $KMnO_4$	NR	$KMnO_4$ is decolorized. (Note: $KMnO_4$ breaks the double bond of ethyne, allowing a series of reactions.)
Bleach	NR	NR
Chlorine gas Cl_2	Color of Cl_2 gas fades slowly if tube is placed in sunlight. (Note: Chlorine replaces a hydrogen atom in the methane or propane molecule.)	Gas decolorizes quickly. (Note: $Cl_2(g)$ reacts with the triple bond of ethyne to form a chlorinated hydrocarbon.)

ANALYSIS

1. Combustion:
   $CH_4 + 2O_2 \rightarrow CO_2 + 2H_2O$ or
   $C_3H_8 + 5O_2 \rightarrow 3CO_2 + 4H_2O$
   Reaction with $KMnO_4$: NR
   Reaction with bleach, NaClO: NR
   Reaction with chlorine:
   $CH_4 + Cl_2 \rightarrow CH_3Cl + HCl$ or
   $C_3H_8 + Cl_2 \rightarrow C_3H_7Cl + HCl$
2. Combustion:
   $C_2H_2 + 3/2O_2 \rightarrow CO_2 + H_2O + C$
   Note: $3/2O_2$ indicates insufficient oxygen for complete combustion (the reason that carbon, in the form of soot, is a product).
   Reaction with $KMnO_4$:
   $8MnO_4^- + 5C_2H_2 + 14H^+ \rightarrow 8Mn^{2+} + 5C_2O_4^{2-} + 12H_2O$
   $2MnO_4^- + 5C_2O_4^{2-} + 16H^+ \rightarrow 2Mn^{2+} + 8H_2O + 10CO_2$
   Reaction with bleach, NaClO: NR
   Reaction with chlorine:
   $NaClO + HCl \rightarrow NaOH + Cl_2$
   $C_2H_2 + 2Cl_2 \rightarrow C_2H_2Cl_4$

CONCLUSIONS

1. A product of combustion, CO_2, can be tested for by bubbling the product gas through limewater. A cloudy or milky appearance confirms the test. Cobalt chloride paper turns blue in the presence of water, the other product of combustion.
2. Each sample could be tested with Cl_2 or Br_2. Rapid decolorization of the halogen is an indication of a multiple bond. Decolorization of $KMnO_4$ solution further confirms a multiple bond.
3. Methane or propane burned with a clean, luminous flame. Ethyne burned with a smoky, sooty flame. Extra oxygen must be provided to make ethyne burn completely.
4. Ethyne was more reactive than methane or propane. Methane and propane are saturated hydrocarbons; ethyne is unsaturated. The multiple bonds of unsaturated hydrocarbons break more easily than the single bonds of saturated hydrocarbons; the multiple bonds are the sites of reaction with other compounds.
5. Answers will vary.

EXTENSION AND APPLICATION

Molecules of related unsaturated and polyunsaturated hydrocarbons tend to be liquids and gases, whereas the related saturated hydrocarbon is a solid. For example, saturated fats are solids, whereas unsaturated and polyunsaturated hydrocarbons of the same carbon chain length are liquids. Regardless of the physical form, saturated hydrocarbons are less reactive. Polyunsaturated hydrocarbons are the most reactive because their multiple double bonds are sites for attack by oxidizing and reducing agents.

Polyunsaturated compounds tend to be liquids rather than solids. They react with halogens to produce halogenated hydrocarbons. Saturated hydrocarbons are relatively inert to halogen attack.

Copyright © Glencoe/McGraw-Hill, a division of The McGraw-Hill Companies, Inc.

Name

Date Class

Analysis of Aspirin

LAB MANUAL

18-2

Aspirin is one of the most familiar organic substances. It is a nonprescription drug that is commonly used to relieve pain and reduce fever. Its chemical name is acetylsalicylic acid. Aspirin is made by the reaction of salicylic acid and acetic acid. The reaction can be shown by this structural equation.

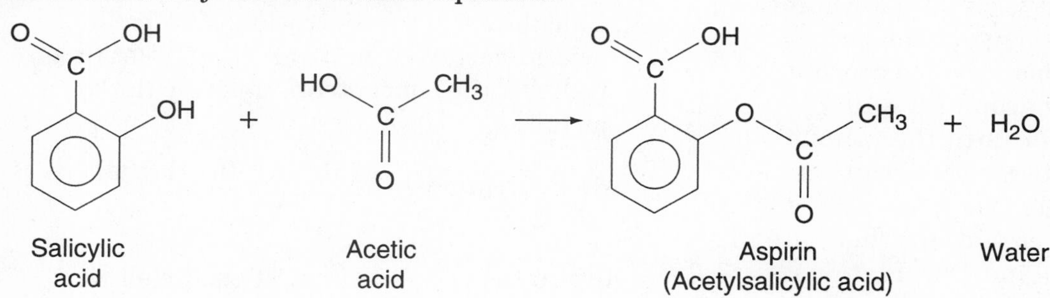

The OH^- group on the salicylic acid molecule reacts with the carboxyl group of the acetic acid to form the ester bond that holds the molecules together.

It is the salicylate part of the molecule that gives aspirin its medicinal properties. Aspirin that has been exposed to heat and moisture may contain unreacted or free salicylic acid. Salicylic acid is not very soluble in water. Therefore, the presence of free salicylic acid in an aspirin product is not desirable.

In this laboratory activity, you will determine the amount of salicylic acid in an aspirin tablet.

OBJECTIVES

- **Prepare** a serial dilution of salicylic acid.
- **Determine** the amount of salicylic acid in an aspirin preparation.

MATERIALS

apron
goggles
96-well microplate
96-well template
microtip pipets (4)
mortar and pestle
glass stirring rod
10-mL graduated cylinder
distilled water
toothpicks

PROCEDURE

Part 1: Preparation of a Salicylic Acid Dilution Series

1. Place the 96-well microplate on a template with the numbered columns away from you and the lettered rows at the left.
2. Use a Microplate Data Form to record all your observations.
3. Use a microtip pipet to obtain one-half a pipetful of standard salicylic acid solution from your teacher. This solution was made by dissolving 0.1 g of salicylic acid in 0.1 L of a mixture of ethanol and water. Therefore, 1 mL of the solution contains 0.001 g of salicylic acid.
4. Add 1 drop of the salicylic acid solution to well A1.
5. Add 2 drops of the salicylic acid solution to well A2.
6. Continue this process across the row, increasing the number of drops in each well, until you have added 10 drops of salicylic acid solution to well A10.
7. Use a clean microtip pipet to add 9 drops of distilled water to well A1.
8. Add 8 drops of distilled water to well A2.
9. Continue this process, reducing the number of drops of water in each well, ending with 1 drop of water in well A9.

10. Do not add any water to well A10.
11. With a clean microtip pipet, dispense 1 drop of iron(III) nitrate solution in each of wells A1–A10. A violet color indicates that free salicyclic acid has formed a complex with iron.
12. Use a toothpick to stir the contents of each well, beginning with well A1.

Part 2: Analysis of Salicylic Acid in an Aspirin Tablet

1. Fill a clean microtip pipet three-quarters full with the ethanol-water mixture provided by your teacher.
2. Counting the drops, add the contents of the pipet drop by drop into a 10-mL graduated cylinder until the volume reads 5 mL. (Remember to read the bottom of the meniscus.)
3. Record the number of drops there are in 5 mL under Data and Observations.
4. Crush an aspirin or a tablet containing aspirin with a mortar and pestle.
5. Add 20 mL of the ethanol-water mixture to the mortar. Stir the mixture with a glass stirring rod.
6. The white solid in the mortar is starch. Starch is used as a filler to ensure accurate dosage of small quantities of aspirin. Allow the starch to settle out.
7. Place 10 drops of the liquid from the mortar in well B5.
8. Add 1 drop of iron(III) nitrate solution to well B5. Stir with a toothpick.
9. Look at the intensity of the color of the liquid in well B5. Compare the color in well B5 with the color of the liquid in each well in row A (the standard row). Record the location of the well containing the solution that is closest in color to the solution in well B5.

DATA AND OBSERVATIONS

1. Drops of ethanol-water mixture in 5 mL:

____

____

2. Location of the well containing the solution that is the closest match in color to the solution in well B5:

____

ANALYSIS

1. Calculate the number of drops of ethanol-water mixture in 1 mL.

____

2. Calculate how many drops of aspirin solution are present when an aspirin is dissolved in 20 mL of ethanol-water mixture.

____

3. Calculate the concentration of salicylic acid in each well in row A. Example: In well A1, the concentration is calculated as follows:

$$\text{concentration} = \frac{1 \text{ drop} \times 0.001 \text{ g/mL}}{\text{analysis 1}}$$

Write the concentration of each well in your Microplate Data Form.

CONCLUSIONS

1. Calculate the number of grams of free salicylic acid in the tablet.

g of free salicylic acid in the tablet =

$$\text{g in well that matched B5} \times \frac{\text{analysis 2}}{10}$$

____

2. Compare your results with those of your classmates. Which preparation seems to be the best—that is, which has the least amount of free salicylate?

____

____

____

____

EXTENSION AND APPLICATION

Aspirin decomposes over long periods of time. Design an experiment that could be used to measure the rate at which aspirin decomposes.

Copyright © Glencoe/McGraw-Hill, a division of The McGraw-Hill Companies, Inc.

Teacher Guide

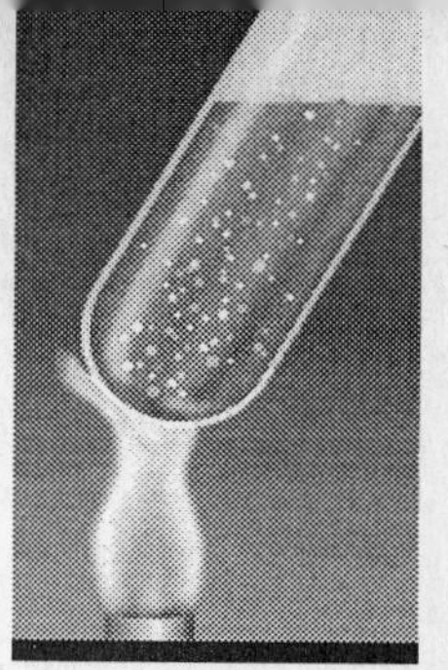

LAB MANUAL

18-2

Analysis of Aspirin

The presence of free salicylic acid can be detected by the addition of Fe^{3+}. Free salicylic acid forms a complex with the iron(III) ion that gives the solution a violet tint. The higher the concentration of salicylate, the more intense is the violet color.

PROCESS SKILLS

1. Students will prepare a series dilution of salicylic acid.
2. Students will prepare a solution of an aspirin tablet in an ethanol-water mixture.
3. Students will observe the reaction between the aspirin solution and an iron salt.
4. By comparing the color of the aspirin solution with those of the dilutions, students can determine the concentration of salicylic acid in the tablet.

PREPARATION HINTS

- Different groups of students can be given different brands of aspirin or aspirin-containing tablets so that they can compare results.
- Solutions for this experiment may be prepared months in advance. Be sure to keep the salicylic acid solution in a tightly stoppered bottle to avoid loss of alcohol. Each mL of the final solution contains 0.001 g of salicylic acid. **CAUTION:** *Salicylic acid is toxic. Ethanol is flammable.*
- In preparing the salt solution, be sure to check the hydration of your stock salt and adjust the mass needed, if necessary.
- For each group, make a copy of the 96-well template on page T8.

MATERIALS

50/50 ethanol-water mixture
0.1*M* $Fe(NO_3)_3$ (40.4 g of $Fe(NO_3)_3 \cdot 9\ H_2O$ in 1 L of solution)
standard solution of salicylic acid (0.1 g of salicylic acid in 100 mL of 50/50 ethanol and water)
aspirin and other tablets containing aspirin

apron
96-well microplate
96-well template
microtip pipets (4)
mortar and pestle
glass stirring rod
distilled water
goggles
10-mL graduated cylinder
toothpicks

PROCEDURE HINTS

- Before proceeding, be sure students have read and understood the purpose, procedure, and safety precautions for this laboratory activity.
- Students can have access to all solutions in limited quantities.
- Use small beakers or vials to distribute standard salicylic acid solutions.
- **Disposal:** Collect all solutions and dispose of them according to **Disposal** J.

DATA AND OBSERVATIONS

1. There are 160 drops in 5 mL of the ethanol-water mixture.
2. The color in well A4 best matches the color in well B5.

ANALYSIS

1. Number of drops in 1 mL: 32 drops/mL
2. Drops of aspirin solution in 20 mL:
   20 mL × 32 drops/mL = 640 drops
3.

		1	2	3	4	5	6	7	8	9	10
Grams $\times 10^{-4}$ of salicylic acid	A	0.3	0.6	0.9	1.2	1.5	1.8	2.1	2.4	2.7	3.0
	B					aspirin solution					

CONCLUSIONS

1. Grams of free salicylic acid in the aspirin tablet:

   1.2×10^{-4} g × 640 drops/10
   = 0.008 g of free salicylic acid

2. Answers will vary.

EXTENSION AND APPLICATION

Students should suggest storing samples of aspirin over long periods of time under different conditions. The tablets could be treated using this lab procedure. The concentration of salicylate could then be plotted as a function of time. Gradually, aspirin decomposes to form acetic acid and salicylic acid.

Name

Date Class

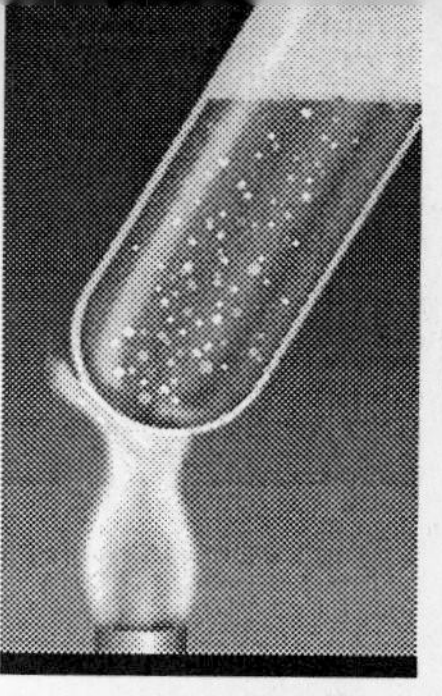

LAB MANUAL

19-1

Biochemical Reactions

Biochemical reactions involve organic molecules that take part in the processes of life. These reactions, the chemistry of life, are studied in biochemistry. Many of the molecules can be classified into three general classes.

Carbohydrates — Carbohydrates are compounds containing hydrogen, oxygen, and carbon only. The breakdown of these molecules provides energy, which is needed for carrying out life processes. The breakdown products are carbon dioxide and water.

Carbohydrates can be classified as monosaccharides, disaccharides, and polysaccharides. Monosaccharides are simple sugar molecules; disaccharides contain two linked monosaccharides. Both monosaccharides and disaccharides are sugars. Polysaccharides contain hundreds or even thousands of monosaccharides bonded together. Polysaccharides include starches and cellulose.

In this lab, you will test for a type of sugar called a reducing sugar—a sugar that can reduce copper(II) oxide to insoluble red copper(I) oxide. Reducing sugars are tested for by use of Benedict's solution. A positive Benedict's test is signalled by a change in color of the Benedict's solution from blue to green, orange, or red, depending on the amount of reducing sugar present. The redder the solution, the greater is the concentration of sugar present.

Starch, a polysaccharide, is tested for by the iodine test. A positive test for starch in a solution is the appearance of a blue-black color when the solution is exposed to iodine.

Lipids—Lipids, like carbohydrates, contain carbon, hydrogen, and oxygen. However, the atoms of lipids are arranged quite differently from the atoms of carbohydrates. Lipids produce carbon dioxide and water when they are broken down, or metabolized, by the body. If lipids are not metabolized, they are stored in the body. The stored solid fat is an energy reserve. Liquid lipids are often referred to as oils.

Lipids can be detected by the translucence test. When a food or some other material containing lipid is applied to a brown, unglazed piece of paper, the paper becomes translucent.

Proteins—Proteins are made up of amino acids, which are linked together to form large molecules. The arrangement, order, and number of amino acid units in a protein are determined by the DNA of the organism that produced the protein. Amino acids, and therefore proteins, contain carbon, hydrogen, oxygen, nitrogen, and, often, sulfur.

Proteins give a positive biuret test. A positive biuret test is signalled by the appearance of a pink, blue, or violet color, depending upon the protein.

In this activity, you will test various substances for the presence of sugar, starch, lipid, and protein.

OBJECTIVES

- **Test** a variety of substances for the presence of sugar, starch, lipid, and protein.
- **Recognize** positive reactions to tests for sugar, starch, lipid, and protein.
- **Identify** substances that contain sugar, starch, lipid, and protein.

MATERIALS

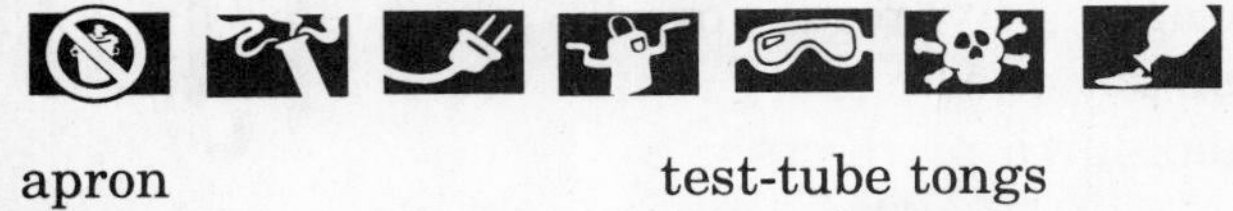

apron
goggles
400-mL beakers (4)
large test tubes (25)
10-mL graduated cylinder
hot plate
test-tube tongs
test-tube racks
unglazed brown paper
thin-stem pipet
marking pencil
distilled water

PROCEDURE

Develop a **hypothesis** about the minimum number of chemical tests that are required to identify the presence of proteins, lipids, and carbohydrates. Write your hypothesis under Data and Observations.

Part 1: Testing for Sugar

1. Prepare a warm water bath. Fill one of the beakers half-full of tap water, and begin heating the water.
2. Label a test tube *HCl*. Place 2 mL of starch solution in the test tube. Add 2 mL of hydrochloric acid to the tube. **CAUTION:** *Hydrochloric acid is very corrosive.*
3. Place the test tube in the water bath. Leave it in the water bath until you have completed all of Parts 1–4 of the Procedure.
4. Label the remaining three beakers: *Sugar*, *Starch*, and *Lipid/Protein*.
5. Label three test tubes for each of the first eight substances listed in the table under Data and Observations. You will have a total of 24 test tubes. Place 3 mL of the appropriate substance in each test tube. For example, place 3 mL of fructose solution in each of three test tubes labeled *Fructose*.
6. Sort out the test tubes so that you have three complete sets of eight sample tubes. Place one set in each of the three labeled beakers.
7. Add 3 mL of Benedict's solution to each of the test tubes in the beaker labeled *Sugar*. **CAUTION:** *Benedict's solution is corrosive.*
8. Add tap water to the beaker until it is half full.
9. Using the beaker as a hot water bath, gradually heat the water to boiling and boil for 2–3 minutes.
10. Record your observations in the data table in this way: Place an *X* in the column labeled *Sugar* to indicate the presence of sugar. If no sugar is present, leave the space blank.
11. Allow the warmed tubes and the beaker to cool.

Part 2: Testing for Starch

1. Add 5 drops of iodine solution to each test tube in the beaker labeled *Starch*. **CAUTION:** *Iodine is corrosive and may irritate the eyes.*
2. Record your results in the data table in the column labeled *Starch*. Use an *X* to indicate a positive test.

Part 3: Testing for Lipid

1. Use a pencil to mark off 8 squares on a piece of unglazed brown paper.
2. Label the squares with the names of the substances in the test tubes. Assign a different substance to each square.
3. With a clean thin-stem pipet, remove 1–2 drops of each substance, and place the drops on the appropriately labeled squares on the brown paper.
4. Allow the spots to dry.
5. Record your results in the data table in the *Lipid* column.

Part 4: Testing for Proteins

1. While waiting for the spots to dry in the test for lipid, test the Lipid/Protein test group with biuret reagent. **CAUTION:** *Biuret reagent is toxic and corrosive.* To each tube, carefully add 3 mL of the biuret reagent. Shake the test tube.
2. Record your observations in the data table in the *Protein* column.

Part 5: Sugar Test Revisited

1. Add 3 mL of Benedict's solution to the warmed starch/HCl solution you prepared in steps 2 and 3 of Part 1.
2. Boil the test tube in the water bath for 2-3 minutes.

3. Record your observations in the data table.
4. Dispose of the contents of the test tubes as your teacher directs.

DATA AND OBSERVATIONS

Hypothesis: ______________________

____

____

Testing Biochemical Compounds				
Substance	Type of Compound			
	Sugar	Starch	Lipid	Protein
Fructose				
Soluble starch				
Albumin				
Maltose				
Dextrin				
Salad oil				
Gelatin				
Distilled water				
Starch/ HCl				

ANALYSIS

1. Classify each of the substances in the data table as a sugar, starch, carbohydrate, lipid, or protein.

____

____

____

____

____

____

____

2. What happened in the tube containing starch and hydrochloric acid?

____

____

____

CONCLUSIONS

1. Compare and contrast the results of the Benedict's test on the starch solution with the Benedict's test on the starch/HCl solution. Account for any difference in results.

____

____

____

____

2. What is the control in all of these tests?

____

3. Was your hypothesis about the number of tests you would need to perform to identify the presence of carbohydrates, lipids, and proteins correct? Explain.

____

____

____

EXTENSION AND APPLICATION

Design a procedure for testing a fast food meal for sugars, starches, lipids, and proteins. With your teacher's approval of your plan, carry it out. Are your tests qualitative or quantitative? Why do you say so?

Teacher Guide

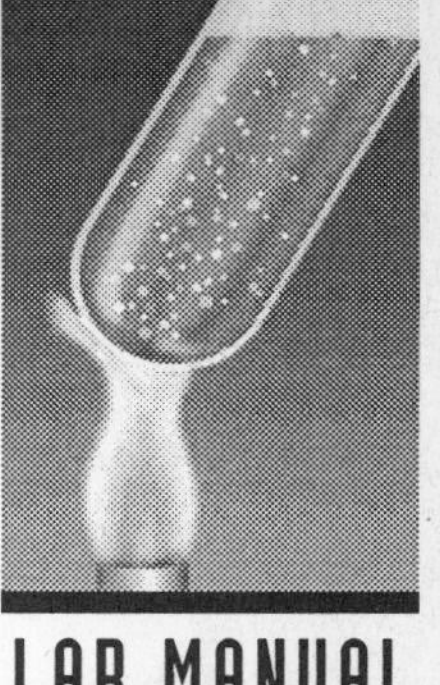

LAB MANUAL

19-1

Biochemical Reactions

Biochemistry is motivating to students. They enjoy using chemical tests to reveal the presence of the molecules referred to in the introduction to the laboratory.

Students will probably recall from biology that there are six classes of nutrients: proteins, lipids, carbohydrates, minerals, vitamins, and water. Yet students may never have thought that only six classes of nutrients are present among all the thousands of foods in a supermarket or all the dishes they can eat in a restaurant or at home.

In this activity, students will test for only three nutrients. But they will carry out four different tests because sugars and starches — both carbohydrates — are tested for separately. The idea of a limited number of nutrient classes and the tests students carry out should lead directly to students' interest in completing the Extension and Application activity. This offers an exceptional opportunity to relate biochemistry to students' own experiences.

Misconception: Students rarely think of what they eat as chemical compounds. This laboratory, including the Extension and Application section, is a good way to show chemistry at work in daily life.

PROCESS SKILLS

1. Students will contrast the results of chemical tests for sugars, starches, lipids, and proteins.
2. Students will use chemical tests to identify the presence of sugars, starches, lipids, and proteins in a variety of substances.

PREPARATION HINTS

- The solutions to be tested cannot be prepared too far in advance. Because they are foods, bacteria and molds grow readily in them. The solutions may be prepared the day before the laboratory period, but they must be stored in a refrigerator overnight.
- Do not use old solutions. Old solutions will not give positive results. Also, an old solution may be a biohazard because it may contain pathogenic bacteria, yeast, or molds.
- All samples, except the oil sample, are solutions of 1 percent nutrient in distilled water. The solutions are prepared by dissolving 10 g of the nutrient in 1 L of distilled water.
- Be sure to use soluble starch in preparing the starch solution. It must be boiled before it will dissolve.
- Oil may be used directly. Salad oil of any type is adequate for this test.
- Prepare several sets of test reagents. The reagent solutions may be prepared as follows.
  - Iodine solution—Add 1 g of iodine to a 50 mL/50 mL mixture of alcohol and water. Place this solution in a dropper bottle.
  - 6*M* HCl—Add 500 mL of concentrated HCl solution to 500 mL of distilled water. Cool the solution as you add the acid to water. *Do not add water to the acid.*
- A grocery bag cut into pieces is ideal for the unglazed brown paper.
- For the Extension and Application section, purchase only one fast food meal per class. A typical meal could include hamburger on a roll, french fries, and a soft drink (non-diet) or milk.

MATERIALS

1 percent solutions of fructose, soluble starch, albumin, maltose, dextrin, gelatin,
biuret reagent (commercial preparation)

Benedict's solution (commercial preparation)
iodine solution
$6M$ HCl
salad oil
unglazed brown paper
apron
goggles
400-mL beakers (4)
18 mm × 150 mm test tubes (25)
10-mL graduated cylinder
hot plate (or burner with ring stand, large ring, and wire gauze)
test-tube tongs
test-tube racks
thin-stem pipet
marking pencil
distilled water

PROCEDURE HINTS

- Before proceeding, be sure students have read and understood the purpose, procedure, and safety precautions for this laboratory activity.
- Students should work in pairs or groups.
- Remind students that iodine stains are extremely difficult to remove.
- Because many test tubes are required in this activity, you may elect to have students prepare each set of nutrients separately. Each lab group would require only 9 test tubes instead of 25. The disadvantage of using the tubes over again for each test is that students will not be able to review the results of the tests after the lab is finished.
- Students should be reminded to wait for the reagents to react with the test solutions. Students should allow at least three minutes for the Benedict's test to show results. Emphasize that patience is needed in this laboratory work.
- Instead of a hot plate, a laboratory burner or an immersion heater may be used to heat the water bath.
- It is a good idea to have students perform the activity suggested in the Extension and Application section as another experiment, if time permits.
- **Troubleshooting:** A problem that may be encountered in this experiment is the age of the nutrient solutions. The nutrients must be fresh.
- **Safety Precaution:** Biuret reagent should be handled very carefully. The NaOH in the reagent is caustic, and the $CuSO_4$, also in the reagent, is poisonous if swallowed.
- **Disposal:** Dispose of the solutions according to the instructions in **Disposal** J.

DATA AND OBSERVATIONS

Students' hypotheses will vary but will likely indicate three tests for the three nutrients. Four tests actually are needed, because detecting sugars and starches requires two different tests.

Testing Biochemical Compounds

Substance	Type of Compound			
	Sugar	Starch	Lipid	Protein
Fructose	X			
Soluble starch		X		
Albumin				X
Maltose	X			
Dextrin		X		
Salad oil			X	
Gelatin				X
Distilled water				
Starch/HCl	X	X		

ANALYSIS

1. See the data table. Sugars and starches are carbohydrates. Salad oil is a lipid, and albumin and gelatin are proteins.
2. The starch and hydrochloric acid gave a positive test for sugar. The starch molecule was broken down to the sugar molecules that were joined to form the starch molecule. If all the starch had not been broken down, a starch test would also be positive.

CONCLUSIONS

1. In the procedure for Part 1, the Benedict's test for sugar in starch was negative. The Benedict's test for sugar in the starch/HCl solution was positive. Some starch in the mixture had been hydrolyzed to monosaccharides by HCl.
2. The control is distilled water.
3. Answers will vary.

EXTENSION AND APPLICATION

A suspension or slurry of each food in the meal could be prepared and tested with each of the reagents used in this experiment. All the tests would be qualitative.

Copyright © Glencoe/McGraw-Hill, a division of The McGraw-Hill Companies, Inc.

Name

Date Class

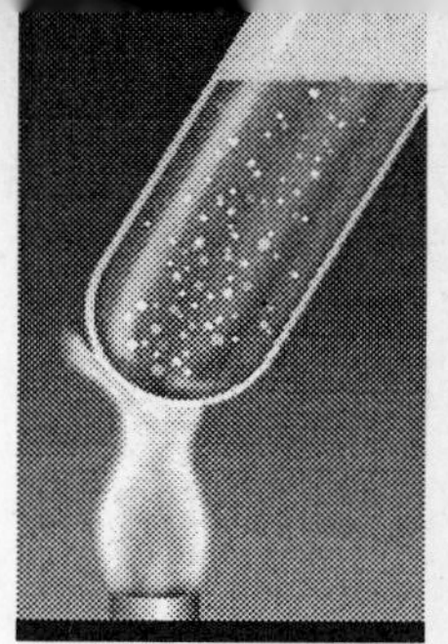

Qualitative Analysis of Food

"Variety is the spice of life" is an old saying that can be applied to how people eat. People do enjoy many kinds of foods. But all foods are made up of six basic nutrients: carbohydrates (starches and sugars), lipids, proteins, vitamins, minerals, and water. It is these nutrients that our bodies use to get energy for life processes, to grow, and to repair themselves.

In this activity, you will run standard tests for nutrients. Then you will carry out the tests on actual foods to determine which nutrients they contain.

OBJECTIVES

- **Observe** the reaction of known nutrients with test reagents.
- **Test** foods for the presence of nutrients.

MATERIALS

apron
goggles
400-mL beaker
thermometer
96-well microplate
96-well template
microtip pipets (6)
hot plate
toothpicks
small strips of brown paper
plastic tray or basin
distilled water
white paper (1 sheet)
thermal mitt

PROCEDURE

1. Fill the beaker with water and begin heating it. When the temperature reaches 50°–60°C, maintain the temperature in this range until the end of the experiment.
2. Place a 96-well microplate on a template with the lettered rows at the left and the numbered columns away from you.
3. Add 5 drops of Benedict's solution to each of wells A1 through A11. **CAUTION:** *Benedict's solution is corrosive.*
4. Add 5 drops of biuret reagent to each of wells B1 through B11. **CAUTION:** *Biuret reagent is corrosive and toxic.*
5. Add 5 drops of iodine solution to each of wells C1 through C11. **CAUTION:** *Iodine is corrosive and may irritate the eyes.*
6. Place a small strip of brown paper in each of wells D1 through D11.
7. Label the Microplate Data Form to show the location of each reagent.
8. Add 5 drops of sugar solution to each of wells A1, B1, C1, and D1.
9. Add 5 drops of starch suspension to each of wells A2, B2, C2, and D2.
10. Add 5 drops of vegetable oil to each of wells A-D in column 3. Add 5 drops of protein solution to each of wells A-D in column 4. Add 5 drops of distilled water to each of wells A-D in column 5. A-D column 5 wells will serve as controls.
11. Label the Microplate Data Form to show the substances added to the wells in each column.
12. Mix the contents of each well with a toothpick. Use a different toothpick for each well.
13. Using the standard tests, you will test real foods (rice water, chicken soup, juice, a soft drink, gelatin, and salad dressing) for the presence of sugar, starch, protein, and lipid. Use your knowledge of nutrients to make a **hypothesis** about the nutrients you will find in each food. Write your hypothesis under Data and Observations.
14. With a microtip pipet, add 5 drops of boiled rice solution to each of wells A6, B6, C6, and D6.
15. Use a clean microtip pipet to add 5 drops of chicken soup to each of wells A7, B7, C7, and D7.
16. Using a clean microtip pipet for each food, add juice, soft drink, gelatin, and

salad dressing to the remaining columns.

17. Label the Microplate Data Form to show the foods added to the wells of columns 6 through 11.
18. Using a different toothpick for each well, mix the contents of the wells in columns 6 through 11.
19. Pour the heated water from step 1 into a tray or basin.
20. Float the microplate in the warm water for approximately 10 minutes.
21. Remove the strips of brown paper from the wells. Place the strips on a clean sheet of white paper.
22. Make observations of all the wells and the strips of brown paper. Record your observations in the Microplate Data Form.

DATA AND OBSERVATIONS

Hypothesis: ______________________

____

____

Record all your observations in the Microplate Data Form.

ANALYSIS

1. Describe a positive test for sugar.

____

____

2. Describe a positive test for starch.

____

____

3. Describe a positive test for lipid.

____

____

4. Describe a positive test for protein.

____

____

5. Describe the changes that occurred in the wells in column 5.

____

____

CONCLUSIONS

1. Which nutrients were present in each of the foods you tested?

____

____

____

____

2. Did your results support your hypothesis? Explain.

____

____

EXTENSION AND APPLICATION

1. The tests you conducted in this activity were qualitative. How are qualitative tests different from quantitative tests?
2. Find out how tests for some minerals and vitamins are conducted.

Teacher Guide

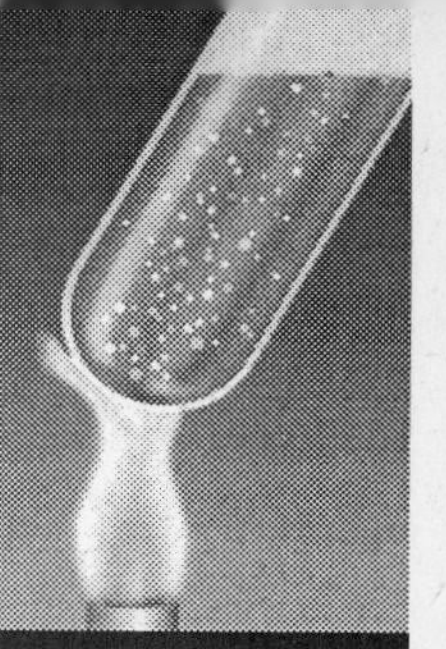

LAB MANUAL

19-2

Qualitative Analysis of Food

The tests used in this laboratory activity can be extended to extracts of any food. For example, a fast food meal consisting of a hamburger, french fries, and a milk shake provides an ample supply of materials for students to analyze. Small quantities of each food can be placed in test tubes and a water extract or slurry of the food can be prepared in each test tube. Students can then use a test matrix similar to the one used in the procedure.

Misconception: Students do not usually consider foods to be collections of chemical compounds. By carrying out this analysis, students can tie the concept of chemical tests to the science of nutrition.

1. Students will carry out standard tests for sugars, starches, proteins, and lipids.
2. Students will test for these nutrients in foods.

PREPARATION HINTS

- Different food samples can be used, depending on availability.
- Be sure the soft drink is not a diet drink.
- Supply the foods to be tested directly from their containers as a motivating device.
- Instead of hot plates, the water can be heated with immersion heaters or over laboratory burners.
- For each group, make a copy of the 96-well template on p. T8.

MATERIALS

Benedict's solution (commercial preparation)
biuret reagent (commercial preparation)
iodine solution (1 g of elemental iodine in 50:50 by volume alcohol/water solution)
brown paper (grocery bag cut into thin strips)
vegetable oil
solutions or suspensions of the following: starch, sugar (glucose), egg albumin or gelatin (1 g of the solid dissolved in 100 mL of distilled water)
other foods: water containing boiled rice, chicken soup, juice, soft drink, salad dressing
apron
goggles
400-mL beaker
thermometer
96-well microplate
96-well template
microtip pipets (6)
toothpicks
hot plate (or burner, ring stand, large iron ring, and wire gauze)
plastic tray or basin
distilled water
white paper (1 sheet)
thermal mitt

- Before proceeding, be sure students have read and understood the purpose, procedure, and safety precautions for this laboratory activity.
- Students may take more than half a period to set up the experiment. Once the reagents and nutrients are mixed, however, the reactions are almost immediate.
- Starch tends to settle out of suspension. Shake the suspension to resuspend the starch.
- **Troubleshooting:** When testing the foods, students must allow enough time for a positive result to develop. In some cases, the concentration of nutrient is low. It is also important for students to heat the microplate as described in the procedure. The Benedict's test is sensitive to heat.
- **Safety Precaution:** Benedict's solution and biuret reagent contain corrosive materials. Make sure students wear goggles and aprons during this laboratory.
- **Disposal:** All solutions and foods should be collected in a basin and disposed of

+ = positive test

Tests for Nutrients

	Glucose 1	Starch 2	Oil 3	Protein 4	Water (control) 5	Rice water 6	Soup 7	Juice 8	Soft drink 9	Gelatin 10	Salad dressing 11	12
Benedict's solution A	+							+	+		+	
Biuret reagent B				+			+			+		
Iodine solution C		+				+	+				+	
Brown paper D			+				+				+	
E												
F												
G												
H												

according to the directions in **Disposal** J. Collect all pieces of brown paper for disposal in the trash. Do not add the paper to the basin in which materials for **Disposal** J are collected.

DATA AND OBSERVATIONS

Students' hypotheses will vary but should include the most obvious nutrients in at least some foods—for example, starch in rice water, protein in chicken soup, sugar in juice, and lipid in salad dressing.

ANALYSIS

1. Benedict's solution changes color from green to yellow and then to orange in the presence of simple sugars.
2. Iodine solution turns blue-black in the presence of starches.
3. The brown paper takes on a translucent appearance in the presence of lipids.
4. Biuret reagent turns pink (or purple) in the presence of proteins.
5. No color changes occurred in the control wells. Brown paper became wet but not translucent.

CONCLUSIONS

1. Rice water contains starch. Chicken soup contains lipid, starch, and protein. Juice and soft drink contain sugar. Gelatin contains protein. Salad dressing contains sugar, starch, and lipid. (Answers may vary if different foods are used. They also may vary depending on the particular contents of prepared foods; check labels.)
2. Answers will vary depending on students' hypotheses.

EXTENSION AND APPLICATION

1. Qualitative tests determine the presence or absence of a particular substance. Quantitative tests determine the amount of a substance present.
2. Chemical tests can detect the presence of particular minerals or vitamins. For example, calcium in foods can be detected by the addition of sodium oxalate to the food sample or extract. Calcium ions combine with the oxalate and form a white precipitate. You can test for the presence of vitamin C by adding indophenol solution to a sample of a food. Indophenol is a blue solution that decolorizes when exposed to vitamin C.

Name

Date Class

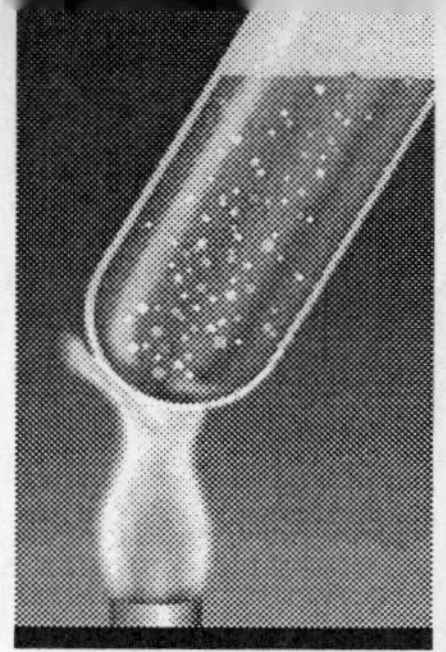

LAB MANUAL

20-1

Energy Changes in Physical and Chemical Processes

Chemical reactions have a tendency to move toward lower energy and greater entropy. The first tendency is for reactions to be exothermic, or release heat. The tendency toward greater entropy means that the components of the system will become more random. A reaction in which the energy of a system decreases and the entropy increases will always be spontaneous. Reactions that are exothermic but have a decrease in entropy can take place spontaneously at low temperatures.

In this activity, you will investigate the flow of energy and the effects of energy on physical and chemical systems.

OBJECTIVES

- **Explore** energy and entropy, using a physical system as a model for chemical reactions.
- **Classify** reactions as exothermic or endothermic and as entropic or more orderly.

MATERIALS

thick rubber band
hot plate
250-mL beakers (2)
1- or 2-kg mass
ringstand and iron ring
24-well microplate
24-well template
ruler
scissors
thermometer
thin-stem pipets (2)
toothpicks

PROCEDURE

Part 1

1. Half-fill a 250-mL beaker with water and place the beaker on a hot plate to heat.
2. While the water is heating, put a thick rubber band around both of your index fingers and hold the fingers apart so that the rubber band is at its full length, *but is not stretched*. Briefly touch the rubber band to your upper lip and sense the temperature of the band.
3. Move your index fingers apart to stretch the rubber band as much as you can. Briefly touch the stretched rubber band to your lip. Notice whether you feel a change in the temperature from your earlier observation. Record your observations in Table 1 under Data and Observations.
4. Slip the rubber band onto the ring attached to the ringstand and let it dangle. Hang a mass weighing 1 to 2 kg from the end of the rubber band, as shown in the diagram.
5. With a ruler, measure the length of the rubber band, and enter the length in Table 1.
6. Place an empty 250-mL beaker below the mass and the rubber band.

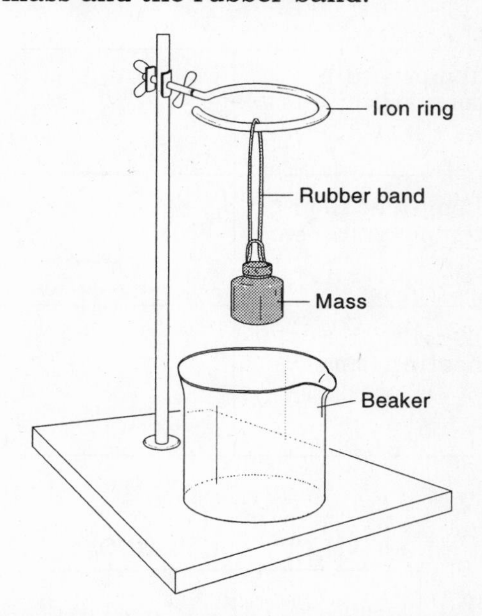

7. Wearing thermal mitts, carefully pour hot water from the beaker on the hot plate over the rubber band, collecting the water in the beaker beneath. Measure the length of the rubber band and record this new length.

Part 2

1. Make a chemical microscoop by cutting the end off the bulb of a thin-stem pipet. (See Laboratory 16-1.)
2. Place ½ microscoop of solid ammonium chloride in well A1 and well B1 of a 24-well microplate. Add ½ microscoop of solid sodium hydrogen carbonate to well A2 and well B2.
3. With the thermometer, measure the temperature of the solid chemicals in wells A1, A2, B1, and B2. Record these data in Table 2.
4. Using the thin-stem pipet, place ½ pipet of water in wells A1 and A2. Stir with a toothpick. Enter any physical changes you observe in Table 2.
5. With the thin-stem pipet, add HCl to wells B1 and B2. Stir with a toothpick. Enter any changes you observe in Table 2.
6. Use the thermometer to measure the temperature of the chemicals as they dissolve in wells A1, A2, B1, and B2. Be sure to rinse your thermometer in cold water between each reading. Record these data in Table 2.

DATA AND OBSERVATIONS

Table 1

A	Temperature comparison	
B	Length at room temperature (cm)	
C	Length after heating (cm)	

Table 2

	1 NH_4Cl	2 $NaHCO_3$
A H_2O		
B HCl		

ANALYSIS

Use your data to analyze the changes you observed in this lab. Specify changes in energy, entropy, and temperature as positive or negative and state whether the change was spontaneous.

Energy

+ = endothermic
− = exothermic

Entropy

+ = more disorder
− = less disorder

Spontaneous

yes = spontaneous
no = not spontaneous

Temperature

any = spontaneous at any temperature
high = spontaneous at high temperatures
low = spontaneous at low temperatures

1. unstretched band → stretched band
2. $NH_4Cl(s) \rightarrow NH_4 + (aq) + Cl^-(aq)$
3. $NaHCO_3(s) \rightarrow Na^+(aq) + HCO_3^-(aq)$
4. $NaHCO_3(s) + HCl \rightarrow$
   $Na^+(aq) + Cl^-(aq) + CO_2(g) + H_2O(l)$

CONCLUSIONS

1. Which reactions took place spontaneously?

____

2. What would be the reverse reaction with the rubber band?

____

____

EXTENSION AND APPLICATION

For the familiar reactions listed below, determine the energy and entropy changes that take place and whether the reactions are spontaneous.

a. Photosynthesis
b. Rusting of a car
c. Formation of a diamond

Teacher Guide

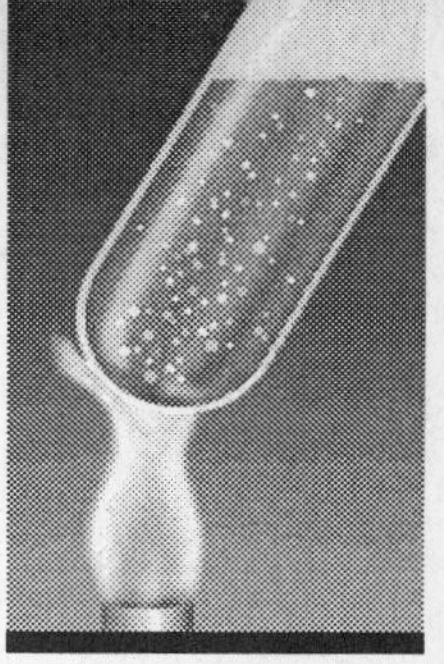

LAB MANUAL

20-1

Energy Changes in Physical and Chemical Processes

When chemical or physical changes occur, heat may be absorbed or released, and entropy, or disorder, may increase or decrease. Changes are spontaneous if the combination of energy and entropy changes are favorable at the given temperature. Negative heats of reaction are favorable—that is, exothermic reactions are more likely to occur spontaneously than are endothermic reactions, assuming all other factors are the same. Increases in entropy are also favorable—that is, reactions in which disorder increases are more likely to be favorable than are reactions in which disorder decreases, assuming all other factors are the same. Temperature conditions also play a role. The overall effect of all these factors determines whether a change is spontaneous or nonspontaneous.

Misconceptions: Students may have difficulty visualizing physical and chemical systems that are spontaneous, exothermic, endothermic, and more or less orderly. This activity will aid their comprehension of these properties.

PROCESS SKILLS

1. Students will observe a physical reaction and will deduce whether it is spontaneous or not.
2. Students will evaluate four chemical reactions and will deduce whether the reactions are spontaneous or not.

PREPARATION HINTS

For each group, make a copy of the 24-well template on page T6.

MATERIALS

ammonium chloride (solid)
sodium hydrogen carbonate (solid)
6*M* HCl (Add 500 mL of concentrated HCl to 500 mL of distilled water.)
thick rubber band
hot plate
250-mL beakers (2)
1- or 2-kg mass
ringstand and metal ring
24-well microplate
24-well template
ruler
scissors
thermometer
thin-stem pipets (2)
toothpicks

PROCEDURE HINTS

- Explain that not all chemical reactions are spontaneous or exothermic, nor do all produce products with higher entropy.
- Troubleshooting: To make it easier to attach the mass to the rubber band, choose objects that have a loop or other projection. A slip knot made in the dangling end of the rubber band can be slipped over the projection of the mass.
- **Disposal:** Dispose of solutions according to instructions in **Disposal** C.

DATA AND OBSERVATIONS

Part 1
Sample data
A. warmer
B. 38.1 cm
C. 34.2 cm
Part 2
See sample data in the table on the next page.

	1 NH_4Cl	2 $NaHCO_3$
A H_2O	Before: 23°C After: 17°C	Before: 23°C After: 26°C
B HCl	Before: 23°C After: 18°C	Before: 23°C After: Bubbles, 18°C

ANALYSIS

1. unstreched rubber band → stretched band
   Energy: – ; Entropy: – ;
Spontaneous: no

2. $NH_4Cl(s) \rightarrow NH_4^+(aq) + Cl^-(aq)$
   Energy: + Entropy: +
Spontaneous: yes Temperature: high

3. $NaHCO_3(s) \rightarrow Na^+(aq) + HCO_3^-(aq)$
   Energy: – Entropy: +
Spontaneous: yes Temperature: any

4. $NaHCO_3(s) + HCl(aq) \rightarrow Na^+(aq) + HCO_3^-(aq) + CO_2(g) + H_2O(l)$
   Energy: + Entropy: +
Spontaneous: yes Temperature: high

CONCLUSIONS

1. Reactions 2, 3, and 4 occur spontaneously.
2. stretched → unstretched
   It would occur spontaneously. Students may have difficulty visualizing physical and chemical systems that are spontaneous, exothermic, endothermic, and more or less orderly. This activity will aid their comprehension of these properties.

EXTENSION AND APPLICATION

a. Photosynthesis
   The reaction is endothermic with a decrease of entropy. Photosynthesis is not spontaneous.

b. Rusting of a car
   The reaction is exothermic and entropy increases. The reaction is spontaneous.

c. Formation of a diamond
   At one atmosphere of pressure the reaction is endothermic and entropy decreases. Formation of a diamond is not spontaneous.

Name

Date Class

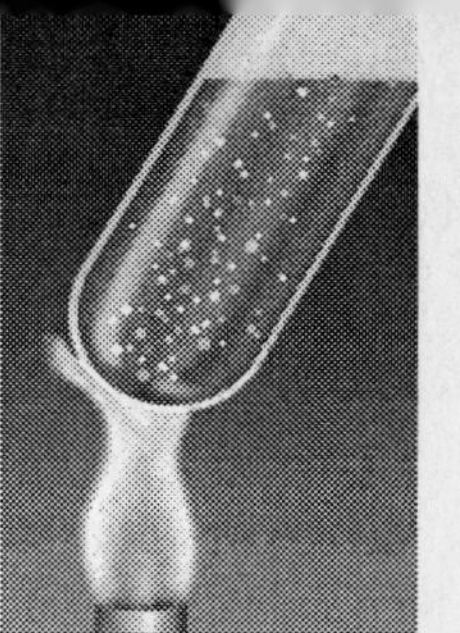

LAB MANUAL

20-2

Measuring the Heat of Reaction

As a chemical reaction progresses, the temperature of the reacting mixture rises or falls according to the amount of heat released or absorbed. The quantity of heat absorbed or given off in a spontaneous chemical reaction is called the heat of reaction. The temperature change can be measured in a calorimeter, a device that isolates the reaction mixture from the surroundings. The temperature change that occurs can then be used to calculate the heat of reaction (ΔH).

If a reaction is exothermic, the temperature rises, showing that heat is released as the reaction takes place. For an exothermic reaction, ΔH has a negative value. If a reaction is endothermic and spontaneous, the temperature falls, showing that heat is absorbed as the reaction occurs. For an endothermic reaction, ΔH has a positive value.

When magnesium reacts with hydrochloric acid,

$$Mg(s) + 2HCl(aq) \rightarrow MgCl_2(aq) + H_2(g)$$

magnesium chloride is the only compound formed, and the heat of reaction is also the heat of formation, ΔH_f. Heat of formation is the amount of heat absorbed or evolved when one mole of compound is formed from its elements. In the formation of magnesium chloride, the magnesium changed while the chloride ion did not change. The heat of formation of H_2 is zero, so the heat of formation of $MgCl_2$ is the same as the heat of formation of Mg^{2+}.

$$MgO(s) \rightarrow Mg^{2+}(aq) + 2e^-$$

In this experiment, you will determine the heat of formation of $Mg^{2+}(aq)$ from MgO(s).

OBJECTIVES

- **Measure** the temperature change for the reaction of Mg with HCl.
- **Calculate** the heat released in the experiment.
- **Calculate** the heat of formation of $Mg^{2+}(aq)$ from Mg(s)

MATERIALS

apron
goggles
ketchup cups (2) with lid (1)
thermometer
10-mL graduated cylinder
metric ruler
rubber band
scissors
paper punch
balance
paper towel

PROCEDURE

1. Construct a plastic calorimeter from two ketchup cups. Place a rubber band around one ketchup cup and then put this cup inside the second ketchup cup. With a paper punch, make one hole in the lid of a cup.
2. Determine the mass of the empty calorimeter and its lid. Record the mass.
3. Place 20 mL of 1.0*M* HCl in the calorimeter. Determine the mass of the calorimeter, lid, and hydrochloric acid. Record the mass. **CAUTION:** *Hydrochloric acid (HCl) is corrosive. Handle this acid with care.*
4. Obtain from your teacher the mass of 1 m of magnesium ribbon. Record the mass.
5. Place a thermometer through the hole in the calorimeter lid. Wait for 1 minute and then record the temperature of the hydrochloric acid to within 0.1°C.

6. Cut a piece of magnesium ribbon 25 mm long from the strip provided.
7. Remove the thermometer from the hole in the top of the calorimeter. Put the magnesium through the hole into the calorimeter and replace the thermometer in the hole. Swirl the calorimeter gently.
8. Observe the temperature as the reaction in the calorimeter proceeds. Record the highest temperature attained to within 0.1°C.
9. Dispose of the reaction mixture as directed by your teacher. Rinse and dry the calorimeter.
10. Repeat steps 3–9. Record your data under Trial 2. Data 1 and 3 will be the same in both trials.

DATA AND OBSERVATIONS

	Trial 1	Trial 2
1. Mass of calorimeter	_____ g	_____ g
2. Mass of calorimeter and HCl solution	_____ g	_____ g
3. Mass of Mg/m	____ g/m	____ g/m
4. Initial temperature of HCl solution	_____ °C	_____ °C
5. Final temperature of reaction mixture	_____ °C	_____ °C

ANALYSIS

Analyze the data for each trial separately.

1. Determine the mass of solution in the calorimeter.

____

2. Calculate the mass of Mg used.

Mass = 0.025 m Mg × data 3

____

3. Calculate the moles of Mg used.

$$\text{Moles of Mg} = \frac{\text{analysis 2}}{\text{molar mass of Mg}}$$

____

4. Find the change in temperature during the reaction.

____

CONCLUSIONS

1. Was this reaction exothermic or endothermic? Explain.

____

2. Calculate the heat released during the reaction. To do this requires making the reasonable assumption that the specific heat of the reaction mixture is the same as the specific heat of water, which is 4.18 J/g • C°.

$$\text{Heat released (J)} = \text{mass of solution} \times \text{temperature change} \times C_W$$

$$= \text{analysis 1} \times \text{analysis 4} \times 4.18\ \text{J/g} \cdot \text{C}°$$

____

3. The heat of reaction for the change $Mg(s) \rightarrow Mg^{2+}(aq) + 2e^-$ is the heat released, in kilojoules, per mole of Mg^{2+} formed. Since the temperature increased, heat was released and the value of ΔH is negative. The value of ΔH for Mg^{2+} is found from the experimental heat of reaction as follows:

$$\Delta H_f \text{ for } Mg^{2+} = -\frac{\text{conclusion } 2 \times 1\text{kJ}/1000\text{J}}{\text{analysis 3}}$$

Calculate ΔH_f for Mg^{2+} for both trials.

____

4. Find the average of your two values of ΔH_f for Mg^{2+}. Compare your average with the accepted value of −466.85 kJ/mol. Is there a reasonably good agreement?

____

EXTENSION AND APPLICATION

Write the complete equation and the net ionic equation for the reaction of Zn and HCl. If the experiment were repeated using this reaction, could the value of ΔH_f for Zn^{2+} be found? Explain.

Teacher Guide

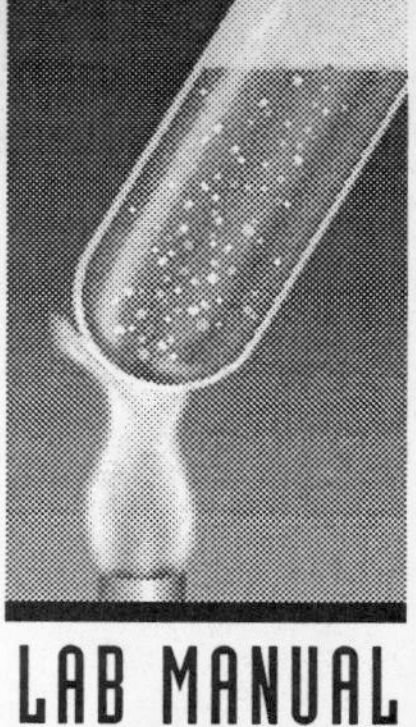

LAB MANUAL

20-2

Measuring the Heat of Reaction

In this experiment, students are determining the heat of formation ΔH of $Mg^{2+}(aq)$ from elemental magnesium. The reaction of magnesium with hydrochloric acid is carried out in a simple, easily made calorimeter.

$$Mg(s) + 2H^{+}(aq) \rightarrow Mg^{2+}(aq) + H_2(g)$$

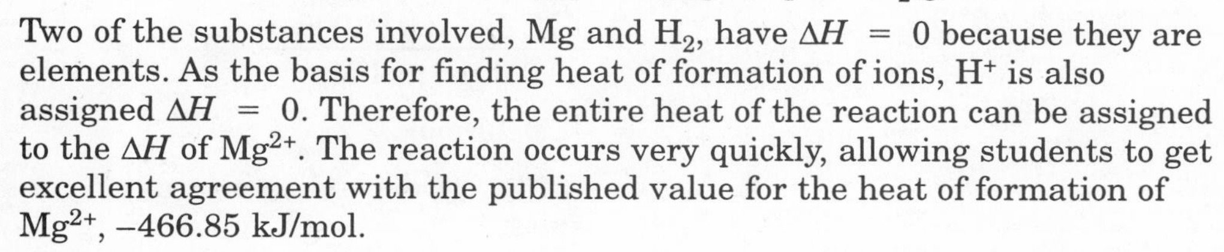

Two of the substances involved, Mg and H_2, have $\Delta H = 0$ because they are elements. As the basis for finding heat of formation of ions, H^+ is also assigned $\Delta H = 0$. Therefore, the entire heat of the reaction can be assigned to the ΔH of Mg^{2+}. The reaction occurs very quickly, allowing students to get excellent agreement with the published value for the heat of formation of Mg^{2+}, –466.85 kJ/mol.

Misconception: The values for constants such as heats of formation do not simply appear in reference books, as students often think. Students should be made aware that the thermodynamic data, as well as other data, in handbooks are the result of years of experimentation by hundreds of researchers.

PROCESS SKILLS

1. Students will use a calorimeter to determine the heat released in the reaction of Mg with HCl.
2. Students will calculate the heat of formation of Mg^{2+}.
3. Students will compare their experimental value of ΔH for Mg^{2+} with the value from a published source.

PREPARATION HINTS

- Store the Mg ribbon in a sealed plastic bag to prevent reaction of the ribbon with moisture and oxygen in the air.
- Determine the mass of the 1 m strip of Mg from which the students will cut their samples. Write the value on the chalkboard for students to add to their data.

MATERIALS

1*M* HCl (add 86 mL of concentrated HCl to enough water to make 1 L of solution. Do not add water to acid.)

Mg ribbon
apron
goggles
ketchup cups (2) with lid (1)
thermometer
10-mL graduated cylinder
rubber band
paper towel
metric ruler
scissors
balance
¼-in. hole paper punch

PROCEDURE HINTS

- Before proceeding, be sure students have read and understood the purpose, procedure, and safety precautions for this laboratory activity.
- **Safety Precaution:** Students should not have free access to the magnesium ribbon. Magnesium ribbon can be set on fire with a match. Magnesium fires cannot be put out with water or CO_2 fire extinguishers. Be certain to monitor students' use of magnesium carefully.
- **Safety Precaution:** Magnesium salts are powerful laxative agents. Handle the resulting solution from this lab with care.
- **Disposal:** Dispose of the solution from this lab according to the directions in **Disposal** J.

DATA AND OBSERVATIONS

Sample student data are given for one trial. Similar data will be obtained for the second trial.

	Trial 1
1. Mass of calorimeter	5.82 g
2. Mass of calorimeter and HCl solution	26.24 g
3. Mass of Mg/m	0.97 g/m
4. Initial temperature of HCl solution	22.5°C
5. Final temperature of reaction mixture	28.0°C

ANALYSIS

1. Mass of solution:
   Mass = 26.24 g – 5.82 g = 20.42 g
2. Mass of Mg used:
   Mass of Mg = 0.025 m Mg × 0.97 g/m
   = 0.024 g Mg
3. Moles of Mg used:
   $$\text{Moles of Mg} = \frac{0.024 \text{ g Mg}}{24 \text{ g Mg/mol Mg}}$$
   = 0.0010 mol Mg
4. 28.0°C – 22.5°C = 5.5°C

CONCLUSIONS

1. The reaction is exothermic, as shown by the fact that the temperature increased during the reaction.
2. Heat released by the reaction:
   Heat released (J) = (20.42 g) × (5.5°C) × (4.18 J/g • C°)
   = 470 J
3. ΔH for Mg^{2+}:
   $$\Delta H = \frac{-470 \text{ J} \times (1\text{kJ}/1000 \text{ J})}{0.0010 \text{ mol Mg}}$$
   = –470 kJ/mol
4. Answers will vary, depending on students' results.

EXTENSION AND APPLICATION

$Zn(s) + 2HCl \rightarrow ZnCl_2(aq) + H_2(g)$

$Zn(s) + 2H^+(aq) \rightarrow Zn^{2+}(aq) + H_2(g)$

This reaction could be used to determine ΔH for Zn^{2+} because the values of ΔH for Zn, H_2, and H^+ are all defined as zero.

Name

Date Class

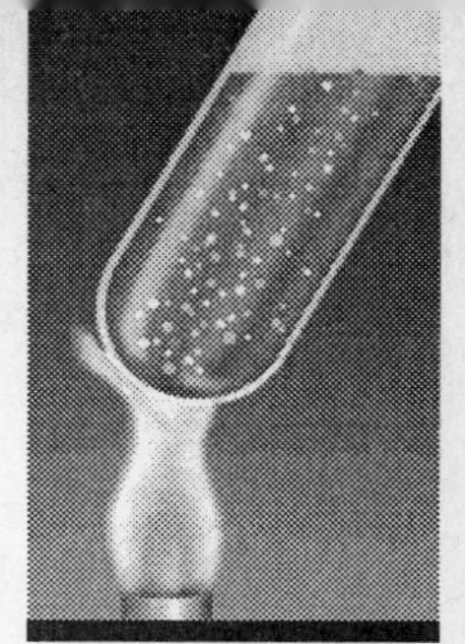

LAB MANUAL

20-3

Heat of Hydration

The products of a chemical reaction differ physically and chemically from the reactants. In addition, there is a difference in energy between the reactants and the products. If, during the course of a reaction, energy is released into the environment, the reaction is said to be exothermic and the products have a lower energy content than the reactants. A reaction that absorbs energy is endothermic and its products have a higher energy content than the reactants. For example, when anhydrous copper(II) sulfate ($CuSO_4$) is exposed to water, the hydrated form of the compound, copper(II) sulfate pentahydrate ($CuSO_4 \cdot 5H_2O$), is formed. This change is accompanied by a change in energy.

In this laboratory activity, you will determine how much energy is involved in the hydration of copper(II) sulfate and whether the change is endothermic or exothermic.

OBJECTIVES

- **Calculate** the heats of solution of anhydrous copper(II) sulfate and hydrated copper(II) sulfate.
- **Determine** the heat of hydration of copper(II) sulfate.

MATERIALS

plastic foam cups (2) and lid (1)
100-mL graduated cylinder
thermometer
balance
evaporating dish
mortar and pestle
glass stirring rod
distilled water
apron
goggles
¼-inch hole paper punch

PROCEDURE

You will determine the heats of solution of anhydrous copper(II) sulfate and hydrated copper(II) sulfate. Using these data you will then determine the heat of hydration of copper(II) sulfate. Form a **hypothesis** about whether the hydration of copper(II) sulfate is endothermic or exothermic. Write your hypothesis under Data and Observations.

Part 1: Heat of Solution of $CuSO_4 \cdot 5H_2O$

1. Insert one plastic foam cup into the other, as shown, to make a calorimeter.
2. Place 50 mL of distilled water in the calorimeter. Record the mass of the water. Remember, 1 mL of water is equal to 1 g of water.

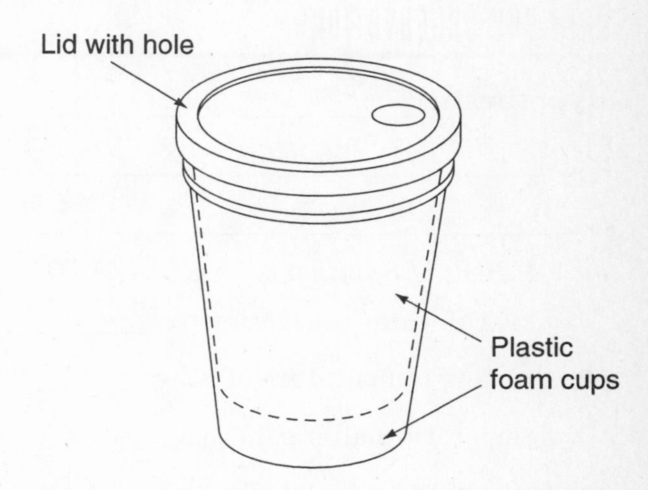

3. With a ¼-inch hole paper punch, make a hole near the rim of the cup lid. Insert a thermometer through the hole into the water in the calorimeter. Record the temperature of the water to the nearest 0.5°C.
4. Measure out 0.02 mole (5.0 g) of copper(II) sulfate pentahydrate in an evaporating dish. (If the compound is in large crystalline form, first crush the solid into a fine powder, using a mortar and pestle.) **CAUTION:** *Copper sulfate is toxic. Wash your hands thoroughly after performing this activity.*

Copyright © Glencoe/McGraw-Hill, a division of The McGraw-Hill Companies, Inc.

5. Add the copper salt to the water in the calorimeter. Stir constantly with a glass stirring rod. Record the highest or lowest temperature of the solution.
6. Discard the solution according to directions from your teacher.
7. Wash and dry the calorimeter and the evaporating dish.

Part 2: Heat of Solution of $CuSO_4$

1. Place 50 mL of distilled water in the calorimeter. Record the mass of the water.
2. Record the temperature of the water to the nearest 0.5°C.
3. Measure out 0.02 mole (3.2 g) of anhydrous copper(II) sulfate in the evaporating dish.
4. Add the copper salt to the water in the calorimeter, stirring constantly with the glass rod. Record the highest or lowest temperature of the solution.
5. Discard the solution according to the directions of your teacher.

DATA AND OBSERVATIONS

Hypothesis: ________________________

____

____

Part 1: Heat of Solution of $CuSO_4 \cdot 5H_2O$

1. Mass of water in calorimeter ________ g
2. Starting temperature of water ______ °C
3. Temperature after solution ________ °C
4. Change in temperature ____________ °C

Part 2: Heat of Solution of $CuSO_4$

5. Mass of water in calorimeter ________ g
6. Starting temperature of water ______ °C
7. Temperature after solution ________ °C
8. Change in temperature ____________ °C

ANALYSIS

1. Calculate the heat of solution of $CuSO_4 \cdot 5H_2O$.

   Heat of solution =
   mass of water × change in temperature × specific heat of water [C_w]
   = data 1 × data 4 × 4.18 J/g • C°

   ____

2. Calculate the heat of solution per mole of $CuSO_4 \cdot 5H_2O$.

   $$\text{Heat of solution/mole} = \frac{\text{analysis 1}}{\text{0.02 mole}}$$

   ____

3. Calculate the heat of solution of $CuSO_4$.

   Heat of solution =
   mass of water × change in temperature × specific heat of water [C_w]
   = data 5 × data 8 × 4.18 J/g • C°

   ____

4. Calculate the heat of solution per mole of $CuSO_4$.

   $$\text{Heat/mole} = \frac{\text{analysis 3}}{\text{0.02 mole}}$$

   ____

5. Now find the heat of hydration of $CuSO_4$.
   Heat of hydration =
   analysis 2 – analysis 4

   ____

CONCLUSIONS

1. Is the dissolving of copper(II) sulfate pentahydrate exothermic or endothermic? How do you know?

   ____

   ____

Copyright © Glencoe/McGraw-Hill, a division of The McGraw-Hill Companies, Inc.

2. Is the dissolving of anhydrous copper(II) sulfate exothermic or endothermic? How do you know?

____

____

____

____

____

____

3. The hydration of copper(II) sulfate is the reverse of dehydration. Do you think the hydration of copper(II) sulfate is endothermic or exothermic? Explain.

____

____

4. How did the results you obtained compare with your hypothesis about the energy change in the hydration of copper(II) sulfate?

____

____

EXTENSION AND APPLICATION

The conversion of copper(II) sulfate pentahydrate to anhydrous copper(II) sulfate is a classic experiment, performed to find the percentage of water in a compound. Find out how dehydration is carried out. How can you predict, using this procedure, whether the hydration reaction is endothermic or exothermic?

Dehydration

$$CuSO_4 \cdot 5H_2O \rightarrow CuSO_4 + 5H_2O$$

Hydration

$$CuSO_4 + 5H_2O \rightarrow CuSO_4 \cdot 5H_2O$$

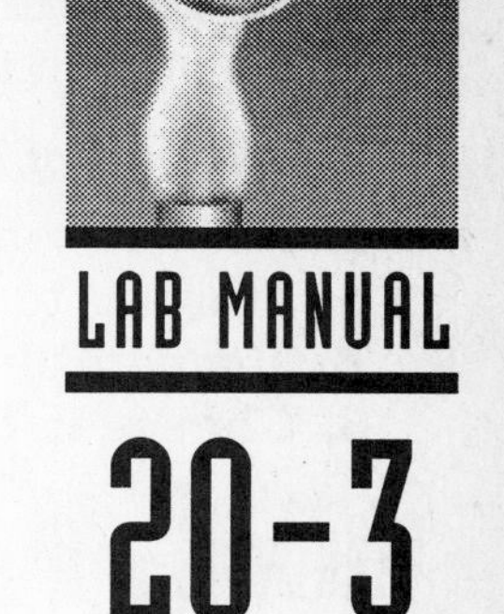

LAB MANUAL

20-3

Heat of Hydration

The conversion of copper(II) sulfate pentahydrate to the anhydrous form is an experiment that has been carried out by chemistry students for almost 100 years. In this laboratory activity, students investigate the reverse reaction and determine the energy involved in the hydration. Students should be reminded that there are two changes occurring, the hydration of copper(II) sulfate and the solution of copper(II) sulfate. By using Hess's law, it is possible to find the heat of solution of only one of the reactions.

Students' results will probably not be very accurate when compared with heats of solution listed in reference tables. Students will, however, be able to determine whether hydration is endothermic or exothermic.

Misconception: Students' associate exothermic reactions with combustion or explosion. This activity shows students that reactions can be mildly exothermic or endothermic.

PROCESS SKILLS

1. Students will measure the temperature changes that occur when hydrated and anhydrous copper(II) sulfate are dissolved in water.
2. Students will calculate the heat of solution for both the hydrated and the anhydrous copper(II) sulfate.
3. Students will determine whether the hydration of copper(II) sulfate is endothermic or exothermic.

PREPARATION HINT

Anhydrous copper sulfate should be obtained from a supply house. Do not use an anhydrous salt prepared by students in a laboratory activity; students might not have driven off enough water of hydration to produce a truly anhydrous salt.

MATERIALS

solid anhydrous copper(II) sulfate
solid copper(II) sulfate pentahydrate
plastic foam cups (2) and lid (1)
100-mL graduated cylinder
thermometer
balance
evaporating dish
mortar and pestle
glass stirring rod
distilled water
apron
goggles
paper punch

PROCEDURE HINTS

- Before proceeding, be sure students have read and understood the purpose, procedure, and safety precautions for this laboratory activity.
- Do not leave the container of anhydrous copper(II) sulfate open to the air. The salt readily picks up moisture. If this should happen, both the mass of the copper salt and the amount of heat produced when the compound is added to water would change.
- **Troubleshooting:** Students should not expect the temperature of the water to change greatly. Changes of only 0.5°C or 0.75°C are not unusual.
- **Safety Precaution:** Use adequate venti-la-tion so that exposure to vapor is minimized. Students with metal allergies should not take part in this experiment.
- **Safety Precaution:** Solid samples should be distributed in small containers. Do not allow students to obtain their own samples.
- **Disposal:** Collect all the salt solutions in a large basin or dishpan. Dispose of the solution according to the instructions provided in **Disposal** J.

DATA AND OBSERVATIONS

Students' hypotheses may vary. The hydration of copper(II) sulfate is exothermic.

Copyright © Glencoe/McGraw-Hill, a division of The McGraw-Hill Companies, Inc.

Part 1. Heat of Solution of $CuSO_4 \cdot 5H_2O$

1. Mass of water in calorimeter 50 g
2. Starting temperature of water 19.5°C
3. Temperature after solution 19.0°C
4. Change in temperature – 0.5°C

Part 2. Heat of Solution of $CuSO_4$

5. Mass of water in calorimeter 50 g
6. Starting temperature 19.5°C
7. Temperature after solution 20.0°C
8. Change in temperature 0.5°C

ANALYSIS

1. Heat of solution of $CuSO_4 \cdot 5H_2O$:

   Heat of solution = mass of water × change in temperature × specific heat of water [C_w]

   = data 1 × data 4 × 4.18 J/g C°

   = 50 g × – 0.5°C × 4.18 J/g • C°

   = –104.5 J

2. Heat of solution per mole of $CuSO_4 \cdot 5H_2O$:

   $$\text{Heat of solution /mole} = \frac{\text{analysis 1}}{0.02\ \text{mole}} = \frac{-104.5\ \text{J}}{0.02\ \text{mole}} = -5225$$

   J/mole

3. Heat of solution of $CuSO_4$:

   Heat of solution = mass of water × change in temperature × specific heat of water [C_w]

   = data 5 × data 8 × 4.18 J/g C°

   = 50 g × 0.5°C × 4.18 J/g • C°

   = 104.5 J

4. Heat of solution per mole of $CuSO_4$:

   $$\text{Heat of solution /mole} = \frac{\text{analysis 3}}{0.02\ \text{mole}} = \frac{-104.5\ \text{J}}{0.02\ \text{mole}}$$

   = –5225 J/mole

5. Heat of hydration of $CuSO_4$:

   Heat of hydration = analysis 2 – analysis 4

   = – 5225 J/mole – (5225 J/mole)

   = – 10 450 J/mole

CONCLUSIONS

1. The dissolving of copper(II) sulfate penta-hydrate is endothermic. The reaction solution became slightly cooler. Heat was absorbed from the environment.
2. The dissolving of anhydrous copper(II) sulfate is exothermic. The reaction solution became slightly warmer. Heat was re-leased to the environment.
3. The hydration of copper(II) sulfate is exothermic. Heat is released when water is added to the compound.
4. Students' answers will vary depending on their hypotheses.

EXTENSION AND APPLICATION

The dehydration of copper(II) sulfate involves heating a sample of the hydrated crystal to drive off the water. Dehydration is therefore an endothermic reaction—it requires heat from the environment. The equation for the dehydration reaction is the reverse of the hydration reaction. Since dehydration is endothermic, hydration must be exothermic.

Copyright © Glencoe/McGraw-Hill, a division of The McGraw-Hill Companies, Inc.

Name

Date Class

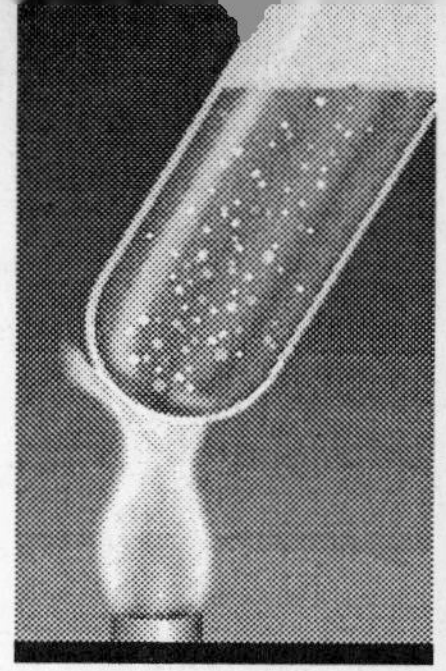

LAB MANUAL

21-1

Radioactive Dating—A Model

Many elements in the periodic table have naturally occurring isotopes. Isotopes are elements that have the same atomic number but differ in atomic mass. The difference in atomic mass is due to a difference in the number of neutrons in the nucleus. For example, there are three isotopes of hydrogen. Notice below that all three isotopes have a single proton in the nucleus, but each has a different number of neutrons. The superscript number gives the atomic mass of the element—the number of protons plus the number of neutrons. The subscript tells the atomic number, or number of protons, of the element. The asterisk next to tritium indicates that this isotope is radioactive.

Hydrogen	**Deuterium**	**Tritium**
$^{1}_{1}H$	$^{2}_{1}H$	$^{3}_{1}H^{*}$

A radioactive isotope is unstable. The nuclei in a sample of the isotope break down at a constant rate, which is usually measured by its half-life. Half-life is the amount of time necessary for one-half of a sample of a radioactive element to decay into another element. Some isotopes have half-lives of a few seconds or less. Other half-lives are millions of years long. Each element has its own half-life. If the half-life of an element in a sample is known and the amount of the element is known, the age of the sample can be estimated.

For example, suppose a piece of charcoal from a fire pit is found in a cave. How long ago did the wood live? Carbon-14 is an isotope found in all organisms. Its half-life is 5730 years. The percentage of carbon-14, compared to other carbon isotopes, is the same in all living organisms. As the C-14 decays, it is replaced by more carbon from the organism's surroundings. However, when an organism dies, it no longer takes in carbon, and the percentage of C-14 decreases over time. If the charcoal sample in the cave holds $\frac{1}{8}$ the amount of carbon-14 found in a piece of living wood today, the carbon-14 in the sample has survived three half-lives. Thus, the sample is 3×5730, or 17 190 years old. The diagram below shows the amount of carbon-14 remaining in the sample after each half-life.

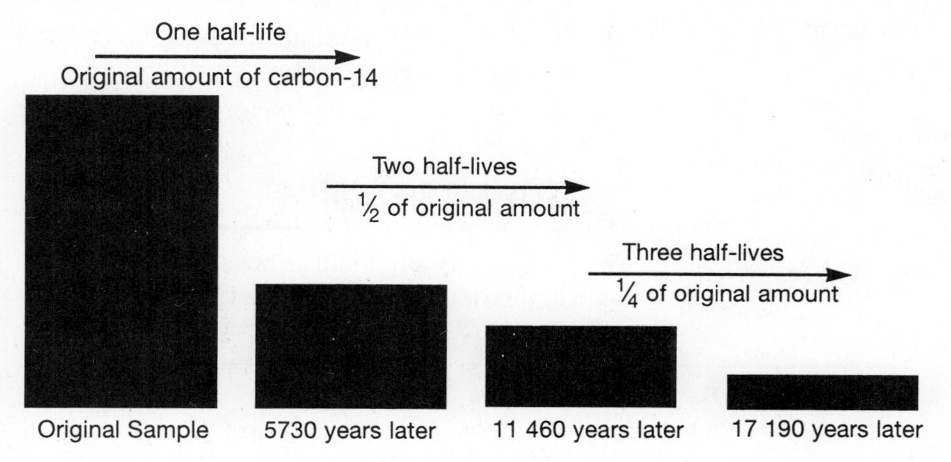

In this laboratory activity, you will simulate the gathering and processing of samples to construct a half-life curve of an imaginary isotope and estimate the age of a sample containing the isotope.

Copyright © Glencoe/McGraw-Hill, a division of The McGraw-Hill Companies, Inc.

OBJECTIVES

- **Simulate** the radioactive decay of an isotope in a sample over time.
- **Make a graph** that shows the isotope's rate of decay (half-life).
- **Infer** from the graph the age of an unknown sample.

MATERIALS

large containers (5) with a mixture of yellow and green split peas in each
film can
balance
white paper

PROCEDURE

1. Examine the mixtures of peas in the containers. The yellow peas represent atoms of a radioactive isotope of the nonradioactive green peas. The mixture in each container represents the same material but at a different age. The material in the container labeled *Sample 1* is zero years old. It has not yet started to decay. The ages of the materials in the other samples are shown in the table under Data and Observations. The age of the material in Sample *X* is unknown.
2. Weigh the empty film can and record the mass in each column of the table.
3. Completely fill the film can with the Sample 1 peas. Pour the peas onto a sheet of white paper.
4. Separate the peas into two groups—those representing atoms of a radioactive isotope (yellow) and those representing atoms of a nonradioactive isotope (green).
5. Return the peas representing the radioactive isotope to the film can. Weigh the can and the peas. Record the mass in the Sample 1 column of the table.
6. Return all peas to the Sample 1 container.
7. Repeat steps 3 through 6, using the remaining samples.

DATA AND OBSERVATIONS

See table below.

ANALYSIS

1. For each sample, calculate the mass of the radioactive isotope. Record these figures in the table.
2. Make a line graph of the data from samples 1 through 4. Graph the age of the sample on the x-axis and the mass of the radioactive isotope on the y-axis. Connect the points with the best fitting curve.
3. Find the age of Sample *X* by marking the mass of the radioactive isotope in the sample on the y-axis of the graph. Draw a horizontal line across the graph until the line intersects the curve. Draw a vertical line from the point of intersection to the x-axis to find the age of the sample.

CONCLUSIONS

1. From the graph that you constructed, what generalization can you make about the amount of a radioactive isotope in a sample?
2. What is the half-life of the radioactive isotope that is represented by the yellow peas?
3. What might be some limiting factors in using radioactive isotopes to determine the ages of samples?

EXTENSION AND APPLICATION

Find out if any physical processes, such as heat or pressure, can change the half-life of an element.

	Sample 1 0 years	Sample 2 5000 years	Sample 3 10 000 years	Sample 4 12 000 years	Sample *X*
1. Mass of empty film can	______ g	______ g	______ g	______ g	______ g
2. Mass of film can plus radioactive isotope	______ g	______ g	______ g	______ g	______ g
3. Mass of radioactive isotope	______ g	______ g	______ g	______ g	______ g

Copyright © Glencoe/McGraw-Hill, a division of The McGraw-Hill Companies, Inc.

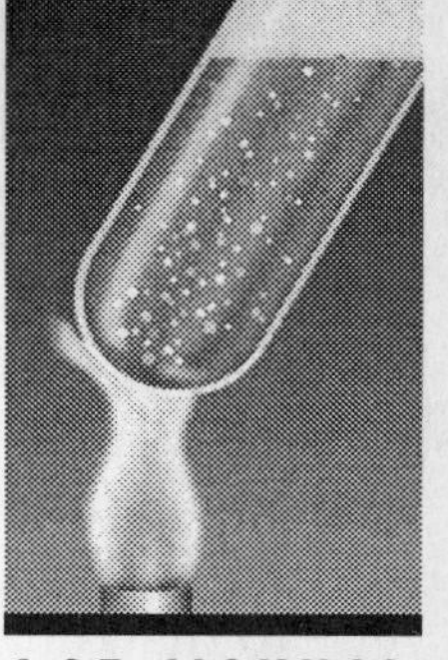

LAB MANUAL

21-1

Radioactive Dating—A Model

Each radioactive isotope has its own characteristic half life. For example, ^{60}Co has a half-life of 5.3 years; ^{131}I has a half-life of 8 days; the half life of ^{90}Sr is 28 years, and ^{238}U has a half-life of 4.5×10^9 years

Although mathematical formulas for calculating the mass of a radioactive material have been developed, the use of a chart can simplify the process for students. For example, using ^{90}Sr with a half-life of 28 years yields the following:

Time	Mass
0 yr	100 g
28 yr	50 g
56 yr	25 g
84 yr	12.5 g

Misconception: Students can usually grasp the idea of half-life. What surprises them is the shape of the graph of their data. The graph is not a straight line with a constant slope. Rather, it is a falling curve, the graph of a natural log function.

PROCESS SKILLS

1. Students will gather, analyze, and interpret data to establish the age of an unknown sample.
2. Students will graph their data.
3. Students will infer from their graph the age of the sample.

MATERIALS

one-pound bags of green peas (5)
one-pound bag of yellow peas
large containers (5)
film can
balance (accurate to 0.01 g)
white paper

PREPARATION HINTS

- Purchase 5 one-pound bags of green peas and 1 one-pound bag of yellow peas. Prepare the following samples in large coffee cans or 1-L beakers and label the containers with the sample numbers. See the table below.
- Be sure samples have been adequately mixed to ensure homogeneity.

PROCEDURE HINTS

- Before proceeding, make sure students have read and understood the purpose, procedure, and safety precautions for this laboratory activity.
- Ask students to compare the ages of the samples, including the unknown, by visually examining the mixtures. Performance of this procedure can be enhanced by placing the pea mixtures in glass beakers.
- **Troubleshooting:** The most common problem in this activity is that the masses of the radioactive isotopes are not in perfect mathematical agreement. Advise students to compare their results with other members of the class and to determine the average of the data.

	Sample 1	Sample 2	Sample 3	Sample 4	Sample X
Green peas:	One pound	One pound	One pound	One pound	One pound
Yellow peas:	120 g	60 g	30 g	15 g	45 g

Copyright © Glencoe/McGraw-Hill, a division of The McGraw-Hill Companies, Inc.

- Be sure students return their samples to the correct containers.
- **Safety Precaution:** Split peas make excellent projectiles. Collect all samples as soon as the data gathering process is completed.
- **Disposal:** Dispose of the peas according to instructions in **Disposal** F.

DATA AND OBSERVATIONS

Sample data:
See table below.

ANALYSIS

1. See table below.
2.

[art: LM 21-1d]

3. The age of Sample X is 7500 years.

	Sample 1 0 years	Sample 2 5000 years	Sample 3 10 000 years	Sample 4 15 000 years	Sample X
1. Mass of empty film can	10.34 g	10.34 g	10.34 g	10.34 g	10.34 g
2. Mass of film can plus radioactive isotope	15.75 g	12.98 g	11.47 g	11.01 g	12.22 g
3. Mass of radioactive isotope	5.41 g	2.64 g	1.13 g	0.67 g	1.88 g

CONCLUSIONS

1. The amount of radioactive isotope in a sample decreases over time.
2. about 5000 years
3. Limiting factors are: (a) not all materials contain radioactive isotopes that can be measured; (b) the sample may be destroyed in analysis; (c) the age of the sample cannot be so many half-lifes that the amount of radioactive material is too small to measure accurately (d) the age of the sample is an estimate with a large error.

EXTENSION AND APPLICATION

1. The rate of decay is a constant for each radioactive isotope. Physical processes such as heat and pressure have no effect on half-life.
2. Students might research cobalt-60 or iodine-131, which are used to treat cancer and have a half-life of about 5 years and 8 days, respectively.

Copyright © Glencoe/McGraw-Hill, a division of The McGraw-Hill Companies, Inc.

Name

Date Class

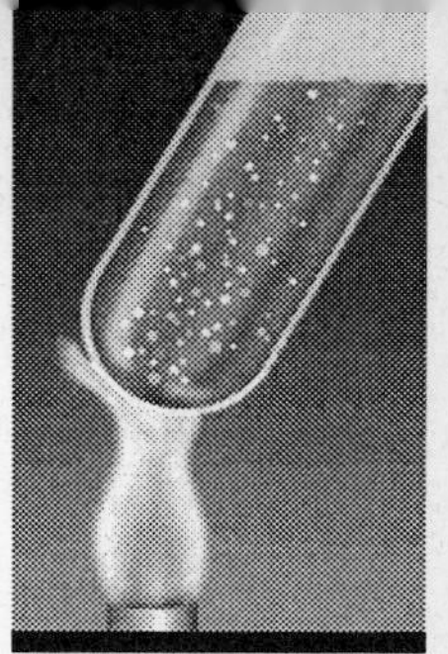

LAB MANUAL

21-2

How Does Microwave Radiation Affect Seeds?

Everyone is concerned about the possible dangers of working with radiation and the effects of radiation on living organisms. Radiant energy can be transferred to tissues of living organisms. This energy can cause chemical bonds to break and cells to function erratically.

Damage to organisms can be classified as either somatic or genetic. Somatic damage causes damage to the organism itself. Large, concentrated doses of radiation may cause sickness or even death. Health problems that are a result of smaller doses of radiation over an extended period of time may not show up for years. Genetic damage causes damage to the genetic makeup of the organism. This damage can be passed on to the organism's offspring.

Seeds are parts of plants. Are seeds affected by the energy associated with radiation? Under the proper conditions the seeds will germinate, or begin to grow, and develop into plants. In this laboratory activity, you will study the effects, if any, that microwave radiation has on the germination of seeds.

OBJECTIVES

- **Observe** and measure the germination results of seeds exposed to different intensities and duration of microwave radiation.
- **Graph** the data to determine the relationship between intensity of exposure and germination and time of exposure and germination.

MATERIALS

microwave oven	glass petri dishes (8)
seeds (160)	paper towels
goggles	scissors
apron	marking pencil

PROCEDURE

1. Form a hypothesis about the effect time of microwave exposure will have on the germination of seeds. Record your hypothesis under Data and Observations.
2. Form a second hypothesis about the effect intensity of microwave exposure will have on the germination of seeds. Record your hypothesis.
3. Cut four circles of paper towels to fit into the petri dishes. Moisten each circle of paper and place one in the bottom of each of four dishes.
4. Distribute 20 seeds evenly on the moist paper in each petri dish. Use a marking pencil to label the dishes *10 seconds, 20 seconds, 30 seconds,* and *Control #1* (not microwaved).
5. Set the microwave oven to medium. Expose the dish marked *10 seconds* to radiation for ten seconds. **CAUTION:** *Follow the manufacturer's instructions for the safe use of a microwave oven.*
6. Repeat step 5 for 20 seconds and 30 seconds for the next two dishes. Do not expose the control dish to microwave radiation.
7. Cut four more circles of paper towels to fit into the petri dishes. Moisten each circle of paper and place one in the bottom of each of four additional dishes.
8. Place twenty seeds evenly on the moist paper in each of the four petri dishes. Use a marking pencil to label the dishes *Low power, Medium power, High power*, and *Control #2* (not microwaved).

Copyright © Glencoe/McGraw-Hill, a division of The McGraw-Hill Companies, Inc.

9. Set the microwave to low power. Expose the dish marked *Low power* to radiation for ten seconds.
10. Repeat step 9 for medium power and high power. Do not expose the control dish to microwave radiation.
11. Place all eight dishes in a secure place, as directed by your teacher. Monitor the seeds in the dishes until germination takes place. When germination is complete, count the number of seeds germinated in each dish. Record your data in the table under Data and Observations.

DATA AND OBSERVATIONS

1. Hypothesis 1: ______________________

____

2. Hypothesis 2: ______________________

____

Data Table:

Dish	Number of seeds present	Number of seeds germinated	Percent of seeds germinated
10 seconds			
20 seconds			
30 seconds			
Control #1			
Low power			
Medium power			
High power			
Control #2			

ANALYSIS

1. Make a bar graph of your results showing the effect of time of microwave radiation on germination. Plot the exposure time in seconds on the x-axis and the percent germination on the y-axis.
2. Make a bar graph of your results showing the effect of radiation intensity on seed germination. Plot the power level on the x-axis and the percent germination on the y-axis.

CONCLUSIONS

1. What is the apparent effect of the length of time of microwave radiation on the germination of seeds?

____

____

____

____

2. What is the apparent effect of radiation intensity on seed germination?

____

____

____

____

3. If a seed fails to germinate as a result of exposure to radiation, is the damage somatic or genetic?

____

____

EXTENSION AND APPLICATION

Study the nuclear power plant disaster in 1986 in Chernobyl, Ukraine. What effect did the radiation have on the rescue workers? How did radiation exposure affect the health and lives of the residents of the area? How did radiation exposure affect plant and animal life of the area?

Copyright © Glencoe/McGraw-Hill, a division of The McGraw-Hill Companies, Inc.

Teacher Guide

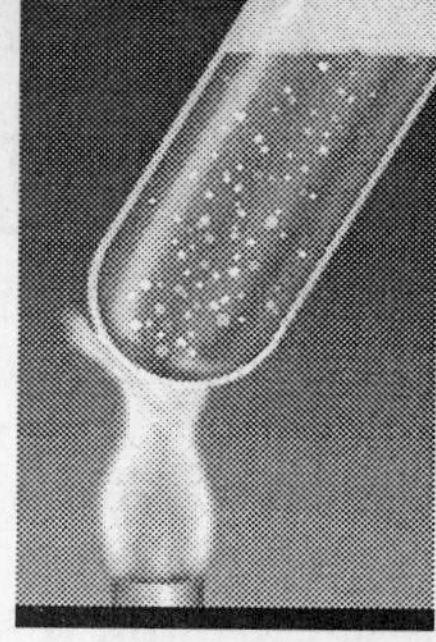

LAB MANUAL

21-2

How Does Microwave Radiation Affect Seeds?

The effects of radiation on living organisms depend on the radiation's ability to penetrate tissue, its penetrating power, and its ability to knock electrons off atoms in the tissue to produce ions, its ionizing power. Some types of radiation, such as alpha particles, have enough mass and energy to cause large amounts of damage to the molecules with which they collide. They have considerable ionizing power. Because alpha particles are so large, however, they can easily be stopped by other molecules, and they don't penetrate very deeply into the tissue. Beta particles, with their smaller mass, have less ionizing power than alpha particles, but they can penetrate deeply into tissue.

Microwaves, like radiowaves and light waves, are part of the electromagnetic spectrum. Microwaves don't have enough energy to produce ions in tissue, but they can cause biological damage because they penetrate tissue and are strongly absorbed by polar molecules such as water.

Misconception: Students often think that exposure to any amount of radiation is lethal or extremely harmful. They forget that they are constantly exposed to radiation from Earth and in the form of cosmic rays. They also receive, electively, radiation from medical diagnoses and therapies, TV tubes, and other commercial consumer products.

PROCESS SKILLS

1. Students will observe the effects of microwave radiation on the germination of seeds and compare the resultant germination to the control.
2. Students will analyze their data by preparing a bar graph which will illustrate the effects of microwaves on seed germination.

PREPARATION HINTS

- Two weeks before beginning the activity, test the seed for germination time and viability.
- Allow time in the schedule as this activity will require postponed observations as well as postponed data recording and analysis. Also be aware of forthcoming vacations and weekends, which may take place when germination needs to be observed and the data recorded. Consider having students take irradiated seeds home for germination observations.

MATERIALS

microwave oven
seeds (160)
  radish or
  bean or
  corn
goggles
apron
glass petri dishes (8)
paper towels
scissors
glass marking pencil

PROCEDURE HINTS

- Germination times vary for different seeds. Radishes will usually germinate in a few days. Beans and corn may require more time. Remember that moisture and temperature have an effect on germination time. Keep the blotting papers in the petri dishes moist.
- **Troubleshooting:** Microwave ovens vary in the intensity of energy they deliver. Make sure that all the seeds are treated in the same appliance.

DATA AND OBSERVATIONS

1. Hypothesis 1: Hypotheses will vary but should indicate that the duration of irradiation either has an effect or has no effect on the germenation of the seeds.
2. Hypothesis 2: Hypotheses will vary but should indicate that the intensity of irradiation either has an effect or has no effect on the germenation of the seeds.

Data Table: Data will vary but might resemble the table below.

ANALYSIS

See the sample graphs.

CONCLUSIONS

1. The data and graph both show that microwave radiation at low power for up to 30 seconds has no apparent effect on seed germination.
2. The data and graph both show that ten seconds of microwave radiation at low or medium power has no apparent effect on the seed germination. Ten seconds of radiation at high power, however, significantly decreases seed germination.
3. The damage is somatic.

EXTENSION AND APPLICATION

Answers and reports will vary but will probably include that workers and wildlife exposed to massive doses became ill and died shortly after the incident. The long-term effect of the disaster is still unknown and under constant study.

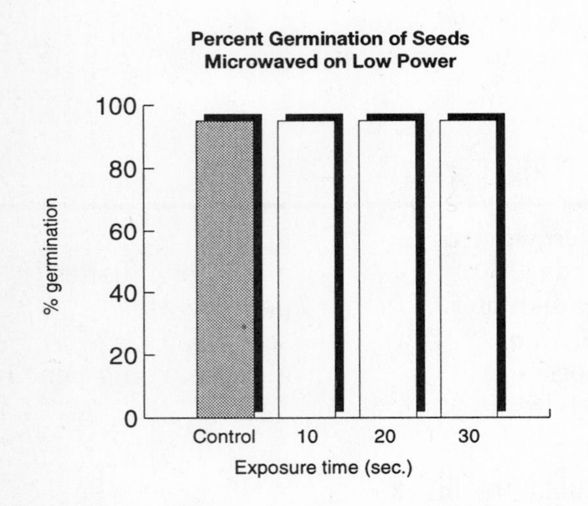

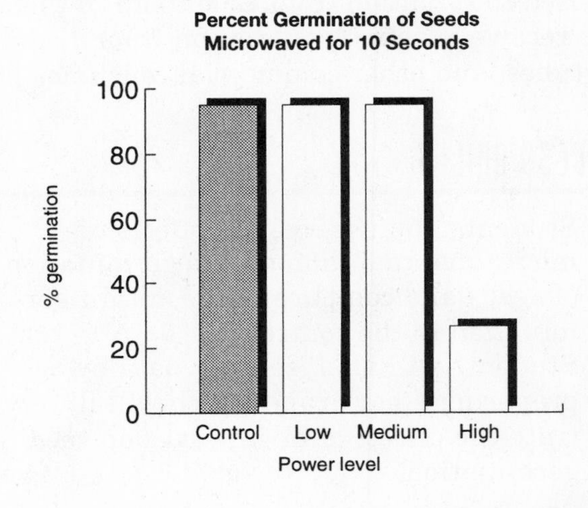

Dish	Number of seeds present	Number of seeds germinated	Percent of seeds germinated
10 seconds	20	19	96
20 seconds	20	19	96
30 seconds	20	19	96
Control #1	20	19	96
Low power	20	19	96
Medium power	20	19	96
High power	20	6	32
Control #2	20	19	96

Copyright © Glencoe/McGraw-Hill, a division of The McGraw-Hill Companies, Inc.